BRITISH GEOLOGICAL SURVEY

B. C. WORSSAM and
R. A. OLD

Geology of the country around Coalville

Memoir for 1:50 000 geological sheet 155
(England and Wales)

CONTRIBUTORS

Stratigraphy
K. Ambrose
M. G. Sumbler
J. Brewster
E. G. Poole
A. Horton
M. R. Henson

Palaeontology
M. A. Calver, M. Mitchell
W. H. C. Ramsbottom, N. J. Riley
A. W. A. Rushton
G. Warrington

Petrology
R. R. Harding

Geophysics
J. D. Cornwell
J. M. Allsop

Water supply
A. R. Lawrence

Economic geology
P. M. Harris
D. E. Highley

LONDON: HER MAJESTY'S STATIONERY OFFICE   1988

088043

© *Crown copyright 1988*

First published 1988

ISBN 0 11 884398 2

Bibliographical reference

WORSSAM, B. C., and OLD, R. A. 1988. Geology of the country around Coalville. *Mem. Br. Geol. Surv.*, Sheet 155.

Authors

B. C. WORSSAM, BSc, MI Geol
19 Warham Road, Otford, Sevenoaks, Kent TN14 5PF
R. A. OLD, BSc, PhD
British Geological Survey, Keyworth

Contributors

J. M. Allsop, BSc, K. Ambrose, BSc,
J. D. Cornwell, MSc, PhD, P. M. Harris, MA, CEng, MIMM, D. E. Highley, BSc, CEng, MIMM, A. Horton, BSc, N. J. Riley, BSc, PhD, A. W. A. Rushton, BA, PhD, M. G. Sumbler, BA, G. Warrington, BSc, PhD
British Geological Survey, Keyworth

A. R. Lawrence, MSc
British Geological Survey, Wallingford,

J. Brewster, BSc, M. A. Calver, MA, PhD, M. Mitchell, MA, R. R. Harding, BSc, DPhil, M. R. Henson, BSc, PhD, E. G. Poole, BSc, W. H. C. Ramsbottom, MA, PhD, formerly *British Geological Survey.*

Other publications of the Survey dealing with this district and adjoining districts

BOOKS

British Regional Geology
Central England, 1969
London and Thames Valley, 1960
The Pennines and adjacent areas, 1954

MAPS

1:625 000
Solid geology (South sheet), 1979
Quaternary geology (South sheet), 1977
Aeromagnetic map of Great Britain (South sheet), 1965

1:250 000
Solid geology
East Midlands Sheet (52°NW 02°W), 1983

Bouguer Gravity Anomaly
Chiltern Sheet (51°N 02°W), Provisional Edition 1980
East Midlands Sheet (52°N 02°W) Provisional Edition 1982

Aeromagnetic Anomaly
Chiltern Sheet (51°N 02°W), 1980
East Midlands Sheet (52°N 02°W), 1980

1:63 360
Sheet 154 (Lichfield), Solid 1926, Drift 1922
Sheet 168 (Birmingham), Solid 1924, Drift 1924
Sheet 169 (Coventry), Solid 1926, Drift 1922
Sheet 170 (Market Harborough), Solid and Drift, 1969

1:50 000
Sheet 140 (Burton upon Trent), Solid and Drift, 1982
Sheet 141 (Loughborough), Solid and Drift, 1976
Sheet 142 (Melton Mowbray), Solid and Drift, 1977
Sheet 155 (Coalville), Solid and Drift, 1982
Sheet 156 (Leicester), Solid and Drift, 1975

Printed in the United Kingdom for Her Majesty's Stationery Office

Dd 0240404 5/88 C20 398 12521

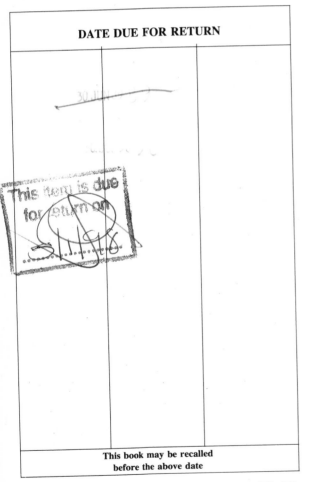

Geology of the country around Coalville

Centrally situated in the Midlands, the district described in this Memoir is largely underlain by Triassic rocks, preserved within the Hinckley Basin. Seismic investigations indicate that the Triassic base lies more than 800 m below OD in the deepest part of the basin. In the north-east the late Precambrian rocks of Charnwood Forest emerge through the Triassic cover to form hills rising to over 200 m OD. In the north, partly concealed by Triassic rocks, are parts of the Leicestershire and South Derbyshire coalfields and in the south-west, the northern part of the Warwickshire Coalfield crops out close to the Cambrian rocks near Atherstone.

The results of a geological survey on the scale of 6 inches to 1 mile, carried out from 1963–1977 and supplemented by deep boreholes at Rotherwood (near Ashby-de-la-Zouch), Twycross and Leicester Forest East, are presented here. Together with much coalfield information made available by British Coal, they enable a modern synthesis of the district's stratigraphy and geological structure to be made.

Additional interest is provided by the glacial deposits which cover a large part of the district, particularly the lower ground. Their emplacement and subsequent dissection by erosion have profoundly influenced present-day topography and drainage.

Although coal reserves within the district are limited, mineral working remains important and its prospects are examined in the chapter on economic geology. The Coal Measures of the South Derbyshire Coalfield include one of the largest accumulations of high-alumina refractory clays in the United Kingdom. The quarrying of igneous rocks as a source of aggregate has regional importance, and there are also extensive workings for brick clays and sand and gravel.

Frontispiece   Banded Charnian tuffs, Bradgate Tuff Formation, Bradgate Park. A 12353

CONTENTS

FIGURES

TABLES

PLATES

SIX-INCH MAPS

Geological National Grid 1:10 560 maps included wholly or in part within the district are listed below with the initials of the surveyors and the dates of survey. The surveyors were K. Ambrose, J. Brewster, B. A. Haines, M. R. Henson, A. Horton, B. Kelk, R. A. Old, E. G. Poole, M. G. Sumbler, K. Taylor and B. C. Worssam. Maps available in printed form are marked P; those available only as dyeline prints are marked D. Those sheets marked * have been surveyed only in part.

SK 20 NE	Austrey D	KT 1964, BCW 1966, '72–73
SE	Polesworth D	KT 1964, BCW 1976
SW	Tamworth P	KT 1964–65
SK 21 NE*	Swadlincote	BCW 1966, '68, 70–71
SE	Netherseal D	BCW 1966
SK 30 NE	Newtown Burgoland D	KA 1975
NW	Norton-Juxta-Twycross D	BCW 1967, '72, KA 1975
SE	Shenton D	JB 1975
SW	Sheepy Parva D	JB 1975
SK 31 NE*	Ashby-de-la-Zouch	BCW 1968, RAO 1975
NW*	Moira	BCW 1967–68
SE	Packington D	BCW 1967, '71, RAO 1975
SW	Measham D	BCW 1967, '71
SK 40 NE	Thornton D	MGS 1975
NW	Nailstone D	MGS 1975
SE	Desford D	BCW 1965–66, '75
SW	Market Bosworth D	AH 1973, JB 1975
SK 41 NE*	Shepshed	EGP 1964–65, RAO 1974
NW*	Whitwick	RAO 1975
SE	Stanton-under-Bardon D	EGP 1964–65, RAO 1974–75
SW	Coalville D	RAO 1975
SK 50 NW	Groby D	EGP 1964–65, MGS, BCW, RAO 1975
SW	Kirby Muxloe D	EGP 1964, BCW 1975
SK 51 NW*	Loughborough	RAO 1974
SW	Woodhouse Eaves D	BK 1973, RAO 1975
SP 29 NE	Merevale D	KT 1963, BCW 1976
NW	Kingsbury P	KT 1964–65
SP 39 NE*	Stoke Golding	KA, MH 1975
NW	Atherstone D	KT 1964–65, MH 1975
SP 49 NE	Earl Shilton D	BCW 1965
NW*	Barwell	AH 1972–73
SP 59 NW	Huncote D	BAH 1961, EGP 1964, RAO, MGS, BCW 1975–77

PREFACE

The district represented on the New Series 1:50 000 Geological Sheet (Coalville) was first geologically surveyed by H. H. Howell and E. Hull, and the results published in 1855 on Old Series 1-inch sheets 63 NW and 71 SW. A memoir 'The Geology of the Leicestershire Coalfield and of the country around Ashby-de-la-Zouch' by E. Hull was published in 1860.

The primary six-inch survey of the district was carried out between 1893 and 1898 by C. Fox-Strangways and W. W. Watts. Six-inch maps covering the coalfields were published in 1905 (Leicestershire and South Derbyshire) and 1923 (Warwickshire). New Series 1-inch Sheet 155 (Atherstone) appeared in 1910. Two related memoirs by C. Fox-Strangways were published: 'The Geology of the country between Atherstone and Charnwood Forest' (1900) and 'The Geology of the Leicestershire and South Derbyshire Coalfield' (1907).

Partial revision of the coalfield areas was accomplished notably by G. Barrow, T. Eastwood, G. H. Mitchell and D. C. Greig, aided by the palaeontological work of C. J. Stubblefield and R. Crookall. Some of this work resulted in Wartime Pamphlet No.22 'The Geology of the Leicestershire and South Derbyshire Coalfield'. by G. H. Mitchell and C. J. Stubblefield in 1941 (revised 1948), and No.25 'The Geology of the Warwickshire Coalfield' by G. H. Mitchell, C. J. Stubblefield and R. Crookall in 1942.

The present resurvey was carried out between 1964 and 1977 by Mr K. Ambrose, Mr J. Brewster, Dr B. A. Hains, Dr M. R. Henson, Mr A. Horton, Dr B. Kelk, Dr R. A. Old, Mr E. G. Poole, Mr M. G. Sumbler, Mr K. Taylor and Mr B. C. Worssam. A list of six-inch maps and the names of the surveyors are given on p.viii. The 1:50 000 map (Solid and Drift) was published in 1982.

Dr R. R. Harding and Mr R. J. Merriman described the petrology of the Charnian rocks and the diorite intrusions. Dr A. W. A. Rushton and Mr R. E. Turner identified the Cambrian fossils, Drs M. A. Calver, W. H. C. Ramsbottom, N. J. Riley and Mr M. Mitchell named the Carboniferous fossils and Dr G. Warrington identified Triassic miospores. The carbonate petrology of the Rotherwood Borehole was undertaken by Mr J. R. Gozzard. Geophysical investigations were described by Dr J. D. Cornwell and Mrs J. M. Allsop. The economic geology was described by Mr P. M. Harris and Mr D. E. Highley, while the section on water supply was contributed by Mr A. R. Lawrence.

Thanks are due to the various quarry and gravel pit operators in the district for allowing access to their excavations; to the National Coal Board for providing detailed plans of underground and opencast workings and for allowing access to borehole cores and opencast workings. Drs M. J. Le Bas and T. D. Ford of Leicester University commented on earlier texts of chapters 2 and 7. Dr J. Moseley of Leicester University provided field and laboratory data for the Charnwood Forest area.

The memoir was edited by Dr A. J. Wadge.

F. G. Larminie, OBE
Director

British Geological Survey
Keyworth
Nottingham NG12 5GG

1st February 1988

CHAPTER 1

Introduction

The district around Coalville (Sheet 155) described in this memoir lies mainly in Leicestershire, but includes parts of Derbyshire, Staffordshire and Warwickshire. Most of the area is given over to agriculture; the largest centre of population is Coalville, a town which has grown up since the early 19th century on the northern part of the Leicestershire Coalfield.

A large proportion of the population of the district is employed in industries related to mineral working. The district includes most of the actively worked area of the Leicestershire Coalfield and parts of the South Derbyshire and Warwickshire coalfields. Coal is, or has been, worked underground and opencast in all three coalfields. Most of the active underground working in the Leicestershire Coalfield is confined to Coal Measures concealed beneath Triassic rocks. Quarrying is carried out on a large scale in the Palaeozoic and Precambrian igneous rocks of Charnwood Forest, south Leicestershire and Warwickshire, and such working has a long local history. For example Francis Goddard, writing in 'Bartlett's Mancetter' (1720), stated that Atherstone 'is paved with marble from a black hardstone dug in the Outwoods as they call them hard by with which it is mostly paved'. The Mercia Mudstone is worked for brickmaking at Ibstock, Heather and Coalville. Pottery clays are worked, along with opencast coal, in open pits in the Middle Coal Measures around Moira and Dordon. Sand and gravel aggregates are produced at Heather, Cadeby and Huncote.

The highest and most varied topography is provided by the Precambrian rocks of Charnwood Forest, where the strike of the beds produces hills with a predominant north-westerly trend. The steep-sided ridges reach 271 m at Bardon Hill and 247 m at Beacon Hill. The rugged hill tops contrast markedly with the smooth-sided valleys cut in younger rocks. The rugged topography of Charnwood Forest gives way westwards to more subdued, undulating farmland underlain by the younger geological formations. Here the land generally lies between 60 m and 165 m above OD.

The Triassic rocks have an overall gentle, south-easterly dip, and the harder sandstones and siltstones within the softer mudstone sequence produce a scarp-and-dip topography. This is particularly well displayed in the west of the district. Carboniferous sandstones also produce strong topographical features, although large tracts of the Coal Measures outcrop have been excavated during opencast coal mining or covered in spoil. The evolution of the drainage of the district is described in Chapter 9.

GEOLOGICAL SEQUENCE

The geological formations described in this memoir are listed on the inside of the front cover.

GEOLOGICAL HISTORY

The Charnian (late Precambrian) rocks of Charnwood Forest are the oldest exposed strata in the district (Figure 1). The Maplewell Group is the result of a major, often violent, volcanic episode. The centre of volcanic activity probably lay near Whitwick, since this is where the agglomerates are coarsest and the dacite lavas are thickest. The succeeding Swithland Greywacke Formation is predominantly a greywacke-mudstone sequence, with sediment derived from a nearby landmass.

The folding of the Charnian may have occurred in latest Precambrian or earliest Cambrian times, although this is not certain. The Charnwood Anticline has a north-westerly 'Charnoid' trend, with a slightly later and oblique cleavage. Soon after the folding, diorites were intruded into the Charnian; the diorites of South Leicestershire may also date from this period, but the nature of the country rocks is uncertain in this part of the district.

The base of the Cambrian is not exposed in this district but is seen just to the south near Nuneaton, where the Hartshill Quartzite rests unconformably upon both the tilted and deeply weathered Precambrian Caldecote Volcanic Series and upon a diorite intrusion.

Mudstones make up the bulk of the Cambrian sequence and are thought to have been deposited in a shallow sea. They contain a rich fauna of trilobites. The Purley Shales (Lower and Middle Cambrian) and Abbey Shales (medial Middle Cambrian) are followed disconformably by the sparingly fossiliferous late Middle Cambrian Mancetter Grits and Shales. The early Upper Cambrian Outwoods Shales, rhythmically banded pale to dark grey and rich in trilobites, are succeeded by the Moor Wood Flags and Shales; the convolute bedding in this part of the sequence is overturned to the north-east, suggesting deposition on a palaeoslope inclined in this direction. The overlying Monks Park Shales form a condensed sequence of black mudstones which passes upwards into the Merevale Shales of Tremadoc age.

As a result of Caledonian (pre-Upper Old Red Sandstone) orogenic movements, the Cambrian rocks were folded and locally cleaved. Some of the cleavage and the folding of the Charnian may be of the same age. The Mountsorrel Granodiorite, and probably the South Leicestershire Diorites, were intruded immediately after these fold movements. Ordovician, Silurian and early Devonian rocks are absent from the district because of the strong Caledonian uplift and erosion. Late Devonian (Upper Old Red Sandstone) strata are exposed just south of the district near Merevale. Spessartitic lamprophyre sills intruded into the Cambrian of Warwickshire and diorite sills intruded into the Cambrian at Merry Lees Drift are also thought to be Caledonian in age.

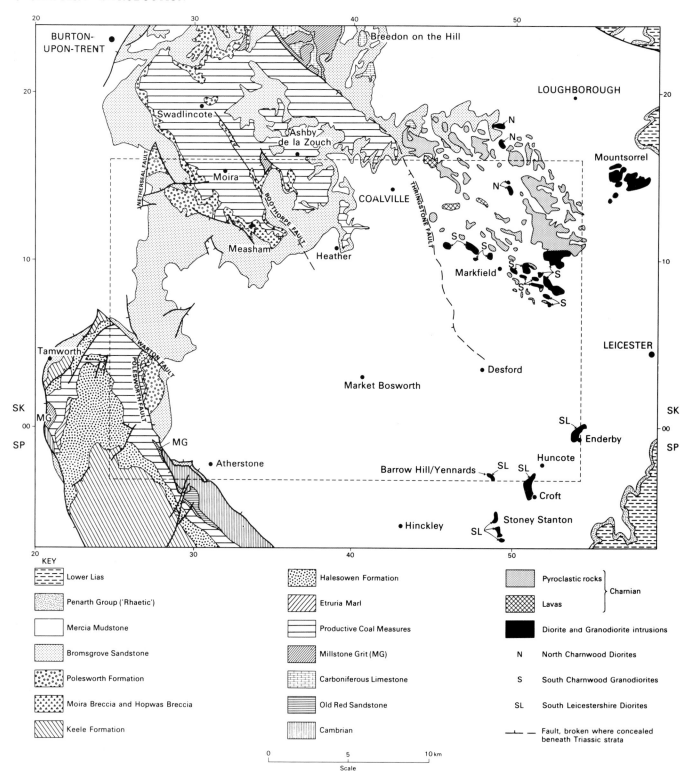

KEY

Lower Lias	
Penarth Group ('Rhaetic')	
Mercia Mudstone	
Bromsgrove Sandstone	
Polesworth Formation	
Moira Breccia and Hopwas Breccia	
Keele Formation	
Halesowen Formation	
Etruria Marl	
Productive Coal Measures	
Millstone Grit (MG)	
Carboniferous Limestone	
Old Red Sandstone	
Cambrian	
Pyroclastic rocks } Charnian	
Lavas	
Diorite and Granodiorite intrusions	
N North Charnwood Diorites	
S South Charnwood Granodiorites	
SL South Leicestershire Diorites	
Fault, broken where concealed beneath Triassic strata	

Scale 0   5   10 km

Figure 1   Geology of the Coalville district and surrounding areas

In Lower Carboniferous times, the district was slowly submerged by seas transgressing from the north, so that sequences are thinner in the south. For example, Viséan limestones are 80 m thick in boreholes at Ashby-de-la-Zouch, but only 8 m thick at Desford. The limestones are dolomitic and were laid down in shallow water, and the underlying mudstones are red and include a laterite, suggesting that the climate was hot and humid. During Namurian times, the deltaic Millstone Grit sequence was laid down; it consists of sandstones, mudstones bearing plants and non-marine shells, and a few thin coals and seatearths. Marine incursions were frequent, although marine beds are mainly of the 'Lingula' facies rather than the fully marine goniatite-bearing facies. Only the higher stages, R_1, R_2 and G_1 are present of the full Namurian sequence but the beds are concordant with the underlying Dinantian, so there is an early Namurian non-sequence across the district.

Similar conditions of deposition persisted whilst the Lower and Middle Coal Measures (Westphalian A, B and lower C) were laid down, but with some important differences. Marine incursions were less frequent, while swamp conditions were more prevalent; forests grew on the swamps and part of their debris contributed to the accumulation of coal seams. Drainage channels through the coal swamps are, in places, filled by sandstone and are preserved as washouts in the coals. Thin bands of volcanic ash, known as tonsteins, in both the Lower and Middle Coal Measures (Eden and others, 1963; Strauss 1971; Worssam and others, 1971) may distantly represent the extensive volcanic activity known from the South Nottinghamshire and Belvoir coalfields or even farther afield.

Towards the end of Middle Coal Measures times, local uplift above the swamp produced the oxidising conditions for the deposition of the red mudstones of the Etruria Marl in South Derbyshire.

The succeeding Halesowen Formation was laid down in deltaic conditions without long-lived coal swamps, so no thick coals occur. The upward decrease in arenaceous sediment and the occurrence of red mudstones may indicate decreased rainfall. Certainly, arid oxidising conditions prevailed during the deposition of the overlying Keele sequence of red marls and sandstones. The Hercynian earth movements at the end of Carboniferous times produced faulting and folding of considerable amplitude, predominantly along north and NNW-trending axes. The Precambrian block of Charnwood Forest was uplifted and no Coal Measures are preserved on the east side of the Thringstone Fault. Even to the west of this fracture, erosion removed all the Carboniferous rocks between the Warwickshire and Leicestershire coalfields, so this tract was not a basin at this time. Subsequently, by basin inversion, this latter area became a depositional basin in Triassic times.

This period was one of continental deposition when conditions were even more arid than in Upper Carboniferous times. Virtually all the rocks are red and were laid down in an oxidising environment, and the rocks immediately beneath the Triassic unconformity are commonly weathered and reddened. Fossils in the Triassic sediments are rare and none of the beds can be given a definite age. Sedimentation began in the Hinckley Basin in the west; succeeding formations overstepped eastwards until even the highest parts of Charnwood Forest were engulfed.

The lowest Triassic deposit, the locally-derived coarse Hopwas or Moira Breccia, is widely variable in age and thickness, and tends to fill local pre-existing hollows in the pre-Triassic land surface. The sandstones and conglomerates of the overlying Polesworth Formation are fluviatile in origin, as are the succeeding Bromsgrove Sandstone and Mercia Mudstone which were deposited in a non-marine basin subject to temporary flooding. Some of the thin sandstones and mudstones in the Mercia Mudstone sequence consist of wind-blown sediment.

The Tertiary period was predominantly one of uplift and erosion during which any Mesozoic formations laid down in the district were removed. Deep weathering of possible Tertiary age has been described from the district (Ford and King 1968, pp.329–330), but there are no Tertiary sediments like those in the southern Pennines (Walsh and others, 1972).

Pleistocene drifts blanket much of the district. Two distinct boulder clays, 'Triassic' and 'Chalky' in provenance, suggest that the ice sheets which deposited them travelled from the north, and from the north-east or east, respectively. They are sufficiently close in their interrelationship to indicate that they were produced by one major glaciation. Sands and gravels were laid down around the margins of the ice, and glacial lake clays formed in an arm of a glacially dammed lake that has been termed Lake Harrison (Shotton, 1953).

Late Pleistocene erosion has destroyed many of the glacial landforms, and glacial debris has contributed largely to the river terrace deposits and alluvium along the present-day drainage system. Extensive post-glacial screes formed in Charnwood Forest as a result of intense frost action, and the widespread deposits of head or solifluction debris also date from post-glacial times.

CHAPTER 2

Precambrian

The Charnian rocks of Charnwood Forest are the most eastern Precambrian rocks cropping out in Britain. These exposures lie at the western end of a buried pre-Triassic ridge of Precambrian rocks that extends eastwards to the Wash (Kent, 1968a, p.146; Wills, 1973) and is part of the Midland Craton (Harris and others, 1975, p.3; Le Bas, 1982; Pharaoh and others, 1987a). The majority of the outcrops lie in the north-east corner of the district, where they are distributed over about 66 km², although the area of Charnian rocks exposed is quite small owing to an extensive mantle of Mercia Mudstone and drift. A few outcrops occur in the adjoining Loughborough (141) and Leicester (156) districts.

The Charnian rocks are folded in the Charnwood Anticline, with a north-west-trending axis and a south-easterly

plunge (Figure 2). Sedimentary structures in the Charnian show that the rocks are not overturned.

Considerable interest was shown in the Charnian during the latter part of the 19th century, culminating in the work of W. W. Watts (not published until 1947), which contains an account of the earlier research and a reference list for the period up to 1945. Watts' major contributions were to unravel the Charnian stratigraphy and to devise the stratigraphical nomenclature used by subsequent workers. The results of research subsequent to Watts are summarised by Evans (1968), Ford (1968) and by Moseley and Ford (1985).

A revised stratigraphical nomenclature for the Charnian succession is given in Table 1. It brings Watts' terminology more into line with current practice (Holland and others,

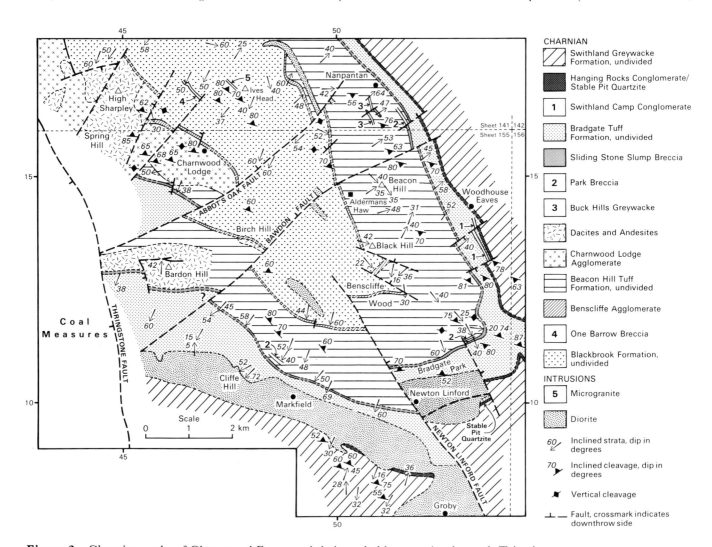

Figure 2 Charnian rocks of Charnwood Forest and their probable extension beneath Triassic cover

1978); some of his original lithological terms were misleading and additional members have now been identified. A formal stratigraphic nomenclature using additional subdivisions has been proposed by Moseley and Ford (1985).

The Charnian can be divided into a lower volcanic group and an upper sedimentary formation; the sequence is more than 3000 m thick and its base is not seen. The Maplewell Group consists predominantly of waterlain pyroclastics which are generally coarser in the Blackbrook Formation, and characteristically laminated and flinty in the overlying Bradgate Tuff and Beacon Hill Tuff formations. In the north-west of Charnwood Forest, around Charnwood Lodge (Figure 2), the Beacon Hill Tuff formation includes lavas and subaerial agglomerates, derived from predominantly calc-alkaline magmas. The lavas are dacites or andesites, and the pyroclastics are characteristically siliceous. The Swithland Greywacke Formation is predominantly sedimentary in origin although the rocks include a small proportion of pyroclastic material, so volcanicity was probably continuing nearby during their deposition. Sutherland and others (1987) conclude that the Charnian rocks formed in an island-arc and an associated back-arc basin. The thickness estimates of each formation shown in Table 1 are derived from measurements of outcrop widths and average dips. Even in the best exposed areas, considerable lengths of unexposed ground occur along each measured section.

A suite of dioritic intrusions post-dates the Charnian rocks, but is still of Precambrian age. The intrusions are described in Chapter 7.

The palaeontological evidence supports a late Precambrian age for the Charnian. The fauna is confined to the Beacon Hill Tuffs and the Bradgate Tuffs. Occurrences in the Blackbrook Formation and the Swithland Greywacke (Boynton, 1978) are referred to as pseudofossils by Ford (1980). Ford (1958, 1968, 1980) reviewed the fauna, and listed the coelenterates *Charnia masoni*, *Charniodiscus concentricus* and *?Cyclomedusa davidi*. These genera have been reconstructed and revised by Jenkins and Gehling (1978). More specimens of these three genera were described by Boynton (1978). A new arthropod, *Pseudovendia charnwoodensis*, was discovered in the Maplewell Group by Boynton and Ford (1979). The most distinctive form is *C. masoni*, which is comparable with other Precambrian frond-like fossils that are constituents of the 'Ediacara faunas' of Australia, USSR, Europe, Africa and N. America. There is general agreement that the Ediacara fauna occurs in the Vendian System, the youngest subdivision of the Precambrian (Cowie and Glaessner, 1975). Twenty samples spanning all the major Charnian subdivisions were collected during the survey in the hope that they might contain microfossils (SAL 5665–5684), but all proved barren of recognisable palynomorphs.

Potassium-argon age determinations by Meneisy and Miller (1963) on Charnian lavas yielded a range of ages which is thought to reflect periods of heating or stress undergone by the rocks since they were first formed. The oldest age was 684 ± 29 Ma obtained from Bardon Hill, and may be taken as the minimum age for the Charnian.

Table 1 Lithostratigraphy of the Charnian Supergroup

Group	Formation and dominant lithology	Members	Watts (1947)
			Brand Series
			Swithland Slates
	Swithland Greywacke Formation (>375 m) greywacke	Stable Pit Quartzite (0–30 m)	Trachose Grit and Quartzite
		Hanging Rocks Conglomerate (0–40 m)	Hanging Rocks Conglomerate
		Swithland Camp Conglomerate (0–5 m)	Hanging Rocks Conglomerate?
			Maplewell Series
			Woodhouse and Bradgate Beds
	Bradgate Tuff Formation (190–530 m) banded tuff	Sliding Stone Slump Breccia (0–10 m)	Slate Agglomerate
Maplewell Group			*Beacon Hill Beds*
		Park Breccia (0–5 m)	—
	Beacon Hill Tuff Formation (1100–1500 m)	Buck Hills Greywacke (0–10 m)	Buck Hills Grit
	banded tuff	Bardon Hill Lavas (0–200 m)	
		Charnwood Lodge Agglomerate (0–900 m)	Bomb Rocks
		Benscliffe Agglomerate (15–75 m)	Felsitic Agglomerate
			Blackbrook Series
	Blackbrook Formation (base not seen) (>1100 m) coarse- to fine-grained tuff	One Barrow Breccia (10 m) (Sheet 141 only)	—

Correlation of the Charnian with other Precambrian out-crops in the Midlands and Welsh Borderlands remains extremely tentative. An Ediacara-type fauna has been reported from South Wales (Cope, 1977), but details of this locality have not yet been published. The nearest Precambrian outcrops to Charnwood Forest are those of the Caldecote Volcanic Series at Nuneaton (Allen, 1957, 1968). These rocks are similar lithologically to the Charnian and are intruded by a granophyric diorite similar to that cutting the Charnian at Markfield (Wills and Shotton, 1934). The Caldecote rocks may therefore be contemporaneous with the Charnian. The frequently-made lithological correlation between the Charnian and Uriconian (e.g. Greig and others, 1968) has been challenged on geochemical grounds by Thorpe (1972), although he considered that they could be of broadly similar age (1974, p.121). Uriconian tuffs from Shropshire yield Rb-Sr ages of 558 ± 16 Ma (Patchett and others, 1980). These authors consider that K-Ar ages of Uriconian rocks by Fitch

and others (1969, p.36) ranging from 632 ± 32 Ma to 677 ± 72 Ma are too great due to 'excess argon problems'.

BLACKBROOK FORMATION

This is the least well exposed of the three formations of the Group. The type area is at Blackbrook Reservoir; the rocks are best exposed to the north of the present district (Watts, 1947, pp.10, 27–28) where Moseley and Ford (1985) have identified various subdivisions within their Blackbrook Group. The area underlain by the Formation on Sheet 155 is less than indicated by Watts, due to a reinterpretation of the fold closures south of Black Hill [506 136]. The succession consists of thin bedded and laminated, coarse- to fine-grained waterlaid tuffs and lesser amounts of greywacke (Moseley and Ford, 1985) and dacite (Pharaoh and Evans, 1987). They are strongly cleaved and generally more deeply

Plate 1 Benscliffe Agglomerate, Pillar Rock, Benscliffe Wood. The massive agglomerate has characteristic, widely spaced, 'pillar' jointing. A 12355

weathered than those of the younger formations, with brown staining common along bedding, joint and cleavage planes. The weathering is probably mainly Triassic. The tuffs differ from those of the overlying Maplewell Group formations in being generally coarser grained, paler grey, softer and not white-weathering, and are compositionally more mature. Towards the top of the formation the tuffs become very similar to those of the overlying formations. The distinctive Benscliffe Agglomerate is a convenient base to the overlying Beacon Hill Tuff.

Thin bedded, pale to medium grey, coarse- to fine-grained, quartzofeldspathic tuffs are exposed in a small quarry at the summit of Charley Knoll [490 158]. Some of the coarse beds have erosional bases and are graded.

Strongly cleaved, pale grey, flinty, rhyolitic tuffs form an elongate knoll at Rock Farm [481 153]. A thin section (E 46494)* reveals phenoclasts of plagioclase and quartz, many

* Numbers with E prefix refer to thin sections in the National Sliced Rock Collections, British Geological Survey, Keyworth, Nottingham.

of which are angular and broken, although some of the quartzes are rounded with embayments identical to those found in lavas. The quartz- feldspar-chlorite mosaic of the groundmass is very uneven in grain size due to recrystallisation, and is crossed by numerous sericite and chlorite-coated cleavage planes. On the evidence of the phenoclasts, it is thought most likely that the rocks are tuffs but the possibility that they are cleaved lavas cannot be ruled out.

The highest beds of the Blackbrook Formation, immediately below the Benscliffe Agglomerate, are exposed in gardens [4945 1180; 4980 1175] forming the eastward continuation of the ridge of Rocky Plantation. They are finely banded, fine-grained, white weathering tuffs very similar to those of the Beacon Hill Tuff. Identical tuffs extend for about 40 m below the Benscliffe Agglomerate at Green Hill [508 131] and show a variety of sedimentary structures such as slumping, undulating bedding and local channelling. Hereabouts they are underlain by 6 to 8 m of dark grey, coarse, quartzofeldspathic tuff resembling the matrix of the Benscliffe Agglomerate. All these beds were placed in the

Plate 2   Laminated, thin bedded tuffs, Beacon Hill Tuff Formation, Beacon Hill. Note the white weathering and steeply dipping cleavage. A 12356

Beacon Hill Tuff by Watts (1947, p.31), who considered the coarse tuff to be part of the Benscliffe Agglomerate. Although this tuff shows the same 'pillar' jointing as the Benscliffe Agglomerate, it is not agglomeratic. Pillar jointing is not confined to the Benscliffe Agglomerate, being well displayed for instance in an agglomeratic tuff well below in the Blackbrook Formation at Lubcloud [479 162].

BEACON HILL TUFF FORMATION

The succession is characterised by thin bedded and laminated, coarse to fine, siliceous, flinty tuffs or hornstones (Plate 2), which are shown to be waterlain by a wide range of sedimentary structures. The monotony of the succession is broken by a number of thin coarser members. One of these, the Benscliffe Agglomerate, persists throughout the Charnian outcrop and is the basal member of the Formation. Other thin, less persistent members occur in the eastern and southern parts of the outcrop (Figure 2). The Charnwood Lodge Agglomerate and the dacite lavas of Bardon Hill and High Sharpley are confined to the Beacon Hill Tuffs on the western limb of the anticline and, unlike the other members, were probably largely subaerial in origin.

The 200 m-thick pile of dacitic and andesitic lava interbedded with the pyroclastic succession at Whitwick and Bardon Hill, and associated with the Charnwood Lodge Agglomerate, suggests that the source of the Charnian pyroclastic debris may have lain in this area. No vents have yet been identified and the characteristic vent lithologies appear to be absent although it is possible that they occur within the Charnwood Lodge Agglomerate. In general, the tuffs of the Beacon Hill and Bradgate sequences are very similar. Watts' assertion (1947, p.13) that the Bradgate Tuff contains more 'grits' or coarse ash beds than the Beacon Hill Tuff has not been confirmed by the present survey; indeed the reverse appears to be true.

Benscliffe Agglomerate

The Benscliffe Agglomerate can be traced right around the Charnwood Anticline. It is a massive, unbedded quartzofeldspathic tuff, sufficiently coarse to be agglomerate and contrasting sharply with the bedded tuffs above and below. Its massive lithology combined with 'pillared' jointing gives rise, on weathering, to piles of huge tumbled blocks. Another peculiarity is the weathering-out of clasts on weathered faces. These characteristics are well-displayed at the type locality—the 'Pillar Rock', Benscliffe Wood [5146 1247] (Plate 1). The coarse lithology of the Benscliffe Agglomerate and its areal extent of at least 25 km² suggest that it was the product of a single powerful eruption. It is 75 m thick near Whitwick, 30 m at Benscliffe Wood and 15 m at Rocky Plantation [493 118].

Watts (1947, p.31) states that a second, and slightly younger agglomerate parallels the outcrop of the Benscliffe Agglomerate westwards from Green Hill. Coarse, agglomeratic tuffs occur at Green Hill and Benscliffe Wood above and below the Benscliffe Agglomerate, but are much thinner and not laterally persistent. The outcrop of the supposed higher agglomerate at Birch Hill was shown, during the construction of the M1 Motorway, to be due to folding (Poole, 1968, pp.141–143).

Thin sections of the agglomerate (75041-2)* show a poorly sorted rock consisting of fresh and altered, whole and shattered plagioclase crystals, trachyte or andesite clasts, and a few quartz grains in a matrix of varying proportions of quartz, albite, epidote, chlorite, white mica, iron oxides and probably titanite (Plate 3c). Epidote in abundant is places, and the chlorite has a marked green to straw pleochroism. Contorted alignments of flakes of mica are common in the matrix of the agglomerate, but it is uncertain whether this is due to compaction or tectonism. The matrix is mostly very fine grained and appears to be completely recrystallised.

On the north-eastern limb of the Charnwood Anticline, coarse, quartzofeldspathic tuffs with dark grey clasts strike north–south at Whittle Hill Farm [4985 1580]. Loose blocks of the same lithology are scattered over the rough ground 0.5 km south of Alderman's Haw [503 141], and strongly cleaved agglomerate occurs along strike at Black Hill [505 135]. The continuity of outcrops of the bed is abruptly terminated by the closure of the Charnwood Anticline and associated faulting at Green Hill and Benscliffe Wood.

At Green Hill [509 130], the pillared outcrops consist of coarse, medium grey, quartzofeldspathic agglomeratic tuff with pink and grey inclusions up to 5 cm across. The crags of agglomerate in Benscliffe Wood and at Brockers Cliffe [505 124] strike approximately east–west, but the continuation of the outcrop through Rocky Plantation has been offset to the south by the Newtown Linford Fault (Figure 2).

The Agglomerate is next seen in the M1 Motorway cutting at Birch Hill (Poole, 1968). Apart from this section it is almost impossible to trace the Benscliffe Agglomerate hereabouts. Boulders of similar rock occur in a small quarry in Fox Covert [4868 1295], but cannot be certainly placed in the Benscliffe Agglomerate. Watts (1947, p.32) recorded its

Plate 3a Beacon Hill Tuff Formation, Beacon Hill [5090 1481]. Width of field 2.9 mm. 75049

Very fine-grained, acid tuff consisting of an aggregate of stongly cleaved quartz and white mica. Lath-shaped and irregular patches (mainly holes) are sometimes filled with quartz mosaic and may be shards or fragments of pumice

Plate 3b Beacon Hill Tuff Formation, Alderman's Haw [5030 1465]. Width of field 2.9 mm. E 41757

Coarse tuff with rounded and angular clasts of quartz, feldspar and fine-grained tuff in a white mica-chlorite matrix

Plate 3c Benscliffe Agglomerate, Benscliffe Wood [5146 1248]. Width of field 2.9 mm. 75042

Coarse aggregate of plagioclase and acid igneous rock fragments. Much altered to chlorite and opaque minerals along cleavage

Plate 3d Park Breccia, Bradgate Park [5310 1149]. Width of field 2.9 mm. 75071

Coarse aggregate of feldspar and rock fragments: granophyre (centre), basic igneous (corner), acid igneous (corner). Feldspar-quartz-epidote matrix

---

* Five-figure numbers in brackets refer to thin sections in the collection of the Geology Department, Leicester University, lent by Dr J. Moseley.

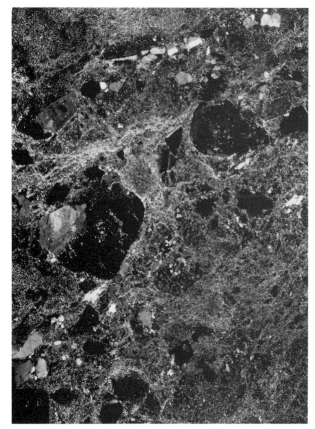

a

b

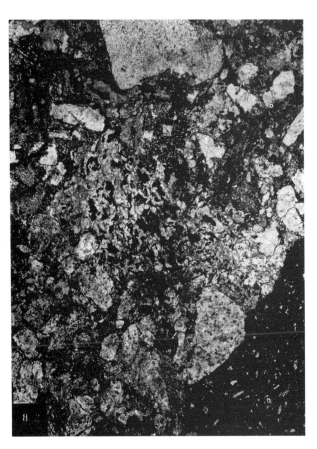

c

d

occurrence in an outcrop, now obscured, west of Poultney Cottage [488 133].

The next outcrop is at Abbot's Oak [464 142] about 2 km north-west of Birch Hill, although scattered blocks of agglomerate occur fairly commonly between. At Abbot's Oak, the rock is an unbedded pale grey quartzofeldspathic tuff with small clasts. No evidence was found here for the upper band of Benscliffe Agglomerate shown on Watts' map.

The outcrop of the Benscliffe Agglomerate next appears 1.5 km to the north-east at Little Hill [476 149] (Moseley and Ford, 1985, p.7). From here to Flat Hill [467 160], it is at its thickest, and can be confidently traced through faulted ground. This thickest development occurs where the Charnwood Lodge Agglomerate appears in the overlying beds, suggesting that a persistent vent lay nearby. At Little Hill and Cat Hill Wood [475 152], the Agglomerate is a dark grey, slightly agglomeratic, coarse, feldspathic tuff. In Cat Hill Wood, the bed forms a steep-sided ridge covered in huge boulders but with little in-situ rock exposed.

To the north-west is the sharp ridge of Collier Hill [470 158], where the coarse Agglomerate contains rounded and angular blocks of pink rhyolite in a coarse, tuffaceous matrix which shows traces of bedding in places. The adjacent outcrops on Flat Hill show coarse, agglomeratic tuffs overlain by agglomerate with angular blocks of rhyolite, derived from the same source as those in the Charnwood Lodge Agglomerate. The outcrops include the Hanging Stone, a spectacular massive, upstanding slab.

Beacon Hill Tuffs

The type locality of these beds is Beacon Hill [509 149] (Moseley and Ford, 1985, p.7) but excellent sections occur at many other localities in Charnwood Forest. Throughout the succession, the tuffs appear almost uniform and were evidently laid down during a long period of pyroclastic activity by a magma source of consistently acidic composition. The banding in the tuffs results from variations in grain size, which suggest the occurrence of a very large number of distinct ash falls. Differential weathering picks out the bedding so that the rocks have a characteristic striped appearance. The finest beds are commonly white-weathering, although the fresh rock is a medium to dark greenish grey throughout.

Sedimentary structures commonly preserved in the finer grained tuffs include graded bedding, ripple marks, small-scale cross-lamination and slumping. The thickness varies from 1500 m on the north-east flank of the Charnwood Anticline to 1100 m at Bradgate Park; it is about the same thickness at Whitwick if the Charnwood Lodge Agglomerate is included.

Thin sections of typical tuffs from the Beacon Hill Tuffs were examined from Beacon Hill (75049), Bradgate Park (75088), Markfield (75078), Benscliffe (75026), and at the Outwoods (75066 and 75067). The rocks are mainly crystal lithic tuffs with variable proportions of angular to rounded quartz, plagioclase and alkali feldspar crystals, and rounded andesite and quartzite grains in a matrix of recrystallised quartz, albite, chlorite, epidote, iron oxides and white mica (Plate 3a). The chlorite is markedly pleochroic from green to pale yellow. At the Outwoods the tuffs, although of similar composition to those described above, are finer and relatively well sorted and bedded (75066). Some rocks contain shards (75049) and show evidence of chlorite-albite-quartz-epidote recrystallisation in the matrix (75067).

At Beacon Hill, tuffs from about the middle of the formation are disposed around a minor east–west-trending syncline in a series of crags (Evans, 1963, pp.71–72). Up to 5 m of tuffs are exposed in any one crag. They show numerous small-scale slump structures and lensing of individual beds.

Tuffs at about the same horizon are exposed on Broombriggs Hill [515 141]. The coarser beds, up to 0.5 m thick, commonly have ripple-marked tops. Several beds have eroded tops, some show graded bedding and there are numerous thin convoluted horizons due to soft sediment deformation.

The 'Alderman's Haw Porphyroid' of Watts (1947, p.69) is part of this pyroclastic succession. Two exposures of this rock, 150 m NE and 500 m N 10°W of Alderman's Haw [5030 1466 and 5012 1498] respectively, resemble strongly cleaved quartz porphyries. A thin section from the former locality (E 47157), however, shows it to be a coarse, crystal-lithic tuff with rounded phenoclasts of quartz and plagioclase, and clasts of quartzite and dacite (Plate 3b). The matrix consists of fine-grained quartz, feldspar and sericite and chlorite. The rock from the latter locality is similar, but strongly sheared (E 47158).

Dip-and-scarp features at the north-eastern corner of Benscliffe Wood are formed by banded tuffs dipping south-west at 16° to 36°. One coarser bed [5138 1282] has eroded at least 20 cm into the bed below. These scarps were thought to be fault lines by Watts (1947, p.31), since he incorrectly correlated several coarse tuffs with the Benscliffe Agglomerate and considered that the latter was repeated by faulting. For example, a coarse 4 m thick quartzofeldspathic tuff cropping out at the top of this sequence [5127 1265] was included in the Benscliffe Agglomerate by Watts (1947, pp.31, 88). The rock shows 'pillared' jointing and has rare large clasts, but its poor bedding and small thickness are atypical of the Benscliffe Agglomerate locally.

The tuffs exposed at Old John Tower [1122 5250] include an impersistent 0.5 m slump breccia similar to the Park Breccia, but probably below it.

An unusual tuff occurs 100 m north-east of White Hill Farm 4831 1180. Bedding is obscure, but one band about 1 m thick contains numerous rounded, siliceous nodules up to several centimetres across, superficially like accretionary lapilli. These are not strongly contrasted to the matrix of the rock in the field. A thin section (E 46495) reveals that the nodules consist of a quartz-chlorite-white mica mosaic with sericite and chlorite concentrated around the rims. The bulk of the rock is also a quartz-chlorite-white mica mosaic with broken quartz-crystals, abundant clinozoisite and rare plagioclase. Clinozoisite is commonly concentrated around the nodules. It thus appears that the nodules originated as concretions rather than accretionary lapilli.

Locally developed agglomerates are interbanded with tuffs at Sandhills Lodge [502 111], Bushey Field Wood [497 111] and Hunt's Hill [524 116]. These beds are too thin and impersistent to justify member status. The agglomerates at the first two localities were considered by Watts to be 'crush breccias' (1947, p.32) and the last to be part of the Sliding

Plate 4 Sliding Stone Slump Breccia, Sliding Stone Enclosure, Bradgate Park.

Clasts of fine-grained tuff, some folded before incorporation, lie in a coarse-grained matrix. A 12351

Stone Slump Breccia (1947, p.39). The 'crushed' appearance of the rocks at Bushey Field Wood and Sandhills Lodge is due to a strong cleavage that disrupts the agglomeratic texture. At Hunt's Hill there are small outcrops and scattered boulders of coarse, strongly cleaved agglomerate with clasts up to 5 cm across. The agglomerate does not resemble the Sliding Stone Slump Breccia in lithology, and its present ascription to a lower horizon eliminates the need for the complicated faulting invoked by Watts to explain its offset relative to the main outcrop of the Sliding Stone Slump Breccia in Bradgate Park.

Charnwood Lodge Agglomerate

These rocks occur as a wedge up to 900 m thick near Whitwick, thinning markedly to the south-east, and apparently not present south-east of the Abbot's Oak Fault. They rest directly on the Benscliffe Agglomerate between Timberwood Hill [472 148] and Flat Hill [466 160], in place of the banded tuffs found to the south and east. The Agglomerate includes the 'bomb rocks' of Watts (1947, pp.15 and 57), and is most spectacularly displayed in the type locality at Charnwood Lodge Nature Reserve [463 157] (Moseley and Ford, 1985, p.11), where it is composed of ellipsoidal blocks of porphyritic dacite 1 cm to 1 m across, set in a subordinate unbedded coarse tuff matrix. The blocks are flattened in the cleavage and can be matched with the dacites of the Whit-

wick area nearby, which are presumably co-eval with them. The dacites vary considerably in the size and relative proportions of the quartz and feldspar phenocrysts. Chemical analysis of the matrix of the agglomerate unexpectedly revealed that it is much less acid in composition than the dacites or the typical Beacon Hill Tuffs (Moseley, 1979). 'Bomb rock' lithology is also found in the grounds of Mount St Bernard Abbey between Charnwood Lodge and Peldar Tor Quarry. The non-intrusive nature of the Agglomerate is clearly shown in one of these exposures [454 158] where it is interbedded with coarse tuffs.

In many exposures, the proportion of tuff matrix is greater than at the type locality. On Flat Hill, for example, the Agglomerate contains fewer and more angular blocks and appears to grade downwards into the Benscliffe Agglomerate. The most south-easterly exposure of the Agglomerate occurs on Timberwood Hill [4727 1487] where rounded and angular fragments of acid lava and coarse tuff reach 0.5 m. The coarse matrix shows traces of bedding with a NW–SE strike. It grades upwards into coarse tuffs containing only scattered large clasts; these have been included in the Beacon Hill Tuffs.

The varied lithology of the blocks of dacite in the Agglomerate and their contrast with the intermediate composition of its matrix, suggest that the agglomerate formed by the explosive disruption of earlier acid lavas. Although the vent of this eruption has not been located, it is unlikely to be far away. Whether explosive vulcanicity producing coarse agglomerates continued throughout the period of formation of the banded Beacon Hill Tuffs, or whether the latter are banked against a slightly earlier mound of agglomerate cannot be decided on the field evidence.

Andesites and dacites

A suite of andesitic and dacitic lavas is found at the north-western end of the Charnwood Anticline and includes the andesites ('good rock') of Bardon Hill and Birch Hill and the dacites of Bardon Hill, Peldar Tor and Spring Hill. Although no vents have been identified it is likely that the source of lavas lay close by.

North-west of the Abbot's Oak Fault there is a close stratigraphical association between the dacite and the Charnwood Lodge Agglomerate suggesting that both were products of the same volcanic centre. The remaining lavas occur between the Abbot's Oak and Bawdon faults. The andesite of Birch Hill lies just above the Benscliffe Agglomerate. The stratigraphical position within the Beacon Hill Tuffs of the lavas at Bardon Hill is not precisely known, but they seem to lie near the top of the sequence.

ANDESITES   The best exposures of andesite are at Bardon Hill, which stands massive and isolated on the eastern outskirts of Coalville. The andesite, 'good rock' in the quarryman's term, won from the huge quarry on its western side has been variously described as a lava, a pyroclastic rock and an intrusion, as discussed by Watts (1947, pp.60–63). Re-examination of the quarry during the survey has shown that the various rock units form part of an interbedded volcanic sequence of lavas and tuffs, dipping east-northeast. In detail, the 'good rock' consists of at least two andesitic to dacitic block lavas whose rubbly exteriors are preserved as breccias (Christiansen and Lipman, 1966; Macdonald, 1972, pp.91–98). The inclusions randomly distributed through the lava may be due to flow-brecciation.

In the south-western corner of the quarry, at the top of the lower face [4533 1300], is a coarse breccia of well bedded flakes of very fine-grained, purple sediment or tuff. The flakes are up to 5 cm across, angular or subrounded and set in a coarse, tuffaceous matrix. Coarse thin-bedded tuffs now seen only as loose blocks were formerly accessible in situ at the top of this face below the roadway (Dr R. J. King, personal communication). These pyroclastics are overlain by an andesitic lava rich in chloritised ferro-magnesian minerals and plagioclase which can be traced for about 50 m along the roadway above the lower face. It is overlain by 1 m of 'breccia' dipping 40° NE. This consists of rounded and angular fragments of the underlying rock set in a dark grey very fine-grained streaky matrix which apparently represents the rubbly top of the lava flow. The breccia is overlain by the andesite described below.

On the northern side of the quarry the andesite has a roughly east-west faulted contact with dacites lying to the north. Watts (1947, pp.68–69) described porphyry (dacite) intruded into andesite, but this occurrence could not be located during the present survey. Subsequent quarrying has revealed an intrusive porphyry dome (Moseley and Ford, 1985, p.15).

On the upper face north of the trigonometric point on Bardon Hill [460 130] a strong east-west trending fault with a low southerly dip separates dacite and andesite. Undisturbed contacts between these rocks occur on either side of the fault. About 1 m to the south of the fault-zone, dacite overlies andesite. The contact is sharp and the dacite shows no sign of chilling. On the other side of the fault and about 12 m above the fault-zone, a 1 m band of andesite is interleaved with the dacite; it dips at 40°NE, apparently in normal stratigraphical sequence. If the upper and lower contacts of the andesite are accepted as showing that it is in normal stratigraphical sequence, then its total thickness is about 380 m.

Although most of the andesite quarried is uniformly brownish grey, closer examination reveals important variations. Angular or rounded blocks are frequently discernible, and these are predominantly acid or intermediate lavas where their lithology can be identified. Commonly, however, the outlines of the blocks are diffuse and only visible because they have a darker or paler shade than the matrix; they are commonly pink in colour. The andesite can be seen to grade downwards into coarse breccia in the upper face 90 m north of Bardon Hill trig. point [4599 1328]. The breccia passes abruptly southwards into a coarse tuff, but the contact can only be traced for a short distance.

Jones (1926) observed that alteration of the andesite has largely obscured the details of its original nature. This is confirmed in three thin sections of this rock. All three are acid or intermediate rocks; the first (E 2661) is a porphyritic andesite (Plate 6b) and the others (E 2658 and E 2663) may be lavas or crystal tuffs, probably the former. All are composed dominantly of feldspar with lesser, variable, amounts of quartz, chlorite, epidote, iron oxide, ?titanite, calcite and white mica. The dominant phenocrysts are euhedra (some cracked or broken) of plagioclase, with variable degrees of alteration mainly to epidote. Probable ferro-magnesian phenocrysts are now represented by chlorite, epidote and iron oxides. Rounded and subhedral quartz phenocrysts in E 2658 have optically continuous overgrowths of groundmass quartz. The groundmasses of the three rocks differ. One (E 2658) has a fine-grained altered igneous texture with feathery-edged plagioclase laths indistinct in an almost isotropic chlorite, which may have been a devitrified glass, and with randomly distributed clots of epidote, ?titanite and opaques. The groundmass of E 2661 is largely alkali feldspar or sodic plagioclase in approximately equant (50 μ) grains with an almost spherulitic texture. Chlorite occurs in small quantities. In E 2663 the original groundmass texture has been obscured by recrystallisation and shearing, and it now comprises chlorite, feldspar, white mica, epidote, opaque oxides and calcite. The disordered texture of this rock could indicate a scoriaceous or tuffaceous origin.

An average of three X-ray fluorescence whole rock analyses from Bardon Hill carried out by J. Moseley of Leicester University indicates an andesitic composition as follows: SiO_2 58.1%, TiO_2 0.48%, Al_2O_3 12.9%, Fe_2O_3 11.2%, MnO 0.16%, MgO 7.7%, CaO 4.7%, Na_2O 2.6%, K_2O 0.32%, P_2O_5 0.03%. The unusually low Al_2O_3 and relatively high Fe_2O_3 and MgO probably reflect the high degree of alteration in these rocks. Further analyses recently published by Pharaoh and others (1987b) also prove an andesitic composition.

A porphyritic quartz-andesite has an elongate outcrop at Birch Hill, where it occurs apparently in the core of a syncline in the Beacon Hill Tuff Formation. No contacts with the adjacent rocks are visible, and Watts considered the 'Birch Hill Porphyry' to be an intrusion (1947, p.69). Although much fresher and lacking in inclusions, the Birch Hill andesite resembles the 'good rock' of Bardon and the two were correlated by Hill and Bonney (1878, p.206) and Bennett and others (1928). This correlation is supported by the appearance of the rock in thin section. It has abundant epidotised, subhedral, plagioclase phenocrysts with minor phenocrysts of augite (altering to hornblende and chlorite) and much corroded quartz. The groundmass is a mosaic of plagioclase, epidote, chlorite and a little quartz (E 48291).

DACITES   Porphyritic dacites occur on the north side of Bardon Hill and at Peldar Tor and Spring Hill, forming outcrops of 'porphyries' which extend northwards into the Loughborough district (Watts, 1947, pp.65–70). The only exposure of a dacite in contact with another Charnian rock has been described above at Bardon Hill, where a dacite lava rests on an andesite lava. Although some of these dacites may be intrusions (Moseley and Ford, 1985), they are more likely to be mainly lavas. The dacites are characterised by numerous elongated inclusions of dacite and andesite aligned in the cleavage; some reach 1 m in length. In places these give the rock a volcaniclastic appearance, but thin sections do not reveal any trace of vitroclastic textures so the rocks are not ignimbrites. The inclusions vary considerably in concentration, but never form discrete beds as they do at the base of the andesite flow at Bardon Hill. They are probably flow-breccias.

Quartz and feldspar phenocrysts are prominent in hand specimens of the dacites: both vary greatly in concentration. The feldspar phenocrysts are usually the larger (up to 1 cm) and more abundant. The very fine-grained groundmass, almost glassy in places, imparts a deep purple-grey colour to the rock. No clear contacts were observed between dacites of varying phenocryst concentration. There is a distinctive variety at Bardon Hill with much smaller phenocrysts in an almost glassy matrix, but its contacts with the more typical dacite are everywhere faulted.

Thin sections show that alteration and cleavage has obscured much of the original igneous texture at Peldar Tor Quarry. Subhedral plagioclase phenocrysts are commonly partially sericitised or epidotised; epidote is also common as discrete grains. Quartz phenocrysts vary from subhedral to rounded and embayed, and many show an optically continuous overgrowth in the groundmass (E 46497) (Plate 6c). The groundmass is a quartz-plagioclase mosaic with flakes of chlorite and white mica in cleavage planes (E 46498). Cleavage has broken many of the phenocrysts giving a rather disrupted texture. The dacites from Bardon Hill Quarry are very similar to those at Peldar Tor, but are generally more strongly cleaved (E 43885–7) (Plate 6d).

Chemical analyses of 'porphyritic lavas' from Spring Hill and Bardon Hill quarries (Thorpe, 1972) show they are basic and intermediate in composition. This is at variance with the field evidence as no major quartz-free phases were identified in the field, and it is possible that these analyses are of inclusions or dykes. The rocks of High Sharpley, in the Loughborough district, which are very similar to those at Peldar Tor and Spring Hill, proved on analysis (Thorpe, 1972, p.274, Pharaoh and others, in press (b)) to be dacites.

Park Breccia

The Park Breccia is probably the lower of two breccias in the Outwoods near Nanpantan in the Loughborough district (Figure 2). The Breccia is also seen 100 m north of Sliding Stone Enclosure, Bradgate Park [531 115], and below the southern abutment to the road bridge carrying the B587 across the A50 at Raunscliffe [4863 1090]. These outcrops lie about 60 m and 40 m respectively below the Sliding Stone Slump Breccia. Although separated by 4.5 km, the two outcrops are considered to be at the same horizon because both display a distinctive lithology. The full thickness of the Park Breccia is nowhere exposed, but probably does not exceed 5 m.

The Breccia in Bradgate Park is crowded with angular, bedded clasts of fine-grained tuff mainly less than 10 cm long, set in as coarse matrix. The exposure on the A50 reveals 3.5 m of similar lithology. Elongated bedded tuff clasts up to 30 cm long occur in several layers in a dark, grey-green coarse tuff matrix. The breccia passes gradually upwards into thin bedded tuffs.

A thin section of the Bradgate Park material (75071) shows that the larger clasts are tuff, and the smaller lithic grains are trachyte, quartzite and cleaved pelite (Plate 3d). The last is of unknown provenance. The bulk of the matrix consists of quartz and orthoclase, albite and oligoclase feldspars, many of which are fragmented. A thin section (75077) from the A50 exposure proved similar, except that trachyte grains and epidote were abundant in the matrix.

The mode of formation of the Park Breccia is discussed below with that of the Sliding Stone Slump Breccia.

BRADGATE TUFF FORMATION

The Bradgate Tuff Formation is divided from the Beacon Hill Tuff Formation by the Sliding Stone Slump Breccia at its base. The full thickness of the formation can be calculated only on the southern and eastern sides of the Charnwood Anticline. Over 530 m of tuffs are estimated to underlie the largely unexposed ground between Raunscliffe and Cliffe Hill Quarry; in Bradgate Park the tuffs are about 500 m thick but they appear to thin rapidly northwards and only 190 m seem to be present at Hangingstone Hills. Because exposure is poor, it is not clear whether the tuffs have actually thinned or whether they have been overstepped by the Swithland Greywacke Formation.

Sliding Stone Slump Breccia

The Sliding Stone Slump Breccia is one of the most important members of the Charnian succession since its appearance is distinctive and it is so persistent laterally that it can be traced right round the Charnwood Anticline. It is characterised by angular, randomly oriented, elongated clasts of fine-grained tuff up to 3 m long. The clasts are frequently contorted or even rolled up (Plate 4). Smaller,

a

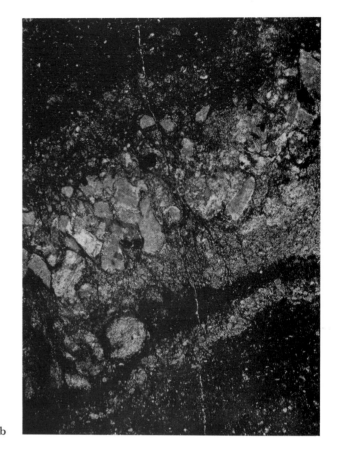

b

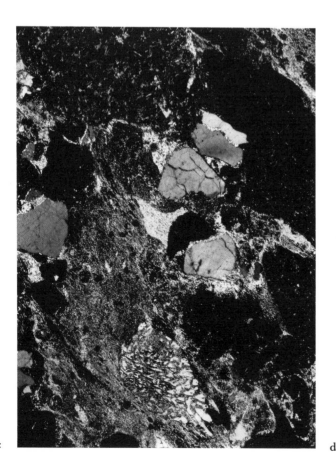

c

d

Plate 5a Sliding Stone Slump Breccia, Warren Hill [5316 1187]. Width of field 2.9 mm. 75090

Ill sorted aggregate of acid lava (right centre), fine tuff (centre), quartz and plagioclase in a fine-grained turgid matrix

Plate 5b Bradgate Tuff Formation, North Quarry, Hangingstone Hills [5223 1556]. Width of field 2.9 mm. 75094

Thin-bedded tuff comprising a quartz-white mica mosaic with coarse band of sericitised plagioclase. Some alignment of grains in cleavage

Plate 5c Swithland Camp Conglomerate, Swithland Wood [5376 1218]. Width of field 2.9 mm. 75133

Coarse, rounded, acid and intermediate lava tuff fragments, granophyre and quartz, all much altered and set in a white mica-chlorite matrix

Plate 5d Stable Pit Quartzite, Deer Park Spinney, Bradgate Park [5375 1053]. Width of field 2.9 mm. 75123

Well sorted quartzite with angular grains of quartz and some acid tuff and rare feldspar

rounded clasts of acid or intermediate lava found at some localities are probably lapilli (Watts, 1947, p.37). The matrix is generally coarse, similar to that of the Park Breccia.

The type locality is the crags just west of the aptly named Sliding Stone Enclosure [5308 1149] at Bradgate, where about 8 m of Breccia is exposed. Overall, the Breccia ranges in thickness from 6 m to 10 m.

Both the Park Breccia and the Sliding Stone Slump Breccia originated through the disruption of partly consolidated bedded tuffs coupled with the influx of coarse pyroclastic material. The outcrop of the Sliding Stone Slump Breccia indicates that the bed extends over an area of at least 70 km². It is visualised that the eruption of the coarse tuffs was accompanied by earthquakes, which disrupted the partially consolidated bedded tuffs already deposited. The clasts thus produced were still plastic, and were incorporated by localised slumping into the coarse tuff and simultaneously contorted and rolled. In the Park Breccia, the tuff clasts are undistorted and were therefore probably better consolidated and less plastic when they were incorporated. The conformity of the clasts to the bedding may have been produced by a relatively strong flow of the deposit in one direction. The most northerly exposure on the east flank of the anticline at the Outwoods [5167 1597] has clasts up to 1.5 m long.

A thin section of the Sliding Stone Breccia from Warren Hill [5316 1187] is composed of clasts of fine-grained tuff in a generally coarser matrix (Plate 5a, 75090). The matrix is poorly sorted and shows crude bedding. Lithic grains and grains of devitrified glass and iron oxides are present, together with a few grains of plagioclase up to 2 mm in diameter. The finer fraction of the matrix is a cloudy, brown intergrowth, probably of epidote, titanite, chlorite and white mica.

A thin section from the Outwoods outcrop (75062) reveals a less altered matrix composed of lithic grains of andesite and quartzite, feldspar and quartz, with a finer fraction of

quartz, chlorite and muscovite. About 8 m of breccia are exposed at the south-west end of Hangingstone Hills [5228 1502]. Some of the clasts are as much as 20 cm long, and show traces of bedding. There are a few lapilli of acid lava. The basal 4 to 5 m contain few clasts.

At Raunscliffe [485 109] the Breccia is 8 to 10 m thick with tuff clasts up to 1 m across. A 20 cm band of fine-grained tuff occurs 3 m above the base.

The Breccia is exposed at The Hollies [478 118] in a number of small outcrops. Unoriented blocks of fine tuff up to 1 m long are set in a coarse tuff. Coarse, bedded tuffs occur immediately below, with two beds of fine agglomerate 30 cm and 50 cm thick respectively. The Breccia from the nearby M1 cuttings has been described by Poole (1968, pp.143–146).

A 4 to 5 m bed of breccia crops out 200 m east of Rise Rocks Farm [4716 1215]. The bed contains numerous angular elongated clasts of tuff up to 20 cm long, some of which are slightly bent. Rough bedding occurs in the top 1.5 m. Several small exposures of breccia occur at the south-eastern end of Warren Hills. That occurring 400 m north-west of Abbot's Oak [4632 1468] shows blocks of tuff up to 3 m across in a coarse matrix. A further 300 m to the north-west, there are very coarse, massive, quartzofeldspathic tuffs with numerous comminuted rock fragments and scattered blocks of fine tuff. Another 100 m along strike, thin bedded coarse tuffs overlain by 3 m of massive tuff containing small angular fragments of tuff are exposed, with scattered blocks of more typical Sliding Stone Slump Breccia with large tuff clasts nearby.

Beds above the Sliding Stone Slump Breccia

Apart from being generally finer grained, these tuffs are indistinguishable from those of the Beacon Hill Tuff Formation. They are named the Hallgate Member by Moseley and Ford (1985). The formation includes most of the known localities of Charnian fossils. They are typically developed to the south of the Sliding Stone Enclosure in Bradgate Park, where large crags include outcrops of the Sliding Stone Slump Breccia. These crags also include some fine examples of cleavage refraction ('ripple cleavage' of Watts, 1947, pp.44–45), for example 100 m south of Sliding Stone Enclosure [5315 1115]. The cleavage planes are refracted at each boundary between tuff bands of differing grain size.

Bradgate Tuffs have been laid bare for 400 m along the north face of Cliffe Hill Quarry [477 106] where they are intruded by Precambrian diorite. The banded tuffs are harder and flintier than those seen elsewhere. It is not clear whether this is due to their fresher condition or to thermal metamorphism, though Evans (1968, p.10) recorded spotting and metasomatism of the tuffs at this locality. At the eastern end of the quarry particular bedding planes have yielded many examples of *Charniodiscus*, most of which are now in Leicester Museum. *Charniodiscus* is normally confined to this formation and has never been found outside the Maplewell Group. All the beds exposed on Billa Barra Hill [466 114] should also probably be placed in the Bradgate Tuff Formation. Coarse tuffs, in beds up to 0.5 m thick, alternate with finer beds showing ripple-marks, erosional features and slump structures. In thin section, they consist of angular and subangular quartz, feldspar, and a few lithic fragments of fine sand grade (75104), in a partly recrystallised matrix of fine-grained epidote, chlorite, albite and white mica. There is some orientation of the long axes of fragments and phyllosilicates in the plane of the cleavage. Watts (1947, p.53) placed the Billa Barra rocks and the banded tuffs at Bardon Lodge [456 119] 'with hesitation' in the Swithland Greywacke Formation. All are now placed in the lower formation, together with the banded tuffs that crop out 200 m north-east of Tythe Farm [458 110]. On the crags around Rise Rocks Farm [469 122] a total thickness of about 150 m of tuffs crops out. The tuffs are predominantly medium- to coarse-grained, with a few slightly agglomeratic beds up to 5 m thick. There are several thin slumped and convoluted bands.

In a quarry at the northern end of Hangingstone Hills, fine-grained, banded tuffs with coarser laminations are also assigned to the Bradgate Tuffs (75094–6). This quarry [5223 1554] is the type locality for *Charnia masoni* (Ford, 1958, 1968). In thin section, the tuffs consist of subangular and angular quartz, oligoclase, orthoclase and lithic grains of fine sand grade, in a matrix of chlorite, white mica, quartz and albite (Plate 5b). In the coarser layers, elongated grains are aligned parallel to the cleavage. The proportions of lithic grains and epidote vary in different beds.

The Hanging Stone itself [5224 1529] is formed from a very coarse, massively bedded, agglomeratic tuff with individual beds up to 6 m thick. Watts (1947, p.46) postulated a strike fault to explain the disappearance of this bed to the north and south, but there is no positive evidence for the existence of this fault, and it is thought more likely that the bed passes beneath Mercia Mudstone to the north and passes laterally into fine-grained tuff to the south.

Poorly sorted but well laminated Bradgate Tuffs occur at Hallgates Reservoir [5536 1155]. The rocks (75100–1) were probably originally dust tuffs with an admixture of mud. Shapes indicative of devitrified shards are quite common, even though the matrix is partly recrystallisated with epidote and chlorite granules forming in a quartz and albite intergrowth.

SWITHLAND GREYWACKE FORMATION

The Swithland Greywacke Formation is exposed only on the eastern and southern flanks of the Charnwood Anticline. In contrast to the older formations, it comprises fine-grained greywackes and pelites with only minor tuffs. A variable cleavage gives rise to slates in places which were formerly of commercial importance. These are the Swithland Slates of earlier writers, and take their name from the slate quarries in Swithland Wood. The predominant colour is deep purplish grey, in contrast to the greenish grey of the tuff formations. The presence of conglomerates and coarse quartzites at the base of the sequence, together with the rapid thinning of the Bradgate Tuffs outcrop northwards from Bradgate Park, suggest that there is an unconformity at the base of the Swithland Greywacke succession. The top of the formation is nowhere exposed but a maximum exposed thickness of 580 m is estimated near Groby. On both flanks of the Charnwood Anticline, Cambrian rocks appear to come on above

Plate 6a Swithland Greywacke Formation, South Quarry, Hangingstone Hills [5255 1508]. Width of field 2.9 mm. 75135

Coarse to fine, bedded greywacke with strong realignment of grains and growth of opaque minerals in cleavage

Plate 6b Andesite, Bardon Quarry [458 132]. Width of field 2.9 mm. E 2661

Subhedral, epidotised, plagioclase phenocrysts in feldspar-chlorite-epidote groundmass. Dark chlorite streaks may be altered ferromagnesian phenocrysts

Plate 6c Dacite, north-eastern face of Peldar Tor Quarry [451 158]. Width of field 2.9 mm. E 46497

Embayed quartz phenocrysts and subhedral plagioclase in quartz-feldspar mosaic. Secondary calcite in quartz and secondary epidote next to feldspar

Plate 6d Dacite, Bardon Quarry [460 133]. Width of field 2.9 mm. E 43885

Phenocrysts of quartz and plagioclase in a groundmass of feldspar, quartz, chlorite and iron oxides. Epidote has replaced some plagioclase cores, and carbonate and chlorite are developed in the groundmass

the Swithland Greywackes. Their stratigraphical relationship is unproved, but presumably the Cambrian beds lie unconformably upon the older rocks. To the west of Charnwood, Tremadocian shales were encountered in the Merrylees Drift (Butterley and Mitchell, 1945, p.706). Slates of possible Cambrian age occur 3 km to the east, on the western side of the Mountsorrel Granodiorite in the Leicester (156) district (Le Bas, 1968, pp.49–51).

The dominant lithology of the Greywackes is seen in thin section to be a medium- to coarse-grained metapelite, commonly phyllitic in texture. In Swithland Wood (75140) the rock consists of quartz, feldspar and some lithic fragments of fine sand grade as framework constituents in a matrix of white mica, chlorite and opaque oxides. At Woodhouse Eaves [5315 1412] the detrital grains range from angular to rounded, and reach 2 mm in diameter (65136). In parts the rock has been intensely cleaved. In the matrix, white mica and chlorite predominate, with chlorite and hematite forming rims to some detrital grains. In the quarry at Hangingstone Hills (75135), bedding and cleavage are both clearly exhibited (Plate 6a). Different proportions of quartz, lithic fragments, and matrix minerals of chlorite, white mica, carbonate and hematite, reveal the bedding while white mica has preferentially developed along the cleavage.

Three members are distinguished in the basal 50 m of the Swithland Greywacke Formation. The Swithland Camp Conglomerate, locally defines the base of the Formation at Brand Hills, but elsewhere is overstepped by the Stable Pit Quartzite or its lateral equivalent, the Hanging Rocks Conglomerate.

Swithland Camp Conglomerate

The member is typically developed on an isolated knoll at the south-east corner of Swithland Camp [537 122] where 5 m of conglomerate are exposed. It also crops out for 600 m just below the Stable Pit Quartzite on Brand Hills. Its distinctive

a

b

c

d

lithology consists of abundant, subrounded, flattened, deep purple-grey siltstone clasts up to 3 cm long in a similarly coloured siltstone matrix; the clasts are now aligned in the near-vertical cleavage, but may have originally possessed a steep imbrication.

A thin section (75133) from Swithland Camp consists largely of elongate, subrounded, siltstone clasts (Plate 5c). Subrounded and angular trachyte, quartz and feldspar grains, from 0.5 mm to 2 mm across, are scattered through the rock. The matrix comprises chlorite, muscovite, white mica, hematite, clay minerals and carbonate. The mineral and lithic grains are quite distinct from the matrix, in contrast to the commonly gradational boundaries between the clasts and matrix, and may be pyroclastic in origin.

The conglomerates are intraformational in origin and may have been produced by subaqueous fragmentation during turbidity flows (cf. Pettijohn, 1975, pp.184–185).

Stable Pit Quartzite

Where the Swithland Camp Conglomerate is absent, the base of the Swithland Greywacke Formation is generally marked by the Stable Pit Quartzite, but locally by the Hanging Rocks Conglomerate. The two members are mutually exclusive and appear to grade laterally into each other.

The quartzite is best exposed at 'Stable Pit', a small quarry in Bradgate Park [5340 0999]. The section here is not easy to interpret; Watts (1947, pp.50–51) described two bands of quartzite separated by a bed of slate. It is now considered that there is only one band repeated by a synclinal fold plunging ESE, with pelite in its core.

On the southern flank of the syncline, some 10 m of coarse, massive and dark purplish-grey quartzite dip at 40°. Steeply dipping 2 to 5 cm thick quartzite bands in the overlying pelites are probably clastic dykes. To the north, the quartzite next appears in Deer Park Spinney [5380 1053] dipping to the south. This outcrop was described by Watts (1947, p.50), but is now overgrown.

Thin sections (75123–5) from Deer Park Spinney are of medium- to coarse-grained arenites and pelites, composed dominantly of well-sorted subrounded quartz grains and variable minor quantities of lithic fragments in a chlorite and hematite- stained clay matrix (Plate 5d). Albite-oligoclase grains are a minor constituent, as is the white mica that is in parts oriented along cleavage and bedding directions. The average diameter of the grains is 0.5 mm, a few grains showing embayments or overgrowths indicating limited pressure solution. In a slide from Stable Pit (75119), quartz and chlorite flakes are oriented in the direction of flexural slip which is indicated in hand specimen by slickensides. Quartz and chlorite also fill sigmoidal tension gashes at right angles to the direction of slip.

Quartzite, dipping east in beds up to 0.5 m thick interleaved with pelite, occurs in a small quarry 100 m south-west of The Brand [5351 1322]. At The Brand, this member approaches a greywacke in composition. A thin section (75127) shows framework grains of sand-grade angular quartz, subrounded pelite, trachyte and welded tuff, in a chlorite and white mica matrix. Prisms of apatite are also present and detrital muscovite is crudely oriented in the plane of the cleavage. Another slide from the same locality (75129) shows

abundant opaque grains, and a porphyritic dacite fragment.

The northernmost outcrop of the quartzite is at Brand Hills where it is separated from the Swithland Camp Conglomerate by about 30 m of greywacke.

To the west of Stable Pit, the grey quartzite crops out at Groby Pool [5224 0842] and westwards along strike as far as the crags at New Plantation [503 090].

Hanging Rocks Conglomerate

The conglomerate is best exposed in crags at the southern end of Hangingstone Hills [5247 1506]. Here, immediately below the greywackes exposed in the nearby quarry, come less strongly cleaved, bedded coarse arenites. Some are conglomeratic with quartz pebbles up to 5 mm in diameter. Below these arenites are coarse conglomerates, with angular and rounded pebbles mainly less than 5 cm across in a coarse matrix (Watts, 1947, p.48). The pebbles are strongly oriented in the cleavage, and are similar in lithology to those from Bradgate Park described below. The whole conglomeratic sequence here is about 40 m thick with 6 m of conglomerate proper at the base. Thin sections of the conglomerate from Hangingstone Hill (75108–9) show coarse matrices, with variable proportions of grains of quartz, plagioclase, devitrified glass shards, white mica, chlorite, opaques, epidote and accessory minerals (mostly rounded zircon). One section (75111) shows pressure pitting and solution of fragments, and probably albitisation of the groundmass. This rock also contains more epidote and hematite than the other thin sections.

The conglomerate also crops out between Coppice Plantation and Cropstone Reservoir in Bradgate Park [542 110], just within the Leicester district. A sample of 127 pebbles from these outcrops was classified by Dr J. Moseley as follows: quartz 25%, pelite 24%, quartzite 18%, trachyte 18%, rhyolite/dacite 11%, quartz-chlorite schist 2%, unidentified 2%. Moseley originally included 2% granite in his total, but this has been re-identified by Dr M. J. Le Bas as porphyritic dacite (personal communication).

Within the rest of the Swithland Greywacke succession there are a few continuous sections. Up to 15 m of fine-grained, dark purple-grey greywackes overlie the Hanging Rocks Conglomerate in a disused slate quarry [5254 1507] at Hangingstone Hills. They are finely banded with sedimentary structures including graded bedding, large ripples and channel-scour structures.

The major slate quarries at Swithland Wood are now flooded and inaccessible. Small exposures nearby show a strong, steeply dipping cleavage oriented between 100° to 120°, which almost completely obscures the bedding.

Coarse- to fine-grained greywackes and pelites are exposed in numerous small knolls between Groby and Markfield. The dips are generally at 30° to 45° SSW, with cleavage dipping NNE roughly normal to bedding. At Little John [5015 0831] and on the adjacent crag to the east, the bedding dips north at 34° and the cleavage south at 72°. Included in this area is the former Groby Slate Works quarry [5080 0825] where 18 m of green and purple-grey greywacke are exposed. In a quarry just to the north of Bradgate Home Farm, 21 m of similar beds are exposed.

CHAPTER 3

Cambrian

Cambrian rocks crop out west of Atherstone in a small triangular area on the eastern flank of the Warwickshire Coalfield and are here taken to include the Tremadoc Series. They form the northern end of the Nuneaton Inlier which includes one of the most complete Cambrian sequences in Great Britain. The whole of the Nuneaton succession, summarised by Allen and Rushton (1968, pp.28–37), is exposed within the present district, apart from the Hartshill Quartzite and the lower half of the overlying Purley Shales (Table 2). To the west the Cambrian is unconformably overlain by Millstone Grit, and to the north and east it is faulted against the Triassic. Just south of the present district the Cambrian is unconformably overlain by the Old Red Sandstone, and itself unconformably overlies the Precambrian Caldecote Volcanic sequence.

The Cambrian rocks of Nuneaton and their biostratigraphy have been described in detail by Taylor and Rushton (1971) and their account of the constituent formations is summarised as follows.

The Purley Shales consist of blocky, purple mudstone with interbedded grey mudstone and a few beds of purple and brown siltstone and sandstone. They pass upwards into the Abbey Shales which are bluish grey mudstones with thin beds of limestone and glauconitic sandstone. The succeeding Mancetter Grits and Shales are grey mudstones with silty and sandy bands, and thin beds of glauconitic, coarse- to fine-grained sandstone. These beds pass upwards into the interbedded grey and dark grey, pyritic mudstones of the Outwoods Shales. There is a gradual upward passage into the Moor Wood Flags and Shales, which consist of beds of convoluted and massive sandstone and siltstone, interbedded with wavy-bedded, micaceous and shaly mudstone. The succeeding Monks Park Shales are dark grey, micaceous mudstone overlain by black mudstone. The Merevale Shales are greyish green mudstone with many irregular, pale buff, dolomitic bands and lenticles. Grey and dark grey mudstones occur in subordinate lamellae and thin bands.

Palaeontological details are given by Rushton (1978, 1979, 1983) and the chronostratigraphy follows Cowie and others (1972).

The stratigraphical relationship between the Cambrian and Charnian rocks of the present district is unknown; nor is the relationship of the Cambrian to the South Leicestershire Diorites directly known although a large block of sediment of probable Cambrian age, enclosed in diorite, was formerly exposed (see Chapter 7). Just east of the present district, hornfelsed shale within the aureole of the Mountsorrel Granodiorite is probably Cambrian in age (Le Bas, 1972; Evans 1979). The Granodiorite gives an Rb-Sr isochron age of 433 ± 17 Ma (Cribb, 1975).

The main period of folding affecting the Cambrian of the Nuneaton Inlier is post-Westphalian in age. This is shown along the flanks of the Warwickshire Coalfield where structures affect the Cambrian and the Westphalian equally. Cambrian strata to the east and north of the Nuneaton Inlier are known from boreholes and mine workings. They differ from the exposed sequence in showing steep pre-Westphalian dips and a variable, generally strong, cleavage. Most of the occurrences were summarised by Le Bas (1968, pp.57–58), Cowie and others (1972) and Evans (1979); additional occurrences are given below. The majority of the dated occurrences are Tremadoc or Merioneth in age. Fossiliferous Lower Cambrian Purley Shales were proved in the Dadlington Borehole [3987 9628] and a breccia of doubtful Lower Cambrian age was penetrated by the Bosworth Wharf Borehole [3916 0323] (Brown, 1889).

DETAILS

In the Rotherwood Borehole [3458 1559], dark grey, thinly bedded, Upper Cambrian mudstones and siltstones with many nodules of siderite or dolomite, and dipping at 60°–70° were proved between the base of the Carboniferous, at 173.90 m and the bottom of the hole at 199.0 m. A strong cleavage lay normal to the bedding in the

Table 2 Cambrian stratigraphy of Nuneaton

Lithostratigraphy	Approximate thickness in Nuneaton Inlier (m)	Chronostratigraphy	
Merevale Shales	>90	Tremadoc Series	Tremadocian
Monks Park Shales	90		
Moor Wood Flags and Shales	20	Merioneth Series	Upper Cambrian
Outwoods Shales	300		
Mancetter Grits and Shales	50		
Abbey Shales	40	St David's Series	Middle Cambrian
Purley Shales	200		
Hartshill Quartzite	250	Comley Series	Lower Cambrian

cores. A specimen of cleaved mudstone yielded a K:Ar whole rock age of 477 ± 19 Ma; this is not the age of formation of the sediment but may be that of the cleavage (Evans, 1979, p.36). An acritarch assemblage identified by R. E. Turner contained the following forms and indicates a Merioneth age: cf. *Dasydiacrodium caudatum*, *Impluviculus milonii*, *Leiosphaeridia sp.*, *Micrhystridium shinetonense*, *Timofeevia pentagonalis*, *T. sp.*, *Veryhachium minutum*, and *Vulcanisphaera turbata*.

In thin section, mudstone from 182.80 m in the borehole (E 49796) is largely clay mica, recrystallised and reoriented at a high angle to the bedding, (Pharaoh and others, 1987a). R. W. Sanderson reports that grains of carbonate about 0.024 mm across are common and may be concentrated along the bedding. Carbonate nodules about 0.4 cm across and composed of aggregates of small (c.0.1 mm) spherules, each with radiating fibrous structure, are also present. XRF analysis shows the carbonate to be a mixture of siderite and dolomite.

Lower Palaeozoic rocks of less definite age were encountered in the Twycross Borehole [3387 0564]. Beneath the base of the Trias at 503.3 m was a microdiorite sill to 506.9 m; this was underlain by barren, purple and greenish grey, silty mudstone of presumed Cambrian age dipping at 30°. The base of the hole was at 508.9 m.

Steeply dipping, tightly folded mudstones with an east–west strike were encountered in two drifts [4680 0591 and 4681 0589] at Merry Lees Colliery (Butterley and Mitchell, 1945). The shales were faulted against Coal Measures along the Thringstone Reverse Fault (see p.109). Fossils in the shales included *Acrotreta* aff. *belti*, *Lingulella* cf. *nicholsoni* and fragments of *Hyolithes*. Stubblefield (in Butterley and Mitchell, 1945, p.706) concluded that the shales are of Tremadoc age.

Presumed Charnian or Cambrian rocks were encountered below the Trias base in the Lindridge No.1 Borehole [4716 0484] (Fox-Strangways, 1907, pp.103, 344); they were logged as 'slaty-looking rock with red marks, dipping at an angle of 70°'.

In the Leicester Forest East Borehole [5245 0283], dark grey mudstone and siltstone dipping at 40° to 60° were proved for 22 m beneath the base of the Trias at 156 m. A strong cleavage was perpendicular to the bedding. A single trilobite, probably *Niobella* (or *Niobe*), suggests a Tremadoc age although an Arenig age cannot be discounted.

Three boreholes near Market Bosworth penetrated grey shales below the Trias, at Bosworth Wharf [3916 0322], Cowpastures [4124 0383] and Kingshill Spinney [3814 0154] (Richardson 1931). A. W. A. Rushton has re-examined specimens from the first two boreholes and compares them lithologically with the Stockingford Shales of Nuneaton. No fossils were recorded, so the age of the beds remains uncertain.

In the Dadlington Borehole [3993 9909], beneath the Trias at a depth of 314 m, grey-green mudstones dipping at 20° to 25° were proved for a further 43 m. They yielded the following fauna identified by A. W. A. Rushton, who correlates it with that from the lower part of the Purley Shales: sponge spicules (cruciform), *Botsfordia* cf.*granulata*, *B.* cf. *pulchra*, *Lingulella spp.*, *Obolella* cf. *atlantica*, *Gracilitheca* cf. *bayonet*, *Agraulos* (s.l.) *sp.*, *Calodiscus* cf. *schucherti*, *Ladadiscus llarenai*, *Serrodiscus* cf. *bellimarginatus*, *Strenuella?* cf. *pustulosa*, *S.* cf. *sabulosa*, *Hipponicharion* cf. *minus* and *Indiana sp.* The age is Lower Cambrian.

The Shuttington Fields Borehole [2642 0610] encountered horizontal grey shales beneath the Trias at 269.7 m. They contained acrotretid brachiopods including *Eurytreta sabrinae*, *Tomaculum problematicum* and burrows (Bulman and Rushton, 1973, p.14); the fauna indicates a Tremadoc age.

A short distance south of the present district, Stockingford Shales occur in a railway cutting at Elmsthorpe [484 959] and in a borehole nearby (Brown, 1889, p.30).   AWAR

CHAPTER 4

Carboniferous: Dinantian and Namurian

No strata of Ordovician, Silurian or Devonian age are preserved in the district and it is not until Asbian or Brigantian times—the later stages in the early Carboniferous (Dinantian)—that the geological record is resumed. The district lies just south of the southern margin of the Widmerpool Gulf, an early Carboniferous depositional basin (Falcon and Kent, 1960), in which the Lower Carboniferous sequence is much thicker and more complete than on its margins. In addition, the marginal and basinal ('block' and 'gulf') sequences show contrasting sedimentary facies. As a result, they can only be correlated in the most general way (Falcon and Kent, 1960; George and others, 1976; Aitkenhead, 1977).

CARBONIFEROUS LIMESTONE (DINANTIAN)

Marine Dinantian sedimentation did not overstep southwards into the district from the Widmerpool Gulf until late Asbian or even Brigantian times. The Dinantian rocks are not exposed and the sequence is known only from three boreholes. In the Rotherwood Borehole [3458 1559], 112.4 m beds, probably all of Brigantian age, rest on upper Cambrian; in the Ellistown Colliery Borehole [4390 1056], Brigantian rocks were encountered at 467.9 m and were un-bottomed at 494 m (Boulton, 1934); 9 km farther south, only 8.2 m of Asbian limestone, resting on diorite, were encountered in the Stocks House Borehole [4680 0212] (Mitchell and Stubblefield, 1948, p.3). No Dinantian rocks underlie the Warwickshire Coalfield within the district; the Namurian rocks around Merevale rest directly on Cambrian. About 2 km west of the district, however, several boreholes prove unbottomed sequences of up to 26 m of Brigantian limestone and sandstone (Taylor and Rushton, 1971, p.52).

The Dinantian sequences found in the district contrast with those of the adjacent Widmerpool Gulf. For example, an unbottomed succession of more than 180 m of limestone and dolomites, ranging in age from Courceyan to late Asbian, is preserved at Breedon on the Hill [406 232], about 8 km north of Coalville (Mitchell and Stubblefield, 1941a; Mortimer and others, 1970) while at Grace Dieu, only 5 km to the south-east of Breedon, limestone deposition on the block did not begin until Brigantian times (Kent, 1968b, pp.80–81). In contrast, a borehole at Hathern, only 11 km north-east of Coalville, proved an unbottomed Dinantian basinal sequence exceeding 350 m, including an 'Anhydrite Series' more than 100 m thick of Courceyan age (Falcon and Kent, 1960; Llewellyn and Stabbins, 1970). In the following account the fossils have been identified by W. H. C. Ramsbottom and M. Mitchell.

Details

In the Rotherwood Borehole (Figure 3), the Carboniferous Limestone succession rests with marked angular unconformity on the underlying Cambrian at a depth of 173.90 m. The basal 4.3 m of the sequence are barren, grey, deltaic sandstone, siltstone and conglomerate. The latter contains Cambrian mudstone clasts up to 8 cm across, fragments of Dinantian limestone and mudstone, and rounded quartz pebbles. Coalified plant debris is abundant. The succeeding 5.1 m of strata are 'red beds' comprising dark brown and ochreous, red, unbedded, barren mudstones, and including a 1.5 m bed with abundant 2 to 3 mm limonite pellets. Such pellets, with onion-skin layering, characterise present-day, lateritic soils and this oxidised sequence probably represents a soil profile developed under hot, humid conditions. The red mudstones are overlain by 14.6 m of predominantly grey, marine mudstones and siltstones, but the shoreline cannot have been far away, since there are abundant plant remains, and clay ironstone bands and nodules are common, and there are also two, thin, red, mudstone bands interbedded with the grey measures. The marine mudstones contain abundant shells and several 'shell reefs' or coquinas are developed. Fossils include chonetoid and spiriferoid fragments, *Cypricardella sp.* and *Phestia attenuata*. A sudden change in sedimentation led to the deposition of 80 m of mainly limestone and dolomite with subordinate clastic rocks. Below a depth of 109 m, the main carbonate is limestone but above it, dolomite predominates. Nodular beds exhibiting a variety of textures are developed in the limestones and more abundantly in the dolomites. Most commonly, they are rounded, unlaminated, carbonate nodules in a darker, laminated, muddy matrix (Plate 7a). The nodules are jumbled and do not exhibit jig-saw fits. The laminated matrix is deflected around the nodules. Comminuted fossils (bivalves, crinoids, corals and foraminifera) are abundant. A rarer texture comprises darker, irregular nodules in a weakly laminated matrix, and off-white, unfossiliferous, porcellanous carbonate patches in irregular clotted aggregates (Plate 7b). At a depth of 112.0 m, a colony of *Syringopora* cuts all three carbonate phases.

The nodular beds probably originated during diagenesis, soon after deposition. There may originally have been planar bedding but it was disrupted by carbonate segregation rather than bioturbation, since there are no burrows. The abundant marine fauna and the absence of roots precludes a pedogenic origin for these beds. Similar nodular beds or quasibreccias occur in the Ballytrea Borehole, Co. Tyrone (Fowler and Robbie, 1961, pp.26–28; plates III to VI) where they were ascribed to prolonged desiccation of lime-mud. The carbonates have been examined petrographically by J. R. Gozzard who reports that most of the limestones are biomicrites or biomicrosparites with micrites below about 136 m. Upwards, the limestones are increasingly interbedded with dolomitic biomicrites and sparry dolomites, and these lithologies predominate in the top 40 m of the succession. In places, remains of calcareous shells are preserved in the dolomite and there are scattered patches of the original limestone. The dolomite is generally vuggy.

Although mainly limestone or dolomite, nearly 25 per cent of the sequence consists of mudstone, siltstone, quartzose sandstone, and minor nodular limestone bands. These non-carbonate beds are commonly burrowed and contain many poorly preserved shells and a few plants. Disseminated pyrite is abundant and there are rare pyrite nodules. Amongst the few identifiable macrofossils in the car-

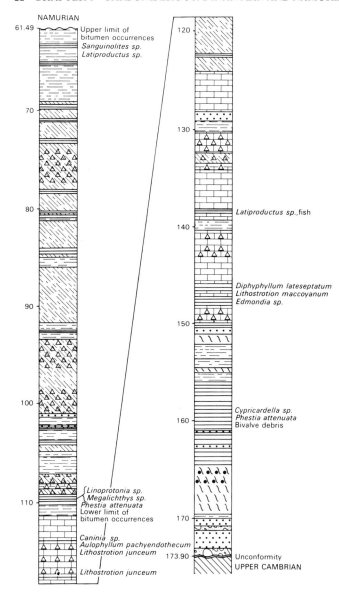

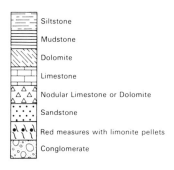

Siltstone

Mudstone

Dolomite

Limestone

Nodular Limestone or Dolomite

Sandstone

Red measures with limonite pellets

Conglomerate

Figure 3   Generalised section of the Carboniferous Limestone in the Rotherwood Borehole

bonates are the following corals: *Aulophyllum pachyendothecum*, *Caninia sp.*, *Diphyphyllum lateseptatum*, *Lithostrotion maccoyanum L. junceum* and *Syringopora sp.* The interbedded clastics have yielded *Edmondia sp.*, *Linoprotonia sp.*, *Phestia attenuata*, *Latiproductus sp.*, *Megalichthys sp.*, (tooth) and a cochliodont tooth. The coral fauna is of Brigantian age.

Carbonate deposition ceased almost as abruptly as it began, and the uppermost 7.5 m of the Dinantian sequence comprise grey siltstones and subordinate mudstones, yielding abundant *Latiproductus sp.*, and a few *Sanguinolites sp.* There is no angular unconformity with the overlying Namurian, but there must be a break since the Dinantian is immediately overlain by mudstones containing an upper Namurian fauna thought to be of R_1 age. An unusual feature of the Dinantian of the Rotherwood Borehole is the abundance of bitumen. Near the top of the succession, it forms casts of *Latiproductus sp.*, and it also occurs in veins with quartz and pyrite, cutting all lithologies down to 111 m, and filling many of the cavities in the dolomite.

In the Ellistown Colliery Borehole, the 26.1 m Dinantian sequence consists predominantly of dolomite in the upper 20 m with mainly limestone below. There are a few beds of sandstone, siltstone and mudstone up to 1.7 m thick.

The Dinantian strata in the Stocks House Borehole are only 8.2 m thick and were recorded by Fox-Strangways (1907, pp.104, 347–348) as follows:

	Thickness m	Depth m
Grey sandy metal and spar	0.9	144.8
Conglomerate (Carboniferous Limestone with *Productus giganteus*)	2.4	147.2
Light blue sandstone	3.1	150.3
Coarse grey sandstone	0.9	151.2
Conglomerate	0.3	151.5
Brown sandstone	0.8	152.3

The 'limestone' is dolomite (p.104), and it is possible that the 'conglomerates' are nodular carbonate beds like those in the Rotherwood Borehole. The specimen of *Productus giganteus* has been assigned by Mr J. Pattison to the *Gigantoproductus maximus* group, and suggests an Asbian (most probably late Asbian) age.   RAO

MILLSTONE GRIT (NAMURIAN)

The Millstone Grit succession in the present district is mainly mudstone including marine bands, sandstone with some beds coarse-grained, and a few thin coals and seatearths.

The sedimentary environment was deltaic in Namurian times but there is no local indication of provenance. The Millstone Grit has a small outcrop in the north around Valley Farm [3445 1545] and another in the south-west where sandstone cropping out in the core of an anticline at Waste Hill [283 983] is assigned to the top of the sequence (Taylor and Rushton, 1971, pp.53–55). The Millstone Grit is also proved in a number of boreholes in the Leicestershire and South Derbyshire coalfields. It is assumed to underlie Coal Measures everywhere in these two coalfields, but its extent beyond the Coal Measures incrop is unknown. Wells at Holly Hayes [4405 1529], described below, indicate an incrop of Millstone Grit beneath the Trias east of the Thringstone Fault. It is not certain that the Millstone Grit is continuous beneath the Warwickshire Coalfield; in the Baddesley Colliery, Stratford No.3 Shaft [2790 9709], Coal

Plate 7a   Nodular, dolomitic limestone at 108.5 m in the Rotherwood Borehole. Bitumen veins occur near the base

Plate 7b   Irregular pale aggregates in nodular limestone at 114.2 m in the Rotherwood Borehole

Measures appear to rest directly on Cambrian mudstone but the contact may be faulted. Details of the Millstone Grit in those parts of the Warwickshire Coalfield west of the present district are given by Taylor and Rushton (1971, pp.53–60).

The local Millstone Grit, like the Carboniferous Limestone, was deposited on the southern margin of the Widmerpool Gulf (Falcon and Kent, 1960). There is a non-sequence at the top of the Brigantian and only the latest three stages of the Namurian are represented: the Kinderscoutian (R_1), Marsdenian (R_2) and Yeadonian (G_1) in ascending sequence. The marine bands which separate and subdivide these stages (Ramsbottom and others, 1978), are all present except the Bilinguis Marine Band (Figure 4), which is generally absent from much of the eastern Midlands (Ramsbottom *in* Taylor and Rushton, 1971, p.59). The Millstone Grit of the present district is scarcely a fifth the

thickness of its correlatives in the Gulf (Frost and Smart, 1979, p.19), and is generally more argillaceous.

The Millstone Grit thins steadily southwards in Leicestershire from over 80 m at Rotherwood near Ashby-de-la-Zouch, to 26 m at Ellistown, and possibly as little as 4 m in the Stocks House Borehole. Detailed lithological correlation is not possible between these borehole sections (Figure 4).

Two factors make it difficult to correlate biostratigraphically between the local Millstone Grit sequences. Firstly, the palaeontological data from the older boreholes is sparse. In particular, the position of the Subcrenatum Marine Band marking the base of the Coal Measures cannot generally be identified with certainty. This is a major problem since the Namurian lithologies closely resemble the early Westphalian sediments. Secondly, goniatite/pectinoid-phase marine bands are replaced southwards by *Lingula* bands whose correlation is uncertain.

Details

The Millstone Grit outcrop near Valley Farm [3445 1545] is faulted against Coal Measures on two sides and overlain unconformably by Moira Breccia to the south-east. Mapping indicates that beds of mudstone or shale, weathering to a stiff grey clay are interbedded with coarse to gritty sandstones with small (5 mm) quartz pebbles.

The Rotherwood Borehole [3458 1559], sited on this outcrop, proved 61.5 m of Millstone Grit overlying Carboniferous Limestone (Figure 4, col.1). The Superbilinguis Marine Band encountered at 27.9 m was 5.0 m thick. In addition to *Bilinguites super-bilinguis*, the black mudstones contained *Lingula mytilloides*, *Posidonia sp.* and an acanthodian spine. Burrows and trails were abundant. The marine band marks an abrupt marine transgression, as it is underlain by a thin coal with a seatearth. The Gracilis Marine Band was encoutered at 51.8 m and is divided by 1.3 m of non-marine beds into upper and lower portions, 1.3 m and 1.7 m thick respectively. Fossils include *Bilinguites* cf. *gracilis*, *Caneyella sp.*, *Dunbarella sp.*, and *L. mytilloides*. A further 3.5 m of marine mudstones formed the base of the Millstone Grit to 61.5 m. They contained *L. mytilloides*, *Orbiculoidea nitida*, *Dunbarella.?* and fish debris.   RAO,BCW

Comparison with the Blackfordby No.1 Borehole [3235 1827] in the Loughborough district suggests that the Rotherwood Borehole started about 20 m below the top of the Millstone Grit sequence, and the boundary with the Coal Measures at outcrop is drawn on this assumption.

The Grange Wood Borehole [2652 1547] penetrated 27.9 m into the Millstone Grit, if the base of the Coal Measures is correctly indentified at 824.03 m. The beds consist of 2.6 m of seatearth overlying grey silty mudstone, overlying a 0.6 m bed of quartz-grit at 838.3 m, in turn overlying dark grey laminated mudstone with traces of non-marine lamellibranchs.

The uppermost 5.7 m of the Millstone Grit, logged as pale grey sandstone with darker silty bands, may have been encountered in the Acresford No.7 Borehole [2960 1297], below a *Lingula* band correlated by Greig and Mitchell (1955, p.61) with the Subcrenatum Marine Band. South of here, the Millstone Grit may be represented in the lowest 9.1 m of the Chilcote boring [2828 1145], consisting of 8.4 m of grey sandstone overlying 0.7 m of conglomerate and coarse sandstone with quartz pebbles.

The Appleby No.1 Borehole [3205 1058] may have entered the Millstone Grit at about 241.9 m (Fox-Strangways, 1907, p.333). 'Dark stone bind and shale' for 4 m at this depth is underlain by mainly argillaceous rocks, with no coal reported, to the bottom of the hole at 297.2 m.

A marine band at 170.7 m in the Horses Lane Borehole [3355 1166], relative to the presumed Norton Coal (Figure 7), is probably at or below the horizon of the Subcrenatum Marine Band (cf. Measham Nos.1, 2, and 5, Figure 4). The 13 m of strata proved beneath the marine band, comprising siltstone and mudstone, may therefore be Millstone Grit. On this correlation the top of the Millstone Grit is about 110 m below the Kilburn Coal. The Snarestone Shaft [3475 1009] (Fox-Strangways, 1907, pp.335–338) penetrated 170 m of strata below the Kilburn at 93.7 m depth, so about 60 m of the mainly argillaceous sequence was probably in the Millstone Grit.

Several wells at Holly Hayes [4405 1529] sunk through the Mercia Mudstone, proved up to 31.4 m of beds beneath which are provisionally assigned to the Millstone Grit. They comprise purple mudstone and seatearth, becoming grey with depth, and lilac and grey sandstone. These beds lie east of the Thringstone Fault. About 2.5 km to the north-west, at Thringstone, undoubted Millstone Grit strata with marine fossils were encountered just east of the Thringstone Fault during trenching carried out by the National Coal Board. Thus a narrow strip of Millstone Grit, and possibly also Lower Coal Measures, seems to occur between the Thringstone

Fault and the Charnian outcrop. It is not known whether the contact between the Charnian and the Millstone Grit is faulted or unconformable. If the latter is the case, the Precambrian of Charnwood must have been an area of rugged relief during Namurian times. The nearest Charnian outcrops at Peldar Tor are only 750 m from Holly Hayes and are at a height of 180 m above OD, nearly 280 m above the base of the deepest well at Holly Hayes, indicating a pre-Namurian slope steeper than 1 in 3. This seems to be too steep to be realistic and it is more likely that the junction is faulted.

A thin section (E 20846) of a sandstone taken at a depth of 98.5 m in one of the Holly Hayes boreholes shows a distinctly bi-modal texture. The larger grains are of highly angular, even splintery, quartz. A few grains of rounded quartz, vein-quartz, Charnian tuff and plagioclase also occur. The larger fraction is scattered throughout, and is seldom concentrated into laminae. The more abundant smaller fraction consists predominantly of angular quartz with a sprinkling of muscovite and many grains of opaque minerals. There is some carbonaceous material. The calcite cement occurs in large 'lustre mottled' plates and has caused some corrosion of quartz gains, adding to their angularity.

In the Ellistown Colliery Borehole [4390 1056], the Millstone Grit succession is apparently 25.6 m thick (Boulton, 1934) (Figure 4, col.5). There is no clear development of the Subcrenatum Marine Band, and the base of the Coal Measures has been taken at the base of a 4.9 m marine band with *L. mytilloides*, *Discina sp.* (= *Orbiculoidea*) and *Naiadites?* at a depth of 442.3 m. This is 10.4 m below the Listeri Marine Band. The base of the Millstone Grit is taken at 467.9 m so as to include the 'Pendleside Beds' of Boulton's log. There is a marine Carboniferous Limestone sequence below. The Millstone Grit succession as recorded by Boulton comprises sandstone, mudstone, seatearth and a thin limestone. Marine fossils occur at five horizons, but without the goniatites found in boreholes elsewhere (Figure 4).

Millstone Grit was encountered in the Measham Boreholes Nos. 1, 2, and 5 [3670 0978, 3639 1192, and 3689 1091], details of which were recorded by Mitchell (1948), and Mitchell and Stubblefield (1948), and graphic and summary logs published by Greig and Mitchell (1955). The fullest section is afforded by the No.1 Borehole (Figure 4, col.4) which starts in Lower Coal Measures. A *Lingula* band encountered at 167.6 m is probably the Subcrenatum Marine Band (Greig and Mitchell, 1955, p.58). The sandstone at 171 m has been tentatively identified as the Rough Rock. Two marine bands encountered at 178.4 m and 183.9 m were identified by Mitchell (1948, p.504) as the Cumbriense and Cancellatum marine bands respectively. A further *Lingula* band occurs near the base of the borehole at 192.6 m.   RAO

In the Osbaston Hollow Borehole [4166 0635], a marine horizon thought to represent the Subcrenatum Marine Band was proved between 154.5 m and 158.5 m (Figure 4, col.7). It was underlain by mudstones, sandstones and seatearths to a depth of 213.4 m. At 202.1 m, marine mudstones 0.30 m thick contained *Lingula mytilloides*, *Productus sp.*, *Aviculopecten?* and a pterioid bivalve. (Mitchell and Stubblefield, 1948, p.5).

In the Desford Colliery No.12 Borehole [4683 0504], 13.9 m of sandstone and shale of the Millstone Grit were proved beneath the Subcrenatum Marine Band. The beds were steeply dipping, and probably represent a stratigraphical thickness of no more than 6 or 7 m.

The Ibstock No.2 Borehole [4653 0453] proved, beneath the presumed Subcrenatum horizon, a 1.5 m coal underlain in turn by seatearth and by 22.8 m of sandstones, some of them coarse. A comparable succession was proved for 4.8 m in Desford No.2 Borehole.   MGS

The Millstone Grit sequence in the Stocks House Borehole [4680 0212] was recorded by Fox-Strangways (1907) as only 3.7 m thick, with its base at 143.9 m. Mitchell and Stubblefield (1948, p.5), however, pointed out that some of the 25 m of beds above 140.2 m, assigned to the Lower Coal Measures, may belong to the Millstone Grit (Figure 4 col.13).   BCW

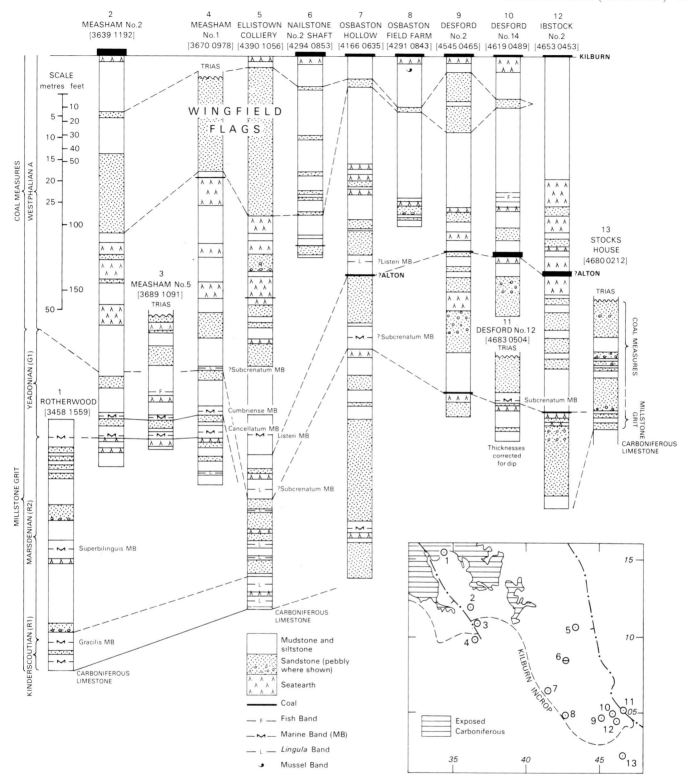

Figure 4 Comparative sections in the Millstone Grit and Lower Coal Measures below the Kilburn in the Leicestershire Coalfield

CHAPTER 5

Carboniferous: Westphalian

The Coalville district includes parts of three coalfields. To the south-west is the north-eastern part of the Warwickshire Coalfield; in the north is the southern half of the South Derbyshire Coalfield, which westwards from Overseal [295 155] is concealed beneath Triassic rocks; and in the centre, from Coalville to Desford, lies the southern part of the Leicestershire Coalfield, again largely concealed beneath Triassic beds. Detailed comparative sections of the Coal Measures in the three coalfields are shown in Figures 5–13. The records of the older shafts and boreholes commonly lack details of seatearths and fossil marker bands, but their absence from the sections does not necessarily mean that they are not present in the sequence.

The Coal Measures consist mainly of grey mudstone, siltstone, sandstone and numerous seams of coal, each resting on a fossil soil or 'seatearth'. The beds occur in cyclic sequences, coal being generally overlain by mudstone, commonly marine, and then by sandstone; in turn this is overlain by siltstone or silty mudstone, on which is developed the seatearth of the next coal. Most of the coal seams are 1 to 2 m thick, though the Main Coal of South Derbyshire maintains a thickness of 3.7 m.

The lowest beds of the Coal Measures are continuous across the Ashby Anticline between the South Derbyshire and Leicestershire coalfields, and many seams correlate from one coalfield to the other. Near Packington [37 14], Leicestershire names have been used for coals in opencast workings on the crest of the Ashby Anticline, so that in terms of local usage the boundary between the coalfields lies on the west flank of the Anticline. In neither field does the detailed succession resemble that of the Warwickshire Coalfield, which is separated from them by tracts of Cambrian rocks beneath the Triassic cover.

The Subcrenatum, Vanderbeckei and Aegiranum marine bands (Ramsbottom and others, 1978) have been identified in Warwickshire and South Derbyshire and the first two are also found in Leicestershire. They allow the three sequences to be broadly correlated, and subdivided since they mark, respectively, the bases of Westphalian A, B and C. Westphalian A corresponds with the Lower Coal Measures but the Middle Coal Measures covers not only Westphalian B but also the lower part of Westphalian C. The base of the Upper Coal Measures is taken at the Cambriense Marine Band in South Derbyshire and at the base of the Halesowen Formation in Warwickshire.

The highest beds in the Leicestershire Coalfield are of late Westphalian B age. The Whitwick Dolerite, a late Carboniferous or Permian sill which intrudes them, has no counterpart in the other coalfields. In the exposed part of the South Derbyshire Coalfield, the Cambriense Marine Band was proved in the Hanginghill Farm Borehole, just north of the district (Figure 10). In the concealed western part of this coalfield, grey measures which possibly lie above Cambriense are succeeded by red mudstones assigned to the Etruria Marl; these are overlain in turn by purplish grey and greenish grey sandstones and mudstones (Halesowen Formation) and by red mudstones (Keele Formation). The age-relationship of these divisions to the lithostratigraphy of the type-area in Staffordshire is uncertain. In the Warwickshire Coalfield, little Etruria Marl is present within the present district, though it occurs a short distance to the west, where the red measures pass laterally northwards and eastwards into grey measures indivisible from Middle Coal Measures.

At the time of their deposition, the Coal Measures of the present district were continuous with those of Nottinghamshire, Derbyshire and Yorkshire as part of the Pennine Province of deposition (Calver, 1969). In the south during Westphalian A to C times, a barrier to deposition was provided by the Wales-Brabant High with its northern limit lying somewhere between Warwick and Banbury. At about the time of the Cambriense Marine Band incursion, a terrestrial facies extended northwards from this area as the red beds of the Etruria Marl in the Warwickshire Coalfield. Later, in Westphalian D times, parts of the Wales-Brabant Island subsided, and grey measures of Halesowen facies extended southwards beyond Oxford. Throughout the Westphalian A to C succession, all divisions of the local Coal Measures thin from north to south; in addition, the faunas of the marine bands become increasingly near-shore, benthic, southwards, and the Pottery Clays facies develops in this direction in the South Derbyshire Coalfield. Tuffs produced by volcanism in Westphalian A times, to the east of the present district, probably led to formation of the tonstein (a 2 to 3 cm layer of kaolinised mudstone) found in the upper part of the Middle Lount seam of Leicestershire, while some, if not all, of the tonsteins in the Pottery Clays coal seams of South Derbyshire probably result from Westphalian C volcanism occurring to the south of the Staffordshire coalfields (Francis, 1969).

WARWICKSHIRE COALFIELD

The succession in the Warwickshire Coalfield is about 500 m thick, and was described by Barrow and others (1919), Eastwood and others (1923, 1925), Mitchell (1942), and Cope and Jones (1970). The seam nomenclature used here is mainly that of the last authors, who demonstrated that the seams between the Smithy and Thin Rider in the present district merge southwards into the Thick Coal of the Coventry area.

The lowest named seam is the Stanhope, beneath which the beds are predominantly argillaceous, lacking the Wingfield Flags sequence of the other two coalfields. The base of the Coal Measures is not accurately determined in Warwickshire since the Subcrenatum Marine Band is absent and its supposed horizon is represented only by a *Lingula* band; nor has the Listeri Marine Band been recorded. Workable coals occur between, and include, the Stumpy and

Four Feet seams. The Vanderbeckei (Seven Feet) and Aegiranum (Nuneaton) marine bands are widely developed but none higher in the succession have so far been recorded. Above the Aegiranum Marine Band, the Middle Coal Measures contain an increasing proportion of sandstone, marking the onset of the conditions which prevailed during deposition of the Upper Coal Measures.

The Halesowen Formation consists of 60 to 120 m of grey and greenish grey, micaceous sandstone, mudstone and seatearth with thin coals, and layers and nodules of *Spirorbis* limestone. Sandstones predominate in the lower part of the sequence and mudstones above. The formation has a broad outcrop in the northern part of the coalfield, the sandstones at crop forming ridges that curve around the coalfield syncline. The succeeding mudstones form low-lying ground beneath the sandstone features of the overlying Keele Formation. The Halesowen Formation also occurs near Warton on the downthrow side of the Polesworth Fault. Within the district the Halesowen Formation is conformable upon the Middle Coal Measures. Farther west, however, along the western margin of the Warwickshire Coalfield between Kingsbury and Dosthill, the base of the Halesowen Formation is unconformable and cuts down below the level of the Bench seam. Similarly in the Bolehall Borehole [2217 0402], all the seams above the Seven Feet are missing beneath the Halesowen Formation (Taylor and Rushton, 1971, p.133).

The Keele Formation is only poorly exposed in the district. It comprises red-brown sandstones and mudstones with thin *Spirorbis* limestones (Fox-Strangways 1900, p.28). Its base is drawn at the base of a persistent red sandstone overlying grey Halesowen mudstones.

Details

LOWER COAL MEASURES (Westphalian A)

The Subcrenatum Marine Band has not been definitely identified in this area (Figure 5), but may be represented by 0.28 m of black mudstone with *Lingula* found at 30.9 m in the Merevale No.2 Borehole [3001 9509], and by 1.5 m of grey mudstone with *Lingula* encountered 2.7 m below the supposed Listeri Marine Band in the Statfold Borehole [2347 0699] in the Lichfield district (Taylor and Rushton, 1971, pp.85 and 137). When the Baddesley No.3 Shaft [2790 9709] was deepened in 1953–55, Cambrian shales were encountered 55 m below the Stanhope with no intervening Coal Measures marine bands, though the bands may have been faulted out by the numerous fractures recorded in the shaft beneath the Stanhope seam.

The measures in this part of the sequence are predominantly mudstones, seatearths and a few thin coals. The Stanhope is too thin to have been worked. It is known mainly from boreholes and shafts, and its outcrop can be traced in only a few places. It ranges in thickness from 0.30 m to 0.7 m. The Stumpy is a persistent, but usually unworked seam about 0.5 m thick. It was, however, worked opencast at Holly Park [288 970] where it was 0.9 m thick. A mussel band in the roof of the Stumpy at Pooley Hall Colliery contains *Carbonicola* cf. *pseudorobusta*, *C.* cf. *rhomboidalis* and *Naiadites flexuosus?*, (Mitchell 1942, p.20) suggesting correlation with the Pseudorobusta Belt of the Communis Chronozone.

The Bench group of coals is variable in the number and thickness of the constituent coals and the intervening partings. Usually two thick seams, the Top Bench and the Bench, are recognised with, in places, several thinner coals—the Bench Thin Seams—in between. The Top Bench and Bench have been worked underground and

opencast. According to the National Coal Board (1957), 'to the south of Birch Coppice Colliery and around Baddesley Colliery . . . the Top Bench Seam lies near the Bench Thin and Bench Seams and the group of seams can be regarded as a composite thick seam'. North from Birch Coppice No.4 Shaft, the Top Bench and Bench seams are separated by as much as 16 m of measures including the

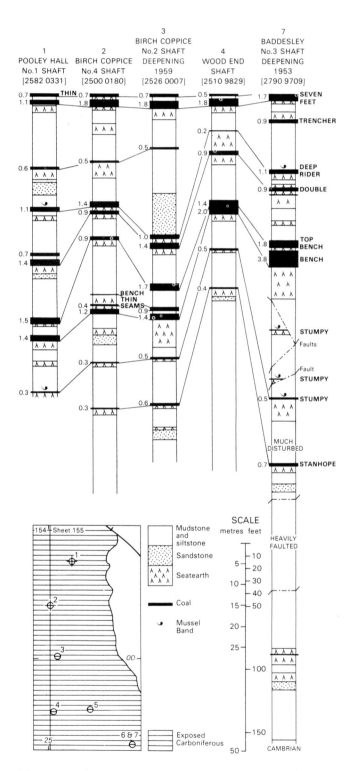

Figure 5 Comparative sections in the Lower Coal Measures of the Warwickshire Coalfield

Bench Thin Seams. An exceptional thickness for the sequence was proved in boreholes at the proposed Orchard St Leonards opencast site [267 001] to read: Top Bench 3.3 m, parting 0.4 m, Bench 3.2 m. A more typical succession was recorded in Birch Coppice No.2 Shaft: Top Bench 1.8 m, measures 3.7 m, Bench Thin Seams (coal 0.3 m, parting 0.03 m, coal 0.2 m, parting 0.08 m, coal 0.5 m), parting 0.4 m, Bench 1.4 m (Figure 5).

The interval between the Bench to the Double is predominantly of mudstone, but thin sandstones are also present, and at the Holly Park opencast site, virtually the whole of the interval was taken up by a 6 m sandstone. The Double has been mined underground and opencast. Its tendency to split is illustrated in Pooley Hall No.1 Shaft (Figure 5), but usually the parting is less than 1 m thick and at the Holly Park site, the Double was a single seam of 1.3 m.

Where the Double and the Deep Rider are close, the intervening beds are predominantly of seatearth, as at Birch Coppice No.4 Shaft [2500 0180], for example, where they are 1.1 m apart. Other lithologies appear where the interval is greater, including thin sandstones proved by exploratory drilling at Orchard St Leonards. A maximum of 7.6 m is reached in Pooley Hall No.1 Shaft where there is a 0.2 m coal about 1 m below the Deep Rider.

The Deep Rider has been worked underground and opencast, and reaches a maximum thickness of 1.6 m at the Orchard St Leonards site. A 0.1 m seam, 3.7 m above the Double Seam in the Birch Coppice No.3 Shaft, is probably the Deep Rider much attenuated. The roof of the Deep Rider commonly contains a mussel band yielding *Anthracosphaerium sp.*, *Carbonicola crista-galli*, *C. os-lancis C. rhomboidalis*, *C. venusta* and the ostracod *Geisina arcuata* indicative of the Crista-galli Belt at the base of the Modiolaris Chronozone.

The Yard seam is thin, and does not persist south of Birch Coppice No.2 Shaft. It is only about 0.5 m thick and has not been worked. It is overlain by a mussel band, which includes *Anthracosia* cf. *lateralis*, *A. regularis* and *Geisina arcuata*, a fauna characteristic of the Regularis Belt, in the early Modiolaris Chronozone. The Trencher is a thin but persistent seam. It fails only in the Birch Coppice No. 4 Shaft where its position is possibly indicated by a 3.1 m seatearth, 4.6 m below the Seven Feet Seam. The Trencher is in places split into two or three; for example at Baddesley Colliery No.3 Shaft, the section is: coal 0.3 m, seatearth 1.7 m, coal 0.4 m, seatearth 0.2 m, coal 0.5 m. At Pooley Hall No. 1 Shaft, the Trencher is a single 0.2 m seam.

The Seven Feet is one of the most economically important seams of the Warwickshire Coalfield and has been worked extensively underground and opencast. North of Pooley Hall Shaft, the Seven Feet and the overlying Thin seam are only about 0.15 m apart and have been worked as one coal. The Seven Feet is typically 1.8 m to 2.1 m thick but reaches 2.5 m at the Holly Park opencast site. A washout in the coal extends east-south-east from Gawby Knob Shafts [2605 0130] (Mitchell, 1942, p.35).

The interval between the Seven Feet and Thin coals consists mainly of seatearth and is very variable in thickness even over short distances. It is generally thicker in the east, reaching a maximum of 6 m at the Abbey opencast site [265 022]. Generally the Thin seam is 0.3 m to 0.8 m thick and has not been worked separately.

MIDDLE COAL MEASURES (Westphalian A and B)

The Vanderbeckei (formerly Seven Feet) Marine Band lies in the roof of the Thin seam (Figure 6) and consists of dark grey mudstone with ironstone nodules. Its full thickness can be established only where it is overlain by mudstones containing non-marine bivalves; elsewhere only its minimum thickness is known. The marine fauna is variable; *Lingula* and arenaceous foraminifera are widespread and locally accessory faunas including benthic bivalves and brachiopods indicate proximity to a shoreline against the Wales-Brabant High. The following fauna is recorded from the Marine Band: *Ammonema sp.*, *Ammovertella inversa*, *Rectocornuspira sp.*, *Spirorbis sp.*, *Lingula mytiloides*, *Orbiculoidea* cf. *nitida*, *Productus carbonarius*, *P.* (*'Pustula'*) *piscariae*, *Spirifer pennystonensis*, *Dunbarella* cf. *papyracea*, *Myalina compressa*, *Anthraconeilo* cf. *laevirostris*, *Schizodus spp.*, *Bucaniopsis sp.*, *Euphemites anthracinus*, *Conularia spp.*, *Geisina spp.*, *Hollinella* cf. *claycrossensis*, *Paraparchites spp.*, *Coelocanthus sp.*, *Listracanthus sp.*, *Megalichthys sp.*, *Pleuroplax affinis*, *Rhabdoderma sp.* and *Rhizodopsis sauroides.*, (Mitchell, 1942, p.21 and later BGS material). The marine faunas are frequently intercalated with non-marine beds, containing the lamellibranch *Anthracosia spp.*, characteristic of this horizon in several British coalfields. The 4 m mudstone bearing *Lingula* recorded at the Orchard St Leonards site is the thickest local occurrence of the Marine Band; normally it is much thinner. At Pooley Hall No.1 Shaft, about 2 m of marine mudstone are overlain by mudstone with non-marine bivalves.

The overlying coals have generally been taken informally to form the Thick Coal group. Formerly restricted to the seams between the Nine Feet and the Thin Rider (Mitchell, 1942, p.7), the term now includes all the seams between the Smithy and the Thin Rider (Cope and Jones, 1970). The seams in the group tend to come together southwards until, in the Coventry district, they combine to form the Thick Coal. The correlation of individual seams within the group is commonly difficult, especially above the Nine Feet.

The Smithy is a variable seam which has been worked to a limited extent at Birch Coppice Colliery and at opencast sites between Dordon and Baddesley. Its maximum thickness is at the Abbey opencast site, as follows: coal 0.8 m, parting 0.1 m, coal 0.6 m. Typically the seam is about 0.8 m thick. At Birch Coppice Colliery (Wood End Pit), the Smithy joins the High Main to give: coal 1.5 m, parting 0.3 m, coal 0.1 m, parting 0.1 m, coal 0.3 m, parting 0.2 m, coal 0.7 m. A north-east-trending washout was found in the Smithy at the Baddesley West Extension South opencast site [269 992]. A mussel band containing *Anthracosia ovum* and its variants, together with *Naiadites sp.*, overlies the Smithy around Polesworth and correlates with the Ovum Belt of the Modiolaris Chronozone.

The High Main has been worked at Birch Coppice Colliery, but elsewhere is too thin for deep mining, being only 0.6 to 0.8 m thick. Some opencast working has taken place between Dordon and Baddesley where the seam is thicker, reaching an exceptional 1.1 m at the Holly Park site. The washout in the Smithy mentioned above also affects the High Main. A sandstone up to 6 m thick is present locally in the roof of the High Main and a mussel band containing *Anthracosia lateralis*, *A. ovum* (and variants) and *Naiadites quadratus* lies above the seam at Birch Coppice Colliery indicating an Ovum Belt correlation.

The Nine Feet (formerly Slate) is not worked underground in the district, but some opencast working has occurred between Dordon and Baddesley, and at Shuttington. The seam is generally 1.1 to 1.4 m thick but is very variable, locally reaching a maximum of 2 m at Holly Park opencast site. It is absent at Baddesley Speedwell Shaft [2663 9841]. A thin parting is present in a few places. A mussel band in the roof of the Nine Feet at Pooley Hall Colliery yielded *Anthracosia lateralis*, *A.* cf. *ovum*, *A. beaniana* and *Naiadites sp.* suggesting an horizon close to the Ovum/Phrygiana Belt boundary, but it has not been recorded elsewhere. A 2 m sandstone overlies the Nine Feet at Baddesley Colliery, and a 1.5 m sandstone was encountered 4.6 m above the Nine Feet at the Alvecote opencast site [252 052].

The Threequarter seam was treated as an upper leaf of the Nine Feet by Mitchell (1942, p.8) and by Cope and Jones (1970, p.588), but it is a distinct seam, well separated from the Nine Feet at all the local opencast sites. It is generally 0.3 m to 0.5 m thick. Small tonnages were won at the Baddesley West [266 996] and Baddesley West Extension South [269 990] opencast sites, where it was called the 'Upper Slate'. The seam is absent in the Birch Coppice No.4 and Speedwell shafts.

A mussel band containing *Anthracosia* cf. *aquilina* group and *Naiadites* cf. *triangularis* group occurs at the base of the seatearth of

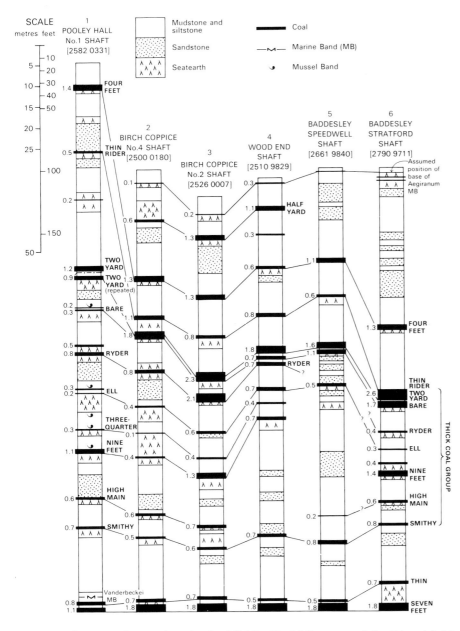

Figure 6   Comparative sections in the Middle Coal Measures of the Warwickshire Coalfield

the Ell seam at Pooley Hall, (Mitchell, 1942, p.20) and a similar band lies within the seatearth 1.5 m below the Ell in a cross-measures drift at Baddesley Colliery [2810 9588]. The seatearth of the Ell is unusually thick here, reaching 5 m.

The Ell seam is too thin to have been mined underground, but it was worked opencast at the Baddesley West, Baddesley West Extension South and Alvecote [251 053] sites. At Alvecote it is 0.7 m thick and near Baddesley, the Ell is 0.9 m thick with a 0.15 m parting 0.3 m above the base. The split is a common feature of the Ell, and reaches 0.5 m in the Pooley Hall No.1 Shaft. A mussel band commonly found in the roof of the Ell yields *Anthracosia* cf. *aquilina*, *A. lateralis, A. phrygiana* and *A. retrotracta* indicating a Phrygiana Belt correlation, late Modiolaris Chronozone.

The Rider, Bare, Two Yard and Thin Rider cannot everywhere be easily distinguished since they vary rapidly in thickness and the upper three seams join and split in a complex way. It is possible that

the Bare identified in opencast workings from Baddesley southwards, may in fact be the Ryder, and the overlying Two Yard may be the Bare and Two Yard joined together as in nearby shafts. All these seams tend to come together southwards (Figure 6). The Ryder has not been worked, although it reached an exceptional 2.1 m in Birch Coppice No.2 Shaft. It is normally much thinner (Cope and Jones, 1970, p.588).

The Bare seam is clearly identified throughout the coalfield except in the opencast sites mentioned above. In Birch Coppice No.4 Shaft and southwards, it is separated from the Two Yard by only a thin parting. It varies from 0.7 m to 1.7 m. When traced southwards the Bare merges with the Two Yard. At Baddesley Stratford Shaft, the interval is only 0.2 m and at Birch Coppice No.3 Shaft the sequence is: Two Yard 1.8 m, parting 0.1 m, Bare 0.7 m, parting 0.2 m, Ryder 0.7 m (Barrow and others, 1919, p.255).

The Two Yard has been worked underground and opencast. It is thinnest in the north; in opencast workings at Alvecote, for example, it is 0.9 m thick. Further south at Baddesley West opencast site it is 1.8 m thick. At Baddesley Stratford it reaches 2.6 m although this is the combined thickness with the overlying Thin Rider. A washout in the Two Yard south of Birch Coppice No.2 Shaft is part of a much larger NNW-trending washout proved in workings to the west of the district. A mussel band occurs about 7 m below the Two Yard at Baddesley Colliery and contains *Anthracosia caledonica*, *A.* cf. *beaniana*, *A. nitida*, *A.* cf. *phrygiana* and *A.* ex. gr. *regularis* suggesting a Caledonica Belt correlation, early Lower Similis-Pulchra Chronozone. Overlying the Two Yard at Baddesley is a higher mussel band with *Anthracosia acutella*, *A.* cf. *aquilina*, *A. atra*, *A. fulva* and *A.* cf. *nitida*. This appears to be the earliest Atra Belt horizon in the district.

The Thin Rider has been worked underground at Birch Coppice Colliery and opencast at Baddesley and Dordon. It is generally 0.9 m to 1.2 m thick, but was only 0.5 m at Alvecote North opencast site. A mussel band characterised by *Anthracosia atra* occurs in the roof of the Thin Rider (Mitchell, 1942, p.9 and table 1), but was recorded in few of the older shaft sections.

The Four Feet is the highest worked seam in the coalfield and has been mined underground and opencast at Baddesley and Dordon. It attains a maximum thickness of 1.5 m at Baddesley West opencast site. A split developes as the seam is traced north. It is first seen at Birch Coppice No.4 Shaft, and north of Polesworth has thickened sufficiently for two seams, the Upper and Lower Four Feet, to be recognised. The combined thickness of the Upper and Lower Four Feet is always less than that of the unsplit Four Feet. At Alvecote North opencast site, they are separated by 3 m of measures. At Birch Coppice Colliery a mussel band occurs in the split and contains *Anthracosia aquilinoides*, *A.* ex gr. *atra*, *A.* cf. *nitida* and *Naiadites sp.*. The Four Feet is overlain by a higher mussel band within the Atra Belt containing *Anthraconaia* cf *pumila*, *Anthracosia acutella* and *Naiadites sp.*.

The Half Yard is the highest named seam in the coalfield but is not worked. It is 1.3 m thick at Birch Coppice No.2 Shaft, but is generally less than 0.5 m thick. It is absent at Baddesley Colliery, where it may be washed out by a thick sandstone at about this horizon. The measures above the Half Yard are predominantly mudstones with a few thin coals.

The Aegiranum (formerly Nuneaton) Marine Band crops out at the Albion Clay Pit, Dordon [2639 0024], as 1.5 m of dark and pale grey mudstone underlain by a thin coal. The fauna includes *Hyalostelia?*, *Orbiculoida* cf. *nitida*, Chonetoid debris, *Productus carbonarius*, *P.* ('*Pustula*') *rimberti*, *Aviculopecten sp.* *Gastrioceras depressum* and *Hindeodella sp.* Mitchell (1942, table 2). The measures between the Marine Band and the basal Halesowen sandstone consist of grey and mottled seatearths and mudstones with brown, ferruginous and greenish sandstones and thin coals, and include beds formerly mapped as part of the Halesowen Formation. The sandstones are gritty and ill sorted in their lower parts with mudstone pellets and white clay fragments. Several shafts penetrate this sequence but these were sunk before marine bands were generally recorded and none of the logs records the Aegiranum Marine Band. A section in Ensor's Clay Pit, Polesworth [2570 0210], in the lower part of the sequence, is:

	Thickness m
Sandstone, grey, yellow weathering	3.1
Conglomerate	1.0
Sandstone, grey	0.8
Sandy shale	about 1.2
Seatearth	about 2.3
Probably seatearth	about 1.5
Clay, pale grey	1.5

Siltstone, pale grey	0.6
Shale, sandy	0.8
Sandstone	0.6
Clay, pale grey, red-streaked and with sphaerosiderite nodules	1.5
Clay, pale grey with sphaerosiderite	0.3
Siltstone, pale grey	0.8
Seatearth, dark grey, with paler grey bands	2.3
Coal—reported	about 0.5
Seatearth, grey and pale grey (much obscured)	4.9
Coal—thin	
Seatearth, grey silty mudstone with numerous ironstone nodules	0.1 seen

Thick sandstones occur immediately above the pit, and about 21 m of pale grey clay with patchy red staining is seen on the steep hillside above [255 022].

The Birch Coppice No.4 Shaft passes through 70 m of beds between the base of the Halesowen Formation and a 0.13 m coal at 133.5 m, probably the horizon of the Aegiranum Marine Band. These measures consist of grey mudstones and seatearths with sandstones and thin coals. Sandstones between 91 m and 114 m in the shaft are probably equivalent to those above the Ensor Clay Pit, and the seatearths and thin coals from 114 m to 134 m probably correlate with the section in the pit.

Thick sandstones equivalent to those above the Ensor Clay Pit were encountered on a building site on Fairfields Hill south of Polesworth [2598 0176]. Variegated, red, purple and yellow clays occurring in trenches above the sandstones are probably a local development of the 'Etruria Marl' lithofacies. The section [2639 0024] at Albion Clay Pit, Dordon, above the Aegiranum, Marine Band, is:

	Thickness m
Sandstone fine-grained	3.1
Seatearth, pale grey with dark bands, faint purplish grey stained, including crimson mottling towards top	about 3.4
Coal	0.5
Seatearth, dark grey	about 0.2
Seatearth, pale grey, with faint purplish grey mottling	about 1.4
Seatearth, dark grey	about 0.3
Gap—probably all grey seatearth with cuboidal weathering—contains sphaerosiderite band	about 4.0
Seatearth, pale grey and dark grey	about 1.1
Coal	0.2
Seatearth, pale grey and greenish grey	1.4
Coal	0.7
Seatearth, dark grey	0.6
Seatearth, pale grey, with cuboidal weathering, brown iron-staining and sphaerosiderite nodules	2.3
Seatearth, medium grey	0.6
Gap—some seatearth and yellow sandstone	6.0 to 9.0
Mudstone with iron stone nodules	0.2

'Coloured' beds recorded at this locality by Barrow and others (1919, p.41) may lie below the Aegiranum Marine Band, as the Etruria Marl lithofacies is known to extend below the Marine Band on the western side of the coalfield.

The sandstone at the top of the Dordon Clay Pit has a gritty 'espley' band at the base, consisting of white clay, greenish mudstone pellets and ill sorted, coarse quartz grains set in a dark brown, ferruginous matrix. It is succeeded by 3 m of greenish grey, false-bedded sandstone of 'Halesowen' type and then by alternations of sandstone, grey mudstone and seatearth. The higher beds of sandstone are coarse and ferruginous, especially at the base. They were partly exposed during road-widening operations at Dordon crossroads [2619 0008], where the section read:

	Thickness m
Sandstone, coarse, brown	0.6
Seatearth, pale grey, sandy, with large ironstone nodules	1.2
Mudstone, smooth, grey, becoming sandy downwards	2.1
Sandstone, hard, brown, with ironstone pebbles, highly ferruginous	about 1.2
Sandy siltstone, grey, with coal streaks	
Coal, dipping at 12°	0.08
Silty mudstone, grey	
Coal—inferior	0.08

In Birch Coppice No.2 Shaft (Fox-Strangways, 1900, pp.55–58), the base of the Halesowen Formation is at 66.5 m, and the supposed horizon of the Aegiranum Marine Band at 152.4 m. Red mudstones in beds up to 4.6 m thick occur between 66.5 m and 94.8 m, but are subordinate to grey measures.

In the Wood End Shaft (Barrow and others, 1919, pp.252–256), the 68 m of strata between the base of the Halesowens, and the supposed horizon of the Aegiranum Marine Band at 229.5 m, are all recorded as Etruria Marl, except the basal 2 m, and comprise mainly red and grey mudstones with a few 'espley' sandstones. This is the only thick sequence of Etruria Marl in the coalfield of the present district and probably indicates a rapid facies change westwards from grey to red measures. A sub-Halesowen unconformity could also have caused the observed lithological change, but such a break has not been proved in this district.

UPPER COAL MEASURES (Westphalian C and D)

At Wilnecote, just west of the Coalville district, the basal sandstone of the Halesowen Formation rests clearly upon the underlying Etruria Marl. As the beds are traced northwards and eastwards around the nose of the coalfield syncline, the Etruria Marl facies passes laterally into grey measures of Halesowen facies, with grey mudstones and lenticular sandstones, and the boundary between the two formations is less clear.

The base of the Halesowen Formation locally is taken at the base of the thick sandstone overlying the red measures at 161.7 m in Wood End Shaft [2510 9829] (Barrow and others, 1919, p.254). This sandstone is probably the '100 ft sandstone' farther south (Eastwood and others, 1923, p.72). Part of this basal sandstone crops out in the disused railway cutting near Hermitage Lane, Birchmoor [2531 0195]:

	Thickness m
Sandstone, greenish grey, fine-grained, argillaceous, with numerous marl pellets; top not exposed	about 1
Mudstone, pale green, soft, with silty micaceous bands and thin bands of siltstone and sandstone; seams of limonite 3–5 cm thick	4.3
Sandstone, greenish grey, fine-grained, slightly coarser towards base, alternately massive and current-bedded; rare thin mudstone pellet bands; base abrupt	6.7

The upper part of the Halesowen Formation, west of the Polesworth Fault, consists of greenish grey mudstones with *Spirorbis* limestones and subordinate sandstones. The mudstones form a broad outcrop with clay soils below the basal Keele sandstone. Exposures are confined to shallow diggings for *Spirorbis* limestone.

Two limestones, neither more than 3 m thick, can be traced west of Baddesley village. They are the Index Limestone below and the Flanders Hall Limestone above (Barrow and others, 1919, pp.50–53; Eastwood and others, 1923, pp.73–76). The Index Limestone lies about 85 m above the base of the Halesowen Formation and its outcrop can be traced as a feature and through sporadic old workings between Baxterley [282 970] and Baddesley Ensor [259 991], where it is marked by scattered fragments of limestone. The Flanders Hall Limestone occurs just below the basal Keele sandstone; a line of pits follows its outcrop for about 1 km through Lower Ridding [267 980] and Big Rough Woods and Ash Spinney [260 985]. Its outcrop is marked by a slight feature and, along the northern edge of the spinney by nodules of dark grey, compact limestone in grey marl. The nodules are not abundant, but the alignment of the nearby pits suggests that they form a definite band. The slight feature continues for some distance west of Ash Spinney but no trace of limestone is seen.

Beds from the upper part of the Halesowen sequence are also exposed at Bramcote Hall [2720 0428] where fine-grained, greenish grey, micaceous sandstone and green, silty, micaceous mudstone dip at 30° NE towards the Warton Fault. A *Spirorbis* limestone, possibly the Flanders Hall Limestone, crops out in a stream [2705 0391] 60 m north-west of Bramcote Pumping Station. All these beds were formerly included in the Keele Formation (Barrow and others, 1919, p.63).

The junction with the Keele Formation used to be exposed in a tramway at Biddle's Wood [2510 9847] where Barrow and others (1919, pp.52–53) recorded a gradual upward transition from grey to red marl. The overlying beds at the base of the Keele Formation were described as an 'alternation of sandstones and marls; the sandstones . . . were all white, but the edges of several have become red by oxidation of the iron and somewhat friable owing to the loss of lime. The marls vary in colour from pale-grey to greenish grey and red. In the lower beds paler colours predominate; upwards red becomes more common and at the top of the cutting all the marls are red'.

Beds near the base of the Keele Formation are also exposed in a quarry [2729 0388] 410 m south of Bramcote Hall. The section reads:

	Thickness m
Sandstone, fine-grained, pale greenish grey and purple mottled	1.5
Conglomerate ('pellet sandstone'), comprising red mudstone pellets less than 1.5 cm diameter	0.8
Sandstone, massive, fine-grained, greenish grey, with purple staining	2.4 seen

The Baxterley *Spirorbis* limestone described by Eastwood and others (1923, p.88) cannot now be located at crop in the type-area [267 969].   RAO, KT.

SOUTH DERBYSHIRE COALFIELD

The sequence between the base of the Coal Measures and the Kilburn seam appears to be thinner in the southern part of the South Derbyshire coalfield than farther north. The succession in the northern boreholes (Figure 7) at Stanhope Bretby and Cadley Hills is substantially thicker than in those within the present district. At Stanhope Bretby (Greig and Mitchell, 1955, pp.54–55), the sequence above the Subcrenatum Marine Band strongly resembles that of the North Derbyshire coalfield, with correlatives of the Soft Beds Flags and the Belperlawn, Holbrook, Second Smalley and First

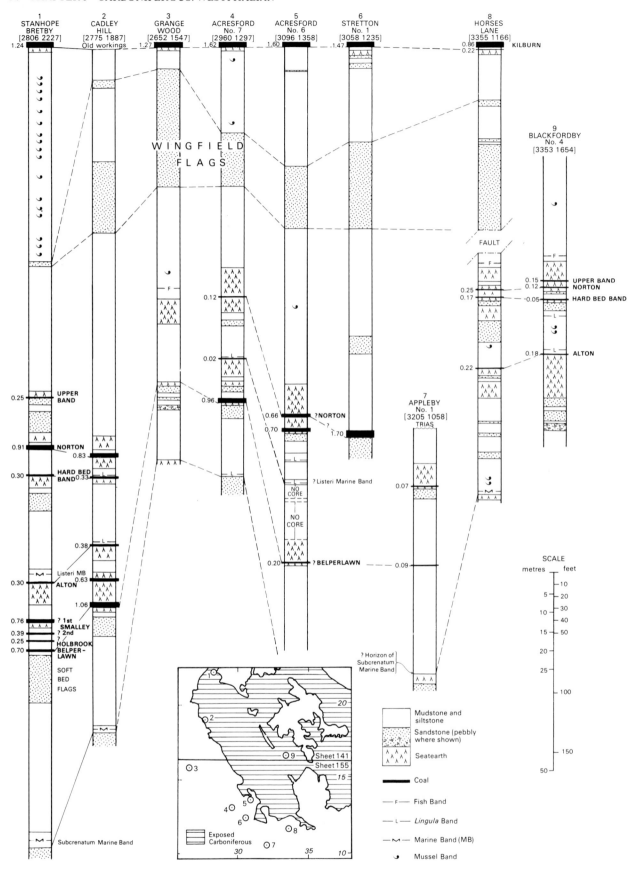

Figure 7   Comparative sections below the Kilburn in the Lower Coal Measures of the South Derbyshire Coalfield

Smalley coals. The thin sandstone just above the Listeri Marine Band may correlate with the Loxley Edge Rock of the Sheffield district.

Within the present district, the sequence above Subcrenatum generally includes a thin coal, probably the Belperlawn, a *Lingula* band perhaps representing the Listeri Marine Band and several thin coals which may correlate with the Hard Bed Band, Norton and Upper Band. The fine-grained sandstones of the Wingfield Flags are generally present in the uppermost 50 m. A hard, ganisteroid, white, fine-grained sandstone with rootlets, lying between the Listeri Marine Band and the Hard Bed Band, crops out

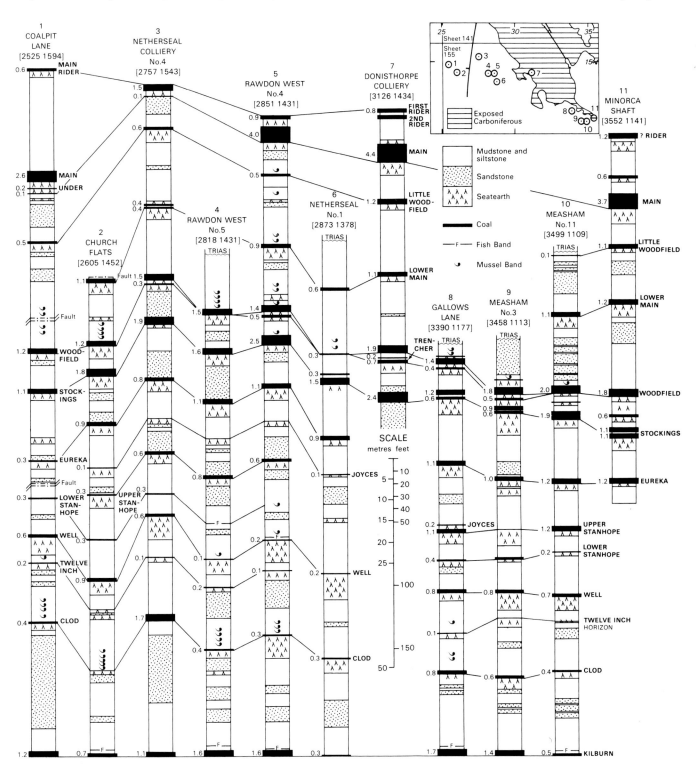

Figure 8. Comparative sections between the Kilburn and Main in the Lower Coal Measures in the South Derbyshire Coalfield

around the Namurian inlier at Valley Farm [3445 1545]. Ganisters have been worked at about this horizon near Sheffield (Eden and others, 1957) but have not been recorded previously from the South Derbyshire coalfield.

Above the Kilburn the Lower Coal Measures include many worked seams, notably the Eureka, Stockings, Woodfield, Lower Main and Main (Figure 8). A tonstein band, recorded from the upper part of the Woodfield seam at some localities corresponds to that in the Middle Lount of Leicestershire. Both these seams are overlain by a mussel band rich in *Carbonicola crista-galli* (Mitchell and Stubblefield, 1948, table I), confirming their correlation within the Cristagalli Belt, earliest Modiolaris Chronozone.

Fossils indicating the Vanderbeckei Marine Band (formerly the Molyneux Marine Band in South Derbyshire) have been found at only a few localities in the coalfield, probably because most of the shafts and boreholes penetrating this horizon date from the nineteenth century, when systematic collections of fossils were rare. The marine band lies in the roof of the Main, or in the roof of the highest Main Rider Coal where this is present. Details of the strata between the Main and Upper Kilburn, the highest coal thick enough to be worked underground, are shown in Figure 9.

Above the Upper Kilburn, the Pottery Clays include numerous thin coals and thick seatearths, many of which are high quality refractory fireclays. The Aegiranum (Mansfield) Marine Band was recognised in these strata by Mitchell and Stubblefield (1948) in Robinson and Dowler's clay pit at Overseal. They gave it the local name of the Overseal Marine Band. They also found two lower *Lingula* bands, which Calver (1968) subsequently correlated with the Maltby (formerly Two Foot) and Haughton marine bands. Recent examination of the Acresford No.6 Borehole has revealed the presence of the Clowne Marine Band and hence the Sutton Marine Band appears to be the only late-Westphalian B marine horizon absent in this part of the South Derbyshire Coalfield. The further discovery, above the Aegiranum, of a number of higher marine bands, up to and including the Cambriense Marine Band, as well as of tonsteins (Worssam and others, 1971; Worssam, 1977), proved that these strata represent a condensed sequence. They correlate with a considerable thickness of Coal Measures of more normal facies in the North Derbyshire and Nottinghamshire coalfields. The thin coals are difficult to correlate individually, and the convention of numbering them as devised by Jago (*in* Worssam and others, 1971; Worssam, 1977), is adopted here for general use in the coalfield (Figure 10). The system, used by the Opencast Executive of the National Coal Board, is based on the sequence exposed in Ensor's No.1 Quarry, Woodville [3060 1765], just north of the present district, where the quarry owners number the coals downwards from 1 to 8. By adding the number 20 to each of these, to allow for the naming of possible higher seams not present at the quarry, and the prefix P (for Pottery), these became the P21 and P28 seams. Tonsteins (as in P27 in Ensor's No.1 Quarry) and marine bands are further aids to identification. The seatearth below P40 is the lowest worked in the pottery clay quarries. To aid the further correlation of borehole sections, the numbers P41 to P43 are given to three thin but persistent coals above the Upper Kilburn.

The positions of the tonsteins known in the district are shown in Figure 10. They have been proved in the P19, P26 and P27 seams, with a possible further occurence in P23 in the Hanginghill Farm Borehole (Worssam, 1977, p.5). The re-numbering of the coals in the Willesley Clay Pit section (Figure 10) makes the highest coal, which includes a tonstein, P19 as suggested by Worssam and others (1971), and not P20 (Worssam, 1977).

Revised numbering of the seams in the Hastings and Grey Shaft (Figure 10) suggests that a 50 mm 'ironstone' logged at 55.2 m, immediately overlying a coal 0.75 m thick, could be the P26 tonstein. The P27 tonstein, proved in the Willesley Clay Pit (Worssam, 1977) may correlate with the Sub-High Main Tonstein of the Sheffield-Nottingham area (Spears, 1970), alternatively named the Supra-Wyrley Yard tonstein (Williamson, 1970).

Scattered outcrops of coarse sandstone in the exposed part of the South Derbyshire Coalfield have for long been assigned to the 'Moira Grits'. Carboniferous plants within the sandstones establish their Coal Measure age, but the coarse-grained sandstones are generally absent from nearby colliery shaft sections. In the nineteenth century, the outcrops were interpreted as outliers of a formerly widespread, gently dipping sheet of sandstone of presumed Upper Coal Measures age, resting uncomformably on more steeply dipping 'Productive' Coal Measures. (Hull, 1860). On the other hand, Fox-Strangways (1907, pp.53–4) and Horwood (*in* Fox-Strangways, 1907, pp. 114–141) regarded the sandstones as part of the Middle Coal Measures, and attributed their absence from the Hastings and Grey Shaft section in particular, to thinning-out of individual coarse sandstones. Detailed mapping of the outcrops, however, has shown that the 'Moira Grits' are not a separate formation. Some of the coarse beds are channel-fill sandstones within the Middle Coal Measures, as for example, the sandstone outlier west of the Bramborough Farm Clay Pit (p.41). Sandstone of 'Moira Grits' type was recently exposed in Willesley Clay Pit (Figure 10) where about 6 m of coarse-grained sandsone rest directly on the P19 coal for a distance of 600 m in the north face [3223 1543 to 3270 1570]. Boreholes drilled to the west of the exposed South Derbyshire Coalfield have proved several coarse, gritty sandstones of limited lateral extent (Figure 9) in the Middle Coal Measures. The term 'Moira Grits' as a lithostratigraphical name is therefore discontinued.

Details

LOWER COAL MEASURES (Westphalian A)

At Valley Farm [3445 1545], the base of the Coal Measures is taken at the base of mudstones resting on a pebbly sandstone 50 m S of the farm. About 20 m higher in the succession, a coarse, yellow sandstone crops out [3450 1525], which may correlate with the pebbly sandstone 200 m west of Valley Farm. This was noted by Fox-Strangways (MS six-inch map) as 'ferruginous sandstone and grey clay' in a stream section [3400 1579], dipping 4° at 280°. A further 70 m west, the stream section shows carbonaceous shale dipping 5° at 260° which probably lies close to the Alton Coal (Figure 7). About 80 m farther west, a hard, greyish white quartzitic silty ganisteroid sandstone with rootlet casts forms a low feature, and at the foot of its dip-slope at Shell Brook [338 164], just north of the present district, Opencast Executive boreholes proved two thin coals identified by Spink (1965, p.46) as probably the Norton and

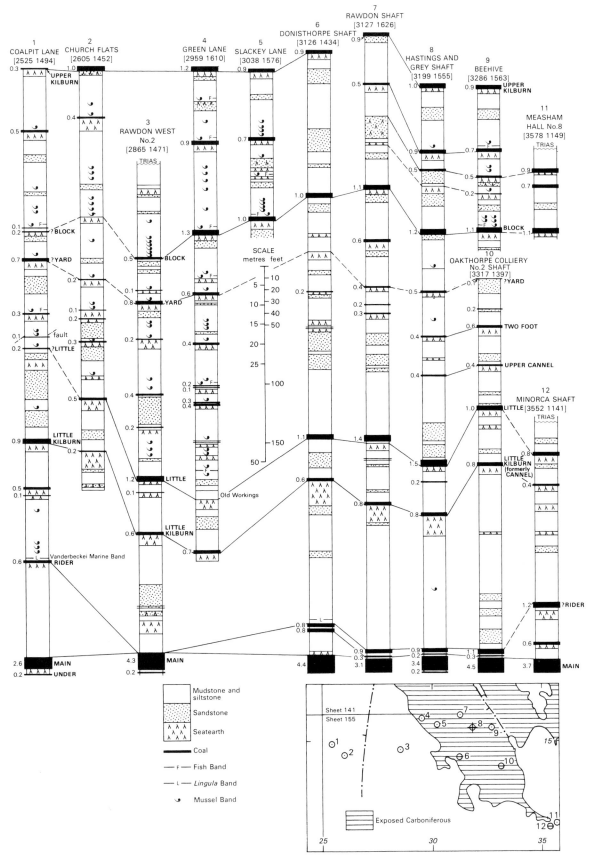

Figure 9 Comparative sections between the Main and Upper Kilburn in the Middle Coal Measures in the South Derbyshire Coalfield

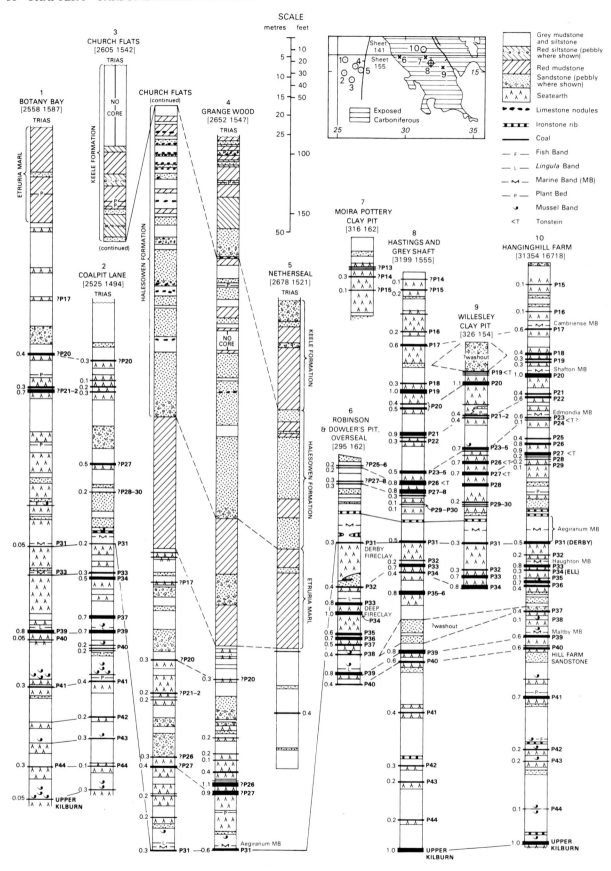

Figure 10 Comparative sections above the Upper Kilburn in the Middle and Upper Coal Measures in the South Derbyshire Coalfield

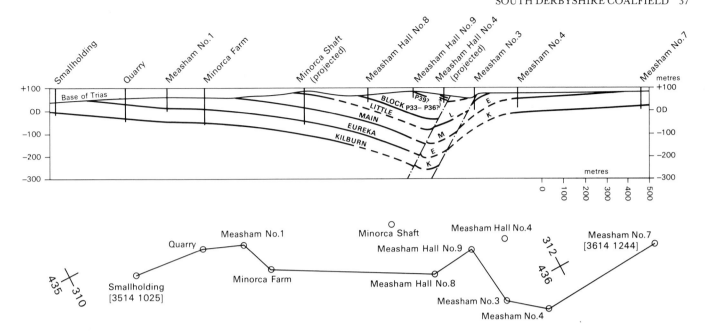

Figure i1 True-scale section across the Boothorpe Fault near Minorca Colliery

'Hard Band'. It is likely that the ganister is part of the seatearth of the Hard Bed Band and corresponds to the 2 m thick 'pale grey, massive' sandstone underlying the seatearth of the Hard Bed Band in the nearbly Blackfordby No.4 Borehole (Figure 7). The borehole further proved a coarse, cross-bedded sandstone with a pebbly and conglomeratic base about 25 m below the ganister which is probably the pebbly sandstone cropping out west of Valley Farm. Unfortunately the borehole penetrated only a further 4 m and did not reach the Coal Measures base.

North of Valley Farm, the ganister dips eastwards in a fault-bounded outcrop [3435 1620] near the railway. Farther east, it dips gently to the west beneath a prominent dip-slope, which is fault-bounded along the Measham to Ashby road. East of this fault, the ganister can be traced for about 0.5 km, to where it is faulted against Wingfield Flags near Rotherwood. The main outcrop of the Wingfield Flags hereabouts extends NNW from the north shore [340 150] of the lake at Willesley, where it is a fine-grained, yellow-brown sandstone.

In the Grange Wood Borehole (Figure 7), no coals were proved below the Kilburn, but a thick seatearth at about 783.1 to 789.5 m may be either the Alton or Norton horizon, and a sandy seatearth underlain by sandstone and grit, at 804.2 to 811.4 m, may mark the Belperlawn horizon. A dark grey mudstone resting on seatearth at 824 m, 104.1 m below the Kilburn is possibly the Subcrenatum Marine Band. In the Acresford No.7 Borehole (Figure 7) a *Lingula* band at 273.1 m was formerly correlated with the Listeri Marine Band (Greig and Mitchell, 1955). Recent examination of the micro-faunas from this interval has revealed the presence of arenaceous foraminifera indicating that the horizon is no older than the Listeri Marine Band, and a lower *Lingula* band, 109 m below the Kilburn, is now presumed to be the Subcrenatum Marine Band.

In the Grange Wood and Church Flats boreholes, about 30 m of Wingfield Flags pass by alternation upwards into the siltstones underlying the Kilburn. W. W. Watts (*in* Fox-Strangways, 1907, p.197) described sandstone from near the bottom of Netherseal No.2 Borehole as having a dolomitic cement, and compared the appearance of one sample to that of Cambrian quartzite from Nuneaton. More recent boreholes suggest, however, that these sandstones, and equivalent strata in Netherseal No.1, lie only

11.5 m below the Kilburn and are therefore part of the Wingfield Flags. They are not pre-Carboniferous, as Fox-Strangways believed (six-inch Geological Map, Sheet Derbyshire 63NW), and as Mitchell and Stubblefield (1948) accepted.

Mussels, including poorly preserved *Carbonicola sp.*, occur in black shale underlying the Kilburn in the Rawdon West Nos.3 and 5, Church Way and Acresford No.7 boreholes. The Acresford No.6 (1949) Borehole (Figure 7) proved 154.8 m of strata below the Kilburn (Greig and Mitchell, 1955, pp.51, 62). A 0.2 m coal at 693 m is correlated with the Belperlawn, but the borehole does not seem to have penetrated to the Subcrenatum Marine Band. The lower of the two marine bands at 667.2 and 673.3 m may be at the Listeri horizon. The 0.71 m coal at 659.6 m may be the Norton (Greig and Mitchell, 1955), but it seems more likely to be the Hard Bed Band with the Norton as the slightly higher coal, at 656.2 m (Figure 7).

Stretton No.1 Borehole (Figure 7) (Greig and Mitchell, 1955 pp.50–51) proved a 1.7 m coal at 288.1 m which may be the Norton, 100 m below the Kilburn, In the Horses Lane Borehole [3355 1166], two thin coals at 118.1 and 120.3 m may be the Norton and Hard Bed Band, and a marine band at 170.7 m, which yielded *Lingula* and marine fossils including a possible goniatite, may be at the Subcrenatum horizon; unfortunately samples were not retained from this horizon.

The Snarestone Shaft [3475 1009] (Fox-Strangways, 1907, pp.335–338) may penetrate the base of the Coal Measures although the log is ambiguous. If this interpretation is correct, the Subcrenatum–Kilburn interval is 129.6 m thick, with the succession, including thin coals and thin sandstones in its lower part, generally resembling that in the Acresford No.7 Borehole. The beds below the base of the Trias at 176.3 m in the Appleby No.1 Borehole (Figure 7), include thin coals and a small amount of 'rock', and generally resemble the sequence below the Wingfield Flags.

The Appleby No.2 Borehole [3176 1166] proved 73.4 m of strata below the base of the Triassic at 145.8 m, including much 'rock', presumably sandstone (Fox-Strangways, 1907, p.334). The presence of only one thin coal suggests that the sequence lies in the Lower Coal Measures.

The Donisthorpe Kilburn No.1 Underground Borehole [3189 1371] proved an exceptional thickness of 42 m of Wingfield Flags below the Kilburn seatearth. The lower part included coarse-grained sandstone of channel type with coal streaks and ironstone pebbles.

The sequence in the Measham No.2 Borehole [3639 1192] (Figure 4) is broadly similar to that at Ellistown Colliery, although thinner. There are 85 m of measures between the Cumbriense Marine Band and the Kilburn seam, but neither the Subcrenatum or Listeri marine bands are present. The Wingfield Flags are 18.9 m of coarse grey sandstone with shale clasts and carbonaceous partings.

The Kilburn Coal averages 1.3 m in thickness, and varies usually between 0.7 and 1.6 m, in boreholes around Neatherseal and west of Overseal. In Grange Wood Borehole, an unusual feature is 2.5 cm of soft bitumen at the base of the seam. The outcrop of the Kilburn west of the Boothorpe Fault is proved mainly by Opencast Executive boreholes (Spink, 1965). The Kilburn is generally 1.0 to 1.5 m thick between Donisthorpe and Measham (Figure 7). Several boreholes have proved the coal beneath Triassic cover between Measham Colliery and Snarestone. In Coronet Borehole [3398 1080], the main part of the seam, 1.68 m thick, is underlain by another 0.75 m of interbedded thin coal and seatearth layers. In Turnover Borehole [3423 0095], the Kilburn is 1.6 m thick, 40.5 m below the base of the Trias.

In boreholes around Overseal and Neatherseal, the Kilburn is overlain by 0.5 m of dark shaly mudstone with fish remains. This band is 1 to 2 m thick in the Measham area, and the overlying 1 to 2 m of mudstone contain *Planolites* and *Cochlichnus*.

The Clod Coal is generally about 0.6 m thick near Neatherseal but reaches 0.97 m in the Rawdon West No.3 Borehole. Farther east, it is 1.08 m thick in the Acresford No.6 Borehole and averages about 0.75 m in the exposed coalfield, but is nowhere worked.

Above the Clod are several thin seams, none workable; they are, in ascending order, the Twelve-Inch, Well (formerly the Three-quarters), Stanhope (lower leaf), Stanhope (upper leaf) and Joyces. Outcrops of these seams, shown on the 6-inch maps to the east of the Boothorpe Fault and north of Willesley Wood, are based on maps by Spink (1965). The Well and Stanhope seams combine to form the Lower Main (formerly the Roaster) of the Leicestershire Coalfield.

A mussel band, generally 6 m thick, immediately overlies the Clod and yields *Carbonicola sp.* ex gr. *pseudorobusta*, *C. martini* and *Curvirimula trapeziforma*, suggesting a correlation with the Pseudorobusta Belt, late Communis Chronozone.

The Twelve Inch is 0.34 m thick in the Donisthorpe Kilburn No.1 Underground Borehole but eastwards is represented by only a few centimetres of coal on a seatearth. The overlying dark grey to black mudstones contain fish remains, and in some sections, mussels including *Curvirimula* cf. *subovata* in Measham No.4 Borehole.

The Well Coal, on its characteristically thick seatearth, is 0.75 to 1 m thick between Pegg's Close [3357 1206] and Measham Colliery, and was formerly worked at Measham Colliery as the 'Stanhope' Coal, being reached by staple pits from the Eureka (Mitchell and Stubblefield 1948, p.15). A mussel band, 1 to 2 m thick, generally overlies the Well and contains *Curvirimula trapeziforma*, *Carbonicola* ex gr. *pseudorobusta* and *Geisina arcuata*. The upper Stanhope was formerly worked west of Minorca Shaft [3552 1141]. East of Measham Colliery the section is: upper Stanhope 0.7 m, measures 9.5 m, lower Stanhope 0.1 m, measures 1.0 m, Well 1.0 m. The Joyces Coal is only 0.1 m thick generally and is absent in places, but the horizon can be recognised locally by a seatearth 1 to 2 m thick and a roof of grey mudstone with worm trails and small ironstone nodules. It does not persist, however, east of Peggs Close.

The Eureka Coal was named in the early 19th century at a colliery in Swadlincote, when it was discovered in an exploration shaft

below the level of coals then being worked (Hull, 1860, p.22). It has since been extensively exploited. In Netherseal Colliery the seam is cut by many washouts running in confluent lines like the tributaries of a river (Fox-Strangways, 1907, p.40). The seam is about 1.2 m thick north of Neatherseal, but thins slightly southwards. In the Rawdon Shaft [3127 1626] it is 1.2 m thick and is 1.5 m in the Hastings and Grey Shaft [3199 1555]. It is 1.45 m thick in the Acresford No.6 Borehole and 1.0 to 1.2 m thick in boreholes between Measham and Measham Colliery. It crops out in a valley at Measham [332 123] and north of Willesley Wood [331 160].

The sandstone or sandy siltstone, commonly forming the roof of the Eureka, is the Eureka Rock (Mitchell and Stubblefield, 1948, p.15). It crops out east of the Boothorpe Fault, to the north of Willesley Wood, where it appears to be lenticular. In the Rawdon West No.5 Borehole it is fine grained, grey, with many silty wisps and carbonaceous and micaceous layers and is 7.8 m thick. South of Measham, the Eureka Rock is absent and the measures consist mostly of silty to sandy mudstone, in places containing numerous plant remains, particularly *Neuropteris*.

The Stockings Coal, the equivalent of the Nether Lount of Leicestershire, was 1.75 m thick in the Oakthorpe Meer opencast site [325 128]. It crops out east of the Boothorpe Fault to the north of Willesley Wood, where it was originally known as the Rafferee Coal (Hull, 1860; Fox-Strangways, 1907). Although of little commercial value in the 19th century, the Stockings has subsequently been extensively worked at Donisthorpe and Measham collieries. Around Neatherseal and west of Overseal, the seam varies between 1.0 m and 2.3 m thick. Farther east, the coal is 3.35 m thick in the Acresford No.6 Borehole, 2.5 m in the Rawdon Shaft and 2.44 m in both the Donisthorpe and the Hastings and Grey shafts. Near Measham, it is about 2 m thick but includes seatearth ('dirt') partings.

The mudstone above the Stockings contains mussels, including *Carbonicola crista-galli*, *C. os-lancis* and *Naiadites sp.*, indicating the Crista-galli Belt, earliest Modiolaris Chronozone. The coal about 0.6 m thick and commonly underlying the seatearth to the Woodfield Coal, is the Trencher.

The Woodfield Coal, named after the colliery near Newhall, north of the present district (Fox-Strangways, 1907. p.40), correlates with the Middle Lount of Leicestershire, and like that seam includes, in places, near its top, a 2.5 cm brown tonstein (Spink, 1965, p.57), described in older records (e.g. Mitchell and Stubblefield, 1948, p.16) as an ironstone. In the western, concealed part of the coalfield, the Woodfield is generally 1.5 to 1.7 m thick and is thickest in the Church Flats Borehole, where the section is: coal 1.88 m, mudstone 0.30 m, coal (Trencher) 0.48 m. In the Coalpit Lane Borehole, a 2.5 cm brown stone band, probably the tonstein, lies 0.25 m from the top of the seam. At Botany Bay the coal is only 0.56 m thick. In the central part of the coalfield, the seam is 1.9 m thick in the Rawdon Shaft and 1.8 m in the Hastings and Grey Shaft. East of the Boothorpe Fault, 0.9 m of coal [3309 1585] dipping 14° WSW in a railway cutting is presumed to be the Woodfield. The Woodfield is 1.5 to 2 m thick in the Donisthorpe-Measham area. A tonstein band 8 mm thick and 1.23 m from the top of the seam was noted at the Oakthorpe Meer opencast site (G. H. Mitchell in MS). The seam was also worked opencast at the small Moira No.1 site [3174 1251].

A persistent mussel band forms the roof of the Woodfield Coal. West of Donisthorpe, it is 6 m thick and ostracods are recorded from it in some boreholes. In some sections it includes a 5 cm coal. Near Measham, the mussel band is only 1.2 to 1.4 m thick and east of the Boothorpe Fault it appears to be replaced by sandstone. The fauna includes *Anthracosphaerium sp.*, *Carbonicola crista-galli*, *C. os-lancis*, *Naiadites flexuosus*, *Geisina arcuata* and *Carbonita sp.* indicating the Crista-galli Belt, earliest Modiolaris Chronozone. Immediately below the Lower Main seatearth in many sections, is a thin cycle of 2 m or less comprising a seatearth, thin coal and thin mussel band with ostracods.

The Lower Main is present throughout the coalfield and correlates with the Upper Lount of Leicestershire. It is thicker and of better quality around Moira and Measham than farther north (Fox-Strangways, 1907, p.41) and within the present district, its former name, Slate Coal, is undeserved. Around Netherseal, the Lower Main is 1.1 m to 1.8 m thick, commonly with a dirt parting in the middle. In the centre of the coalfield, the Lower Main is 1.2 m thick at Rawdon and 1.1 m in the Hastings and Grey Shaft. Southwards, it is 1.4 m thick in the Acresford No.6 borehole, 1.1 m in the Donisthorpe shaft and 1.2 m at Measham. Around Neatherseal, its roof is grey mudstone, overlain by 1 m of black shale with mussels and ostracods.

The Little Woodfield Coal was formerly known as the Toad around Moira, probably because its surface is commonly spotted with pale brown impressions of mussels (Fox-Strangways, 1907, p.42). It crops out in a faulted inlier 0.5 km south-west of Oakthorpe and was worked in the Oakthorpe Meer opencast site towards Measham. Old workings in the seam were uncovered at this site. Boreholes north and west of Netherseal show a thin but persistent seam, 0.5 m to 1.17 m thick. It thickens towards the centre of the coalfield, being 1 m in the Reservoir Pit shaft [3008 1645], 1.2 m at Rawdon and 1.1 m in the Hastings and Grey Shaft, and reaches similar thicknesses in the Donisthorpe to Measham area.

Fox-Strangways (in MS, notebook VI, p.142) recorded the following section in a roadside well at 'Oakthorpe Tollgate' [3282 1283]: Coal smut [Main Coal], 2.74 m; fireclay 0.3 m; 'stone' 0.46 m; bind 5.49 m; coal [Little Woodfield], seen for 0.3 m.

Within the present district, the Main Coal comprises two layers, the Nether and Over Coal, which farther north around Swadlincote are separate seams. In some places, one or more Main Rider seams separate off from the Over Coal and lie at varying distances above the Main. In the present district, only the Over Coal has been generally worked, while farther north, the Nether Coal was wrought instead.. The Nether Coal, left untouched in the old workings, has been worked on a large scale in recent years from Donisthorpe Colliery. The Main is equivalent to the Upper Main and High Main of Leicestershire (Figure 12). Hull (1860) reported that the cellar of the Gate Inn at Oakthorpe [3258 1294] was said to be excavated in the Main.

The Main around Netherseal and west of Overseal has a thin lower leaf in most sections. In the Coalpit Lane Borehole, this is separated by 1.2 m of seatearth from the Main and is distinguished as the Under Main (0.15 m thick). In some boreholes there is a distinct Main Rider coal, which is overlain by the Vanderbeckei Marine Band. At Coalpit Lane the Main Rider, 0.56 m thick, is 24.2 m above the Main, which is 2.62 m thick; the interval consists mainly of siltstone and silty mudstone. In other sections, the Main is generally about 4 m thick and is 5.5 m in the Acresford No.6 borehole. The Reservoir Pit, Rawdon and Hastings and Grey shafts proved Rider and Second Rider coals close above the Main (Over) Coal, separated from it by less than 1 m of seatearth. In the Measham Shaft 3.7 m of coal was proved, without Rider or Under seams. Mitchell and Stubblefield (1948, p.17) recorded that, in the Measham working, the Main is traversed by a number of 'washouts', in some places up to 100 m across, filled with grey sandy mudstone locally with plant remains. The upper part of the seam commonly contains bands of fine sandstone usually 20 mm or so thick but occasionally reaching 150 mm.

According to Hull (1860, p.28), at the Bath Pit [3097 1560] at Moira, the Second Rider is 1.2 m thick and rests directly on the Main, producing a combined thickness of nearly 5.5 m. The section proved in the Ashby Road Borehole [3403 1329] is: coal (Rider), 1.4 m; seatearth, 0.66 m; coal, 0.48 m; old workings, 0.74 m; coal, 2.01 m (Main Over and Nether). The Main Coal was 3.7 m thick in an excavation [3319 1267] on the south side of the Oakthorpe-Measham road (Fox-Strangways, 1907, p.45), and in Measham Colliery. It has been worked around the Minorca shaft from its incrop to the Boothorpe Fault. A 1.2 m coal, 17 m above the base of the Main in local sections, is presumed to be a Main Rider.

North-east of Measham, are numerous disused mine-shafts, probably sunk to the Main. Hull (1860, p.29) stated that in this area, then known as Measham Field, the Main was 20 to 25 yards (18 to 22 m) deep, the beds dipping gently to the south and west. He also stated that the Over Coal had, at that time, almost been exhausted under Measham Field, although the Nether remained for the most part undisturbed.

Between 1976 and 1978, air appears to have entered disused workings in the Main causing the coal to ignite near its outcrop, south-west of Park Farm [around 340 134] and on the north side of the road to Donisthorpe [3354 1262], recalling Hull's (1860, p.29) comment on the Main Coal, 'About 80 years since, Oakthorpe Hill was on fire all along the basset and the combustion continued for several weeks, defying all attempts to extinguish it'.

MIDDLE COAL MEASURES (Westphalian B and C)

The Vanderbeckei Marine Band, which marks the base of the Middle Coal Measures and of Westphalian B, overlies the Main or the highest of its Rider coals. The marine band is represented by the presence of *Lingula mytilloides* in the roof shales of the Rider at Donisthorpe Colliery and in a borehole near Potkiln Cottages [3377 1271] (Mitchell and Stubblefield, 1948, pp.17, 28–29). In the Ashby Road Borehole [3403 1329] the marine band contained *Lingula mytilloides*, *Dunbarella sp.*, *Myalina compressa*, turreted gastropods and arenaceous foraminifera, together with intercalations of non-marine faunas including *Anthracosia* ex gr. *lateralis*, *Anthracosia* ex gr. *phrygiana*, *Curvirimula* cf. *subovata* and *Geisina arcuata*. It occurred just above the Rider and was 12.6 m thick. In the Park Farm Borehole [3404 1379], the thickness proved was 5.87 m. The presumed Rider Coal in the Minorca Shaft of Measham Colliery is overlain by 6.7 m of 'blue bind' which may include the Vanderbeckei Marine Band, though no fossils were recorded. The marine band is commonly overlain by a mussel band, as in Botany Bay and Rawdon West No.4, dominated by *Anthracosia ovum* (and variants) and *Naiadites quadratus*, indicative of the Ovum Belt within the Modiolaris Chronozone.

The sandstone above the Main Coal has a wide outcrop at Oakthorpe but dies out east of the village. Just west of Park Farm, its outcrop is narrow where the sandstone dips steeply away from the Boothorpe Fault. Some elongate hollows on this outcrop [341 138] may be overgrown quarries for building stone. Higher in the succession is a thin but persistent coal, the Little Kilburn, formerly known as the Cannel (Hull, 1860, p.32; Fox-Strangways, 1907, p.46). It crops out in places between Oakthorpe and the north of Measham where it is overlain by an impersistent sandstone. The Little Kilburn is present in most boreholes around Netherseal and Overseal and in these sections the overlying sandstone is 5 to 8 m thick.

The overlying Little seam was worked as 1.1 m of coal in the Steam Mill Bridge opencast site [320 135], and is correlated with a 0.8 m seam at 18.7 m depth in the Minorca Shaft. The Little has been extensively worked in Rawdon Colliery, maintaining a thickness of 1.4 to 1.5 m. Around Overseal and Netherseal the Little thins westwards, being 1.24 m thick in the Acresford Plantation Borehole, 1.27 m in Rawdon West No.2 and 0.91 m in No.1. In the Rawdon West No.3 Borehole, it consists of three leaves, the thickest being 0.74 m. In the Coalpit Lane borehole, however, if correctly correlated, it is only 0.20 m thick. In the Grange Wood and Botany Bay holes only a few centimetres of coal occurred at this horizon and the seam is apparently absent in the Netherseal Colliery No.4 Borehole. The Little is generally overlain by a mussel band containing *Anthracosia aquilina* and *Naiadites triangularis*.

The Upper Cannel was formerly worked on a small scale at the Daisy Pit [3350 1372], a shaft of the Oakthorpe Colliery. The seam there comprised 0.61 to 0.66 m of coal overlying 0.61 to 0.91 m of cannel (Fox-Strangways 1907, p.47–8), 13.6 m above the Little. Nearby, the Upper Cannel was exposed in the Gin-Barn-Pit open-cast site, west of the Measham to Ashby road. Mitchell and Stubblefield (1948, p.18) recorded here an upper coal 0.99 m thick, resting in places on a 0.65 m dirty, pyritous seam and in others separated from it by as much as 5 m of measures. In places, the lower bed was washed out while the upper bed persisted. In Saltersford Brook [3275 1378], the upper coal of the Upper Cannel lies close above a lower dull pyritous seam:

	Thickness m
Sandstone, grey, fine, hard, ferruginous	0.6
Siltstone, pale grey, hard, ferruginous	0.6
Silty mudstone, grey, hard, ferruginous	1.23
Coal	0.91
Coaly shale, with traces of rootlets	0.38
Coal, dull, with bright streaks	0.43
Coal, bright, some pyrite on joints	0.36
Coal, dull, argillaceous	0.33
Seatearth, mudstone, pale grey	1.8
Coal	0.05
Seatearth, mudstone, grey	to 0.6

The name Two Foot Coal was given during the primary survey to a seam mapped by Hull near Oakthorpe Colliery. This is probably the seam, 0.61 m thick at 15.2 m depth, in the Oakthorpe Colliery shaft section (Figure 9) (Fox-Strangways 1907, pp.48, 241). It lies 21.5 m above the Little Coal, and seems to have been worked to only a small extent. Spink (1965, p.43) correlates it with the Threequarters (Jack Head) Coal of Leicestershire. West of Oakthorpe Colliery, it is uncertain which of the thin coals between the Little and the Block correlate with the Upper Cannel and Two Foot (Figure 9).

Slightly higher in the sequence is the Yard Coal. Mitchell and Stubblefield (1948, p.18) described it as a thin coal rarely of any economic value, but locally important to the north of the present district. In Acresford No.6 Borehole [3096 1358], a diverse non-marine fauna overlies the Yard and includes *Anthracosia aquilina* trans. *lateralis*, *A. beaniana*, *A. phrygiana*, *A. retrotracta*, *Naiadites sp* and *Spirorbis sp* representing a fauna within the Phrygiana Belt, late Modiolaris Chronozone. In the Rawdon West No.2 Borehole, a 0.53 m coal with mussels in its roof shales, 11.4 m below the Block, has been identified as the Yard. A thin unnamed coal lies just above it, but may die out westwards before the Coalpit Lane Borehole is reached (Figure 9). Farther east, the Yard is present in the Green Lane Borehole but its correlation beyond is uncertain.

Hull (1860, p.33) said the Block generally had a friable and jointed nature; in sinkings it was usually found to be charged with water. It was worked opencast in 1942 along most of its outcrop between Donisthorpe and Willesley, where it was up to 1.52 m thick (Mitchell and Stubblefield, 1948, p.17). Its thickness in trial boreholes at Willesley [338 149] was 1.47 m (Spink, 1965, p.64). A 1.1 m seam encountered in Measham Hall No.8 Borehole is the Block (Figure 9). The seam is thin around Netherseal and not everywhere present. It is typically overlain by a mussel band, up to 5 m thick, containing *Anthracosia atra*, *A. concinna*, *A. nitida*, *Naiadites alatus* and *Naiadites subtruncatus*, indicating an horizon within the Atra Belt, Lower Similis-Pulchra Chronozone. In the log of Measham Hall No.8 Borehole (Fox-Strangways, 1907, p.252), 2.1 m of 'black bind', at 2.2 m above the Block, and 0.3 m of 'shale' immediately overlying the coal are lithologies that might contain the Block's mussel band but no fossils were collected from the borehole. Between Donisthorpe and Willesley, a sandstone overlying the Block crops out as a prominent escarpment for about 1 km [322 142 to 329 124], but dies out to the east.

In the Beehive [3286 1563], Hanging Hill [3141 1728] and Whitborough Farm [3177 1643] boreholes, three thin coals occur between the Block and Upper Kilburn. The middle of these seams is underlain hereabouts by a sandstone with irregular bedding and much *Stigmaria*.

The Upper Kilburn (formerly Dicky Gobler) maintains its thickness of about 1 m over much of its extent. It is thicker near Overseal, being 1.07 m at Poplars Farm and 1.14 m at Green Lane, but a little farther west, in the Rawdon West No.1 Borehole, the Upper Kilburn is only 0.23 m thick. West of the Netherseal Fault, the Upper Kilburn appears to be only 0.3 m at Coalpit Lane, although core recovery of the seam in this borehole may have been incomplete.

The seam is exposed in the railway cutting [3155 1406] 350 m SE of Donisthorpe Colliery, though only the crest of the small faulted anticline seen by Fox-Strangways (1907, Figure 5, p.49) is now visible. The section is: grey shaly mudstone to 0.3; black shale with fish scales and mussels 0.2; coal 1.1 to 1.2; grey mudstone-seatearth. The Upper Kilburn was worked at outcrop in 1942 [323 145 to 334 146] north of the Donisthorpe to Willesley road. Spink (1965, p.64) traced its continuation towards the Boothorpe Fault near Willesley [337 148] from NCB Opencast Executive trial borings, in which the seam was 0.9 to 1.1 m thick. The Upper Kilburn is overlain by a mussel band which contains *Anthraconaia sp.*, *Anthracosia* cf. *aquilinoides*, *A. fulva*, *A* cf. *lateralis*, *A.* cf. *nitida* and *Naiadites spp.*, indicating the Atra Belt, Lower Similis-Pulchra Chronozone.

Above the Upper Kilburn, four thin, but persistent, coals are numbered P44 to P41 (Figure 10). They lie at a similar horizon to the Excelsior Coal of the Leicestershire Coalfield, and may be splits of that seam (cf. Spink, 1965, fig. 3). All four seams are present in the Donisthorpe Shaft, where P41 is only 9.75 m below ground surface. Mussels occur in the roof shales of the Upper Kilburn and of P42 in the Acresford No.6 Borehole. The Hill Farm Sandstone (Worssam, 1977, p.4), between the P41 and P40 coals, is present in most sections across the coalfield. It forms a prominent feature around Hill Farm [332 148] and extends northwards through Willesley Wood. On the western side of the Moira Main Fault, its outcrop can be traced for about 1 km westward from Rawdon Colliery [3127 1626], and the sandstone was formerly dug from a small quarry [3076 1603], now buried beneath a tip-heap, (Fox-Strangways, in MS). A hill-top outlier of the sandstone occurs to the south [311 158]. Farther west and south lenticular outcrops of the sandstone occur at Gorsey Leys [302 155], near Short Heath [313 148], and just west of Donisthorpe Colliery [310 144].

In the concealed coalfield, the P44 to P41 coals are present in all of the boreholes near Overseal except Botany Bay, where P42 and P43 are represented only by shales with plants and coaly streaks. A persistent mussel band 3 to 4 m thick, overlies the Upper Kilburn and mussels may also occur in dark mudstones above the other seams. Of these P41 was the thickest at 0.45 m, recorded in the BGS Hanginghill Farm Borehole [3135 1672], and the overlying mussel band included cf. *Anthraconaia pulchella* and *Anthracosia sp.*.

The Hanginghill Farm Borehole provides a basis for correlation (Figure 10) of the sequence above the P40 Coal. This succession of Pottery Clay coals has been extensively worked in opencast diggings in the north of the district, at Robinson and Dowler's Pit now filled, 0.5 km north-east of Overseal Church; the Haywood Works Extension Pit [306 162]; the small, disused Bramborough Farm Pit [320 149]; the extensive Willesley Clay Pit [326 154]; and a small working in Willesley Wood [311 155].

Robinson and Dowler's Pit, in 1939, exposed beds from the P39 Coal to just above the Aegiranum Marine Band (Mitchell and Stubblefield, 1948, p.20, fig. 4). A composite section of beds exposed between 1968 and 1971 is shown in Figure 10. Beds above P38 were then exposed on the south-west face [295 162], while P39 and P40 were exposed on the northern face of the pit [2915 1675].

The P39 Coal was overlain by the Maltby Marine Band, consisting of black, pyritous mudstone with *Lingula*. Mitchell and Stubblefield (1948, p.20, table 1) recorded the roof of this coal [2990 1605] as 8 cm of hard, dark shale with *Lingula mytilloides*, *Megalichthys* and 'fucoid markings', succeeded by 0.3 m of grey, soapy shale containing stunted mussels (names revised and brought up to date); *Curvirimula sp.*, *Anthracosia* cf. *acutella*, *Anthracosia spp.* and *Naiadites spp.* The P38 Coal, 0.23 m thick, was overlain by 1.37 m of brownish-black shaly mudstone, with thin-shelled mussels and *Lepidodendron* in a band of black shale about 0.3 m from its base, and thin shells (?*Naiadites*) 3 to 5 cm below the junction with the overlying P37 seatearth. The section in the pit up to the P32 Coal generally resembled that in the Hanginghill Farm Borehole (Worssam, 1977) though the coals were thicker in the pit. The seatearths of P36, P34 and P33 were high-alumina clays, with 30 to 32 per cent Al_2O_3 (information from Messrs Robinson and Dowler).

The Haughton Marine Band, above P33, was 2.7 m of dark grey, thinly bedded mudstone with scattered ovoid ironstone nodules elongated along the bedding; 'double tube' trace fossils were numerous about 1.5 m above the base. Mitchell and Stubblefield (1948, p.20, fig. 4 and MS) recorded *Lingula* from this horizon in Robinson and Dowler's pit and *Orbiculoidea* cf. *nitida* from Sutton's pit [307 167]. The P32 coal was overlain by strongly cross-bedded, fine- to medium-grained sandstone with coaly streaks, but sandstone is not known at this horizon elsewhere in the coalfield. Messrs Robinson and Dowler report the P31 seatearth to be a siliceous clay with 60 to 80 per cent SiO_2. An extensive fauna from the Aegiranum Marine Band, listed by Mitchell and Stubblefield (1948, table 2) includes *Hyalostelia sp.*, *Lingula mytilloides*, *Orbiculoidea* cf. *nitida*, *Chonetinella flemingi* var. *alata*, *Crurithyris carbonaria*, *Dictyoclostus* cf. *craigmarkensis*, cf. *Linoproductus anthrax*, *Rugosochonetes skipseyi*, *Aviculopecten* cf. *delepinei*, *Parallelodon* cf. *reticulatus*, *Pernopecten carboniferus*, *Posidonia sulcata*, *Schizodus* cf. *antiquus*, *Solemya* cf. *primaeva*, *Streblochondria* cf. *fibrillosum*, *Donetzoceras aegiranum*, cf. *Politoceras jacksoni*, *Metacoceras cornutum*, ostracods and conodonts. The highest beds exposed in the pit included four thin coals, which seem to correlate with P25–28.

The highest sandstone in Robinson and Dowler's Pit (Figure 10) crops out for about 500 m to the south-east, where it is unconformably overlain by Moira Breccia. The outcrop of the Aegiranum Marine Band, parallel to that of this sandstone, must be similarly overstepped by the Breccia, for the Poplars Farm Borehole [2996 1514] started in Moira Breccia, entered the Coal Measures at 13.1 m and proved the Maltby Marine Band at 16.3 m.

The Haywood Works Extension Pit is at the southern end of opencast workings extending northwards for 1.5 km to Albert. In 1973 a rectangular pit [306 162] exposed beds between the P39 and P36 coals, overlain by about 6 m of made ground. This was the fill of an earlier working at the site, probably part of the Haywood's Pit of Mitchell and Stubblefield (1948), in which P35 up to P33 at ground level had been worked. The section correlates in detail with that in Robinson and Dowler's Pit. In Haywood's Pit, the P34, P33 and P32 coals can be recognised. The quarry manager in 1973, Mr R. Griffin, reported that in the earlier workings, the most valuable fireclays had been the Deep Fireclay, beneath P33, containing 37–39 per cent Al_2O_3, and the P34 clay with 32–33 per cent Al_2O_3. Although it immediately underlay P35, P36 Coal had not been worth digging owing to its poor quality—22 per cent ash and only 8000 B.t.u./lb. The best of the Pottery Clays coals had been P39, with 4 per cent ash and 11 000 B.t.u./lb. P39 in the Hanging Hill Farm Borehole gave values of 5.9 per cent and 12 120 B.t.u./lb respectively; (Worssam, 1977, p.14). The beds seen in 1973 dipped gently south-eastwards to a north-trending fault at the east end of the Pit [3068 1615] throwing down 4.5 m to the west. On the up-

throw side of the fault P39 cropped out [3073 1623] and was overlain by the Maltby Marine Band. In the Rawdon Colliery Shaft (Fox-Strangways, 1907, p.209; Worssam, 1977, Figure 2) P39 is 0.66 m thick, at 7.1 m below surface.

The Bramborough Farm Clay Pit [320 149] showed the following section in 1967:

	Thickness m
Seatearth and coaly shale, interbedded	1.42
P30 Coal	0.15
Coaly shale; *Neuropteris* and stems	0.56
Seatearth, mudstone, sandy	0.25
Sandstone, fine- to medium-grained, with cross-bedding; rootlets in top 0.2 m and slump-structure in lower part, forming bottom-set of cross-bedding units	0.4 to 1.20
Seatearth, dark grey	0.91
Ironstone (Main Stone)	0.13
(Aegiranum Marine Band) Mudstone, dark grey shaly with ironstone nodules	4.95
P31 Coal	0.36
Seatearth (Derby Fireclay)	about 1.98
P32 Coal	0.23
Seatearth	0.13
Sandstone, fine-grained, with rootlets	0.38
Seatearth, mudstone, medium to dark grey	1.22
P33 Coal	0.71
Seatearth, pale brownish grey; bottom 18 cm darker (Deep Fireclay)	2.01
P34 (Ell) Coal < 0.20 m to	0.61
Seatearth, mudstone, pale yellowish brown; sphaerosideritic ironstone in middle part of bed	about 1.52
Coal, inferior	0.15
Seatearth, up to	0.15
Coaly shale, black	0.91
P35–6 Coal	0.46
Coaly shale	0.13
Coal and coaly shale, interbedded in 5–10 cm layers	0.46
Seatearth, mudstone, brownish grey; a sandstone streak near top and ironstone in lower part, seen for	2.1

The Haughton Marine Band, which should lie above P33, was unrepresented. The Pit was bounded on its west side by a north-trending fault. The west face just beyond the fault showed:

	Thickness m
Weathered grey clay, about	0.3
Obscured (probably mainly sandstone)	about 2.7
Sandstone, grey, fine-grained with irregular lenticular bedding; casts of plant stems 0.1 m wide and 0.3 to 0.6 m long	to 1.2
Mudstone, medium grey, silty to sandy; no rootlets; large rounded ironstone masses 0.3 to 0.6 m in diameter, about 1.2 m below top of bed	about 3.0

The sandstone in the face extends westwards to cap a flat-topped hill, formerly considered an outlier of Moira Grit (Worssam and others, 1971, fig. 2; Worssam, 1977, fig. 1). The sandstone is too fine grained to be Moira Grit, however, and correlates best with one of the sandstones between P36 and P37 in the Hanginghill Farm Borehole (Figure 10). The lowest of these beds in the borehole, 1.1 m thick, contains much coaly detritus and some mudstone flakes in its bottom 0.18 m, and rests so sharply on the P37 coal as to suggest a channel-sandstone. The same bed may be responsible for the washout of the P37 and P38 coals in the Hastings and Grey Shaft (Figure 10). Farther north, sandstone at about this horizon crops out west of the Moira Main Fault, 100 m west of Rawdon Colliery Shaft. It dies out westwards and could represent the westerly limit of a north- or NNW-trending washout.

The Willesley Clay Pit is now back-filled, but Figure 10 shows a composite section of beds exposed between 1967 and 1972. The numbering for the coals adopted here supercedes that given previously (Worssam, 1977). The section in the Pit correlates with boreholes drilled by the National Coal Board Opencast Executive during 1976–77 in Willesley Wood [311 155], west of the Clay Pit. The lowest three seams are P34, P33 and P32 and the Haughton Marine Band was again absent from above P33. The P31 (Derby) Coal and the Aegiranum Marine Band were typically developed. The sandstone overlying the marine band, about 7 m above P31, was up to 0.9 m thick and contained rounded or balled-up masses, 0.3 to 0.6 m across, projecting down into mudstone and suggesting soft-sediment deformation or slumping. The sandstone was overlain by seatearths with thin coals and coaly shales at the P29 or P30 (Soup Kitchen) horizon. Above this came four 0.6 to 0.75 m thick coals, P28, P27, P26 and P23–24. Both the P27 and P26 were overlain by a tonstein (Worssam, 1977). The P25 Coal, just below P23–24 was only 0.2 m thick; another thin seam 0.1 m thick, between P26 and P25 is unnumbered.

A 6 to 9 m thick, thinly-bedded siltstone overlying P23–24 displayed large-scale trough cross-bedding, with many foresets suggesting sediment derivation from the north-west, ending abruptly without bottom-sets on the planar top of P23–24. The base of the bed was erosive since it cut out about 1 m of dark grey mudstone with fish remains and mussel fragments, within a short distance in the west face of the pit [3224 1530]. The dark mudstone may correlate with the Edmondia Marine Band of the Hanginghill Farm Borehole.

On the eastern face of the Pit [3290 1540], the section was:

	Thickness m
Sandstone, pale grey, weathered orange-brown, fine- to medium-grained, with *Calamites* and strap-like stems, seen for	0.7
Mudstone, dark grey, shaly, with *Neuropteris* and other leaflets in lowest 0.6 m	2.4
P21 Coal (inferior)	0.25
Mudstone	0.05
P22 Coal	0.41

In the western part of the Pit [3230 1550] the P21 Coal was overlain by only 0.25 m of shaly mudstone with *Neuropteris*, which then passed upwards into a thick seatearth, locally known as Smith's Strong Clay. Northwards, the sandstone in the above section crops out from the railway [3290 1580] for 400 m to the NNW.

The section in the north-eastern part of the Pit [3232 1556] read:

	Thickness m
Sandstone, coarse, pale grey to yellowish brown	about 6.0
Sandstone, coarse, pale grey, with coaly streaks	0.2
P19 Coal, lower part cross-bedded in part of face, in lenses cutting into underlying seatearth	up to 0.23
Seatearth, mudstone, silty, laminated brownish and dark grey, carbonaceous, with poorly preserved rootlets and *Stigmaria*	0.05
Coal, inferior	0.08
Tonstein, pale brownish grey, hard, fine-grained, with dark grey streaks	0.03 to 0.04

Seatearth, mudstone, mottled dark grey and brownish, soft, with numerous listric surfaces	0.30
Coal	0.25
Seatearth, mudstone, silty, medium to dark grey	0.91
Seatearth, mudstone, greyish black, with rootlets	0.61
P20 Coal	1.14

The tonstein was described by R. K. Harrison (*in* Worssam, 1977, p.5, where it is incorrectly assigned to P20). The sandstone at the top of the section, formerly held to be Moira Grit, is now regarded as a local washout sandstone.

The coals of the Hastings and Grey Shaft are numbered in Figure 10 to correlate with the Willesley Clay Pit. The former numbering by Worssam (1977, fig. 2) was incorrect.

In the Acresford No.6 Borehole, the Maltby Marine Band overlying P39 at 118.70 m is developed as a grey pyritous mudstone containing *Anthraconaia sp.* juv. *Edmondia sp.* juv. and arenaceous foraminifera. Above P38, between 113.54 and 114.22 m, a dark *Lingula*-bearing mudstone (recorded by Greig and Mitchell, 1955, p.66) occurs with an association of arenaceous foraminifera and possible estheriids in its upper part, correlating with the Clown Marine Band. It is overlain by a non-marine fauna including *Anthracosia concinna*, *Naiadites sp.* and poorly preserved *Carbonita sp.*. Beneath P33, a brown clay seatearth 3.1 m thick correlates with the Deep Fireclay. The Haughton Marine Band overlying P33 contains pentagonal crinoid ossicles, *Lingula mytilloides*, *Productus sp.* and fish remains. The seatearth (Derby Fireclay) below the Aegiranum Marine Band is also well developed. The Aegiranum Marine Band is 6.25 m thick and is capped by a 150 mm thick ironstone band, but the overlying sequence up to the Trias base is not readily matched with other sections.

The Measham Hall No.9 Borehole [3576 1168] (Fox-Strangways, 1907, p.253) proved, beneath the Trias, 5.1 m of 'blue bind' and 'dark bind' overlying a 2.39 m coal, separated by 1.83 m of 'black smut' from an underlying seam 0.15 m thick. Projecting known dips (Figure 11), these coals are likely to be in the Pottery Clays and may be P39 and P40. The apparent thickness of the upper coal may be due to a locally steep dip. Measham Hall No.4 Borehole [3577 1181] (Fox-Strangways, 1907, p.251) is down-dip of No.9 (Figure 11) and the beds encountered in it should be higher in the succession. The four coals, in two groups of two, proved in the borehole may be the P33 to P36 coals, and a considerable and hitherto unrecognised thickness of the Pottery Clays adjoins the Boothorpe Fault.

The Pottery Clays succession, including the Maltby Marine Band, is proved beneath Triassic cover in boreholes west of the Netherseal Fault (Figure 10). P40 is a thin, split seam, while P39 is well developed and about 0.6 m thick. Between the Maltby Marine Band and P31, both coals and marine bands are less well developed in the boreholes than at outcrop. At Botany Bay, this interval is largely seatearth, but it includes the Haughton Marine Band with *Lingula* and *Productus*, overlying a 2.5 cm thick P33.

Foraminifera recorded, with some uncertainty, above a coal at this horizon at Church Flats may mark the Haughton Marine Band. Comparison of the sections suggests that the P35–P36 coals die out westwards. All the boreholes show the Derby Fireclay seatearth below P31, but the Sutton Marine Band and P32 are absent.

The Aegiranum Marine Band overlying P31 is readily recognisable in the boreholes. It is overlain by seatearths with very thin, impersistent P30–P28 coals. Soft sediment deformation and slumping of the bedding, in a sandstone 6 m above the marine band in the Coalpit Lane Borehole, recalls similar sedimentary structures in this sandstone elsewhere.

The correlation of the coals higher than P28 is tentative (Figure 10). The presumed P20 is the highest well developed coal, lying 9.0 m below the Etruria Marl at Grange Wood, 29.3 m below it at

Church Flats and 34.7 m below it at Botany Bay. P17 and the overlying Cambriense Marine Band may be represented by dark grey mudstone with coal streaks overlying purple-stained slickensided mudstone, in Botany Bay and by similar beds in Church Flats.

The sandstones in the Pottery Clays sequence proved in the boreholes around Netherseal are generally very coarse to gritty and 4.5 to 6 m thick.

Upper Coal Measures (Westphalian C and D)

An outlier of grey Upper Coal Measures occurs around the Hastings and Grey Shaft at Moira and the Cambriense Marine Band probably overlies P17 at 17 m in the shaft. At outcrop, the Cambriense Marine Band has been mapped just below the thin sandstone between the P16 and P15 coals in the Hanginghill Farm Borehole.

Another outcrop of grey Upper Coal Measures occurs west of Donisthorpe, although the exact horizon of the beds at surface hereabouts is uncertain. At the Hoo opencast site [303 145], 28 m of strata include two coals, each 1.1 m thick, and 1.5 m apart at the base of the section, and a third seam, 0.64 m thick, at 18 m above the base. The lowest coal seam, lying 207 m above the Upper Kilburn, may be 40 to 45 m above the Cambriense Marine Band.

In the Moira Pottery Clay Pit [316 162] (Figure 10) 15 m of seatearth include a coal, correlated tentatively with P15. The succeeding coal, P14, is the only seam good enough to work. The highest seam, P13, is composite, as follows:

	Thickness m
Carbonaceous shale, black, with 5 cm coal lenticles	0.08
Coal (inferior)	0.05
Shale, hard, black to brownish black, with dark grey, soft, clayey streaks in lower part; includes ?fish	0.23
Ironstone, silty, hard, brownish black	0.08
Cannel, brownish black, with conchoidal fracture	0.08
Carbonaceous shale, soft, black	0.05
Seatearth, mudstone, medium grey, with rootlets	0.08
Coal, inferior	0.13

A similar section to that at the Moira Pottery Pit, and presumably from the same part of the sequence, was measured by G. H. Mitchell in 1939 in the now-filled Newfield Pit [322 159].

Above the grey Upper Coal Measures, the Etruria Marl consists predominantly of reddish brown, unbedded mudstone mottled with a wide range of subsidiary colours, especially purple, yellow, green and blue-grey. It is not exposed at the surface but thicknesses of the Marl have been encountered in the following boreholes: Botany Bay, 29.0 m, overlain by Trias; Caldwell No.2, about 44.7 m, overlain by Trias; Grange Wood, 34.1 m; Church Flats, 35.0 m; and Netherseal 26.3 m. A coarse 7.5 m sandstone lies in the middle of the Etruria Marl at Grange Wood. The base of the Etruria Marl seems to be sharp. In the Botany Bay and Church Flats boreholes, it is underlain by red-stained Coal Measures with seatearths but no coals; the absence of coals is probably due to penecontemporaneous oxidation.

The Halesowen Formation is 69.1 m thick in the Grange Wood Borehole. The lowest 45 m consist almost entirely of sandstone, mostly purple and locally very coarse. The upper part consists of interbedded purple and green-grey sandstones, siltstones and laminated marls. The only limestone in the section is a 100 mm band at the top of the succession.

In the Church Flats Borehole, the basal 52 m of the Halesowen Formation is mainly purple sandstone, at some levels coarse and gritty, and with bands of siltstone with limestone nodules. The upper 29 m are mainly dark red-brown mudstones and siltstones; containing small concretionary nodules of cream, grey and purple

limestone (calcrete). Rounded pebbles of the limestone, derived by intraformational erosion, form locally derived bands of conglomerate or pellet-rock, usually in the basal part of the sandstones.

In the Netherseal Borehole, the Halesowen Formation consists mostly of red and grey sandstone, 37 m thick. The succeeding 28.8 m of red-brown mudstones and pellet-conglomerates may belong, in part, to the Halesowens rather than to the Keele Formation, to which they were assigned by G. Barrow (MS notes, 1919), who logged the borehole.

The Halesowen Formation was recorded from the Caldwell No.2 Borehole (Greig and Mitchell 1955, pp.52, 62) but the beds concerned are now regarded, after re-examination of the specimens, as basal Triassic sandstones.

The Keele Formation is recorded only from the Grange Wood, Church Flats and Netherseal boreholes. At Grange Wood, 30.4 m of mainly dark red-brown mudstone and siltstone overlies a 7.3 m sandstone. Some plant stems of *Lepidodendron* type were noted in an 0.6 m micaceous siltstone. Coarse sandstones with quartz pebbles occurred in the highest 13 m, interbedded with red-brown mudstones.

In the Church Flats Borehole, 46.7 m of Keele Formation were encounterd, below the Trias base, and consisted predominantly of dark red-brown mudstone with subsidiary siltstone, sandstone and mudstone-pellet beds. Green reduction spots with black centres, 'fish-eyes', are common in the mudstones. BCW

LEICESTERSHIRE COALFIELD

The Leicestershire Coalfield succession has been described by Fox-Strangways (1907), and Mitchell and Stubblefield (1948). Spink (1965) dealt mainly with the exposed coalfield. The lowest seam in the sequence is the Kilburn; it can be recognised throughout the coalfield but is commonly too thin to have been worked. The underlying measures include the locally thick Wingfield Flags sandstone sequence (Figure 4). Although the beds below the Kilburn crop out north of Packington, full details of the succession are known only from boreholes. Spink (1965, pp.47–51) showed that the Heath End Coal (Fox-Strangways, 1907, p.23; Mitchell and Stubblefield, 1948, p.7) is a cannel-coal development of the Kilburn.

In the succession above the Kilburn (Figure 12), a widely developed tonstein occurs within the Middle Lount seam (Strauss 1971, p.1526). Where the seam is split, the tonstein lies above its lower leaf.

The Vanderbeckei Marine Band (Figure 13) was first recognised at Bagworth Colliery [443 081] by Mitchell and Stubblefield (1941b, p.12) and named the Bagworth Marine Band (1948, p.12). Since then, it has been proved extensively just above the High Main. The highest worked seam in the coalfield is the Excelsior. The maximum proven thickness of measures above this seam and below the sub-Triassic unconformity is 46 m in the Bardon Hill Station Borehole [4389 1256]. A marine horizon correlating with the Haughton Marine Band occurs in this sequence in the Battleflat, Hugglescote Grange and Bardon Hill Station boreholes.

Reddening to a depth of 1 m or so is usual below the base of the Trias, but in an area near Newbold Verdon, at the southern end of the coalfield, the Coal Measures are reddened to much greater depths and many seams are missing

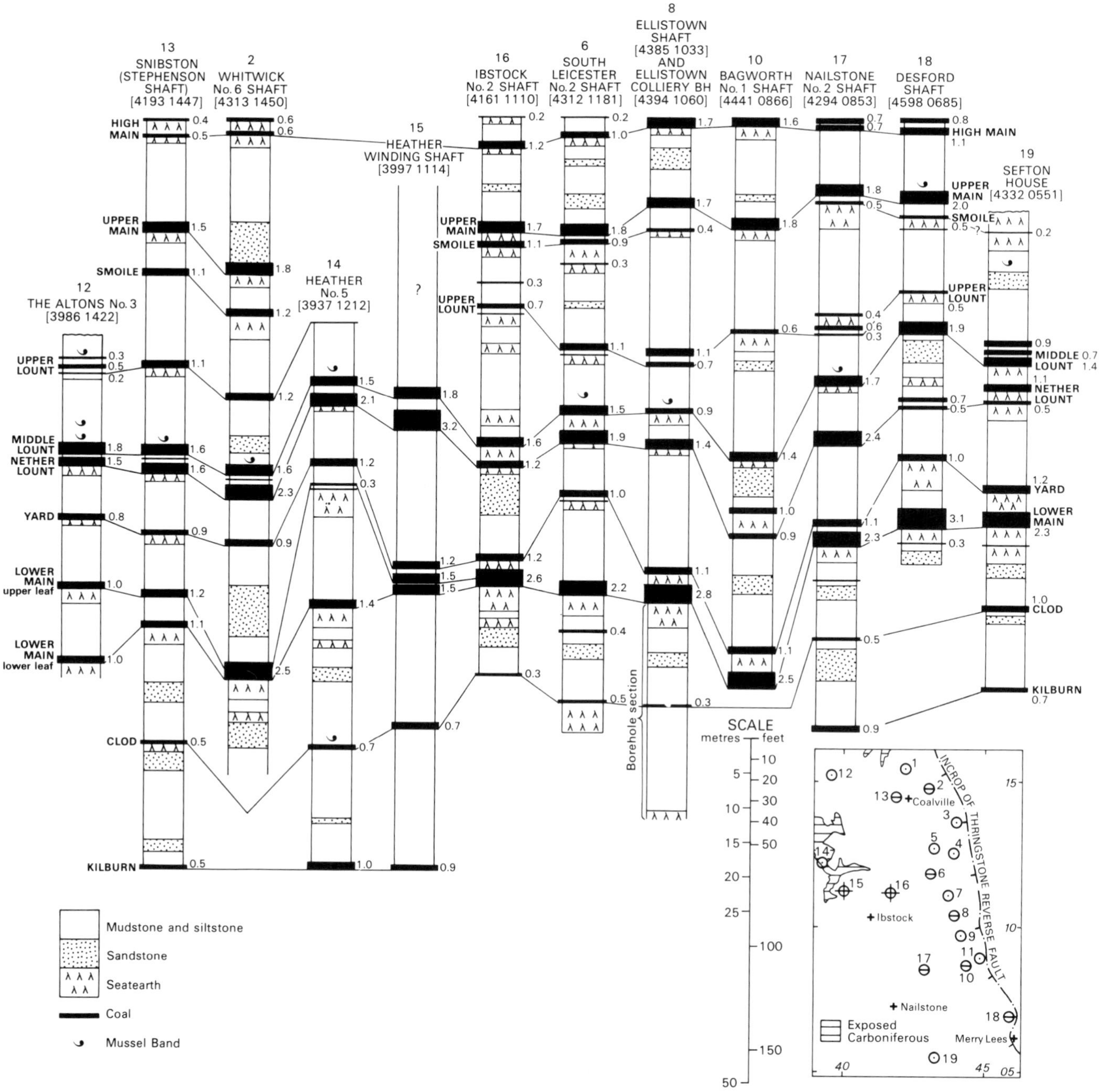

Figure 12   Comparative sections above the Kilburn in the Lower Coal Measures in the Leicestershire Coalfield

owing to oxidation below the pre-Triassic land surface; similar reddening occurs near Coalville (Horton and Hains, 1972).

Details

LOWER COAL MEASURES (Westphalian A)

The thick sandstone capping Windmill Hill [360 160], is probably part of the Wingfield Flags. In a disused quarry on the east side of the hill [3630 1603], 3 m of buff, soft, flaggy sandstone overlies 2 m of buff, pink-stained, more massive, false-bedded sandstone with frequent small ironstone nodules and a 6 cm band of purple-brown sandy siltstone. The sandstone is faulted down to the east where it crops out southwards to Packington. It is well exposed in a quarry [3605 1485] at Packington as 4.5 m of coarse, even-grained, pale yellow-brown sandstone. It is made up of massive 1 m to 1.5 m beds alternating with flaggy, micaceous, false-bedded sandstone. Ironstone nodules are scattered throughout. In a stream 100 m west of the quarry, the basal 2.5 m of this sandstone rests on grey mudstone, the lowest 0.4 m containing abundant plants and a 1 cm coal smut.

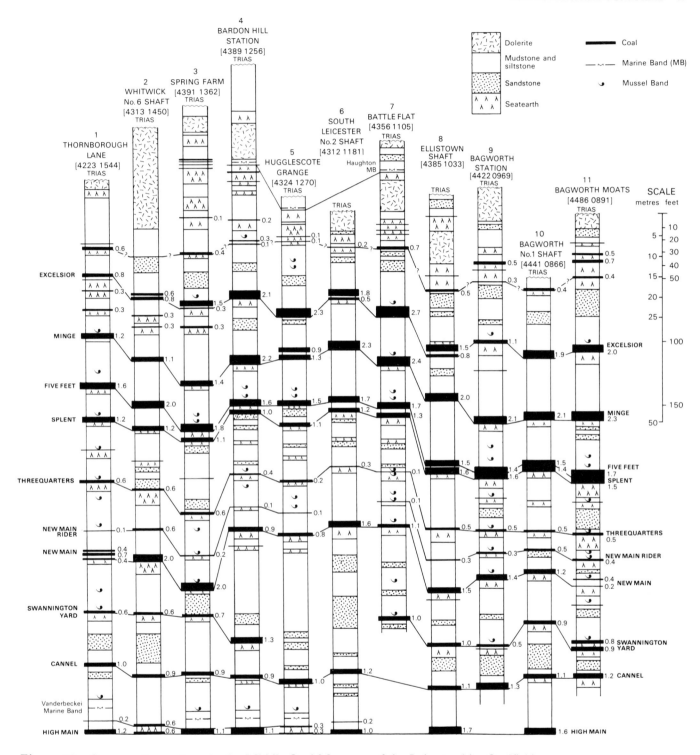

Figure 13   Comparative sections in the Middle Coal Measures of the Leicestershire Coalfield

The Wingfield Flags are proved in several wells at Heather Waterworks [3940 1073], where 25.6 m of sandstone have been proved with the base of the bed at 38.4 m below the Kilburn.

In the Ellistown Colliery Borehole (Figure 4, col.5), the base of the Coal Measures is taken as the base of the 4.9 m marine band, presumably the Subcrenatum Marine Band, 105 m below the Kilburn. In addition to *?Naiadites* and *Discina* (= *Orbiculoidea*) recorded by Boulton (1934, p.325), Stubblefield has identified *Lingula mytilloides*, *Anthraconeilo* cf. *laevirostris* and a coelocanth scale

from the band. A second marine band, possibly as much as 9 m thick, occurs 10.4 m higher in the sequence. The fauna recorded by Boulton and amplified by Stubblefield shows it to be the Listeri Marine Band and includes *Gastrioceras listeri*, *Pterinopecten* (= Dunbarella) *papyraceus*, *L. mytilloides*, *Orbiculoidea* cf. *nitida*, *Posidonia sp.* and a nautiloid. Above the marine band come alternating mudstones, seatearths and sandstones topped by an almost unbroken sandstone sequence, 36 m thick, correlating with the Wingfield Flags.

In the Osbaston Hollow Borehole [4166 0635] (Figure 4), *Gastrioceras* and fish remains including a *Diplodus* tooth were encountered at 154.5 m and *L. mytilloides* at 158.8 m (Mitchell and Stubblefield, 1948, p.5). These horizons, about 70 m below the Kilburn, are possibly parts of the Subcrenatum Marine Band. A 0.41 m coal, 13 m above the marine band, and possibly correlating with the Alton Coal, is overlain by other marine mudstones yielding *L. mytilloides* which probably represent the Listeri Marine Band.

Marine mudstones in the Desford Colliery No.12 Borehole [4683 0504], contain a rich fauna (Mitchell and Stubblefield, 1948, p.7), including *Gastrioceras*, cf. *subcrenatum* together with *Lingula mytilloides*, *Orbiculoidea sp.*, *Dictyoclostus sp.*, *Parallelodon reticulatus* and fish debris, almost certainly marking the Subcrenatum Marine Band.

A mussel band is commonly developed below the seatearth of the Kilburn south of Ibstock. In the Osbaston Field Farm Borehole [4291 0483], 33 m of grey, fissile mudstone, in parts sandy or micaceous and becoming sandy at the base, underlies the Kilburn and includes a black fissile mudstone, 3 m below the base of the seam, containing *Lepidodendron lycopoctoides* and *Carbonicola sp.*

The Stocks House Borehole entered reddened beds, presumably of Lower Coal Measure age, with *Lingula mytilloides* and scales of *?Rhizodopsis* (Gibson, 1905; Mitchell and Stubblefield, 1948, p.5) between the Trias base at 113.4 m and the Millstone Grit at 140.2 m.

In the Ibstock No.2 and Desford No.2 boreholes, the base of the Coal Measures is taken at a dark mudstone, not sampled but presumed to represent the Subcrenatum Marine Band.

The Kilburn, 0.5 to 0.7 m thick, was worked opencast at the Workhouse site [369 169] just north of the present district. The cannel development in the Kilburn at Heather Colliery Meadow Pit [3950 1091] (Fox-Strangways, 1907, p.292) is reminiscent of the 'Heath End Coal' (Spink, 1965, p.50). The Heather Colliery log is:

	Thickness m
Kilburn Coal	1.17
Seatearth	0.25
Black mudstone	0.26
Coal, cannel	0.81

North of Ibstock the Kilburn varies only from 0.9 to 1.2 m. In Stephenson Shaft [4193 1447], it is only 0.5 m thick, but rapidly increases to 1.3 m within 23 m along a north-western heading from the shaft; the variation in thickness is ascribed to a partial washout. The Kilburn is absent in the Ellistown Shaft, but is 0.9 m thick in the Meadow Row Borehole [4127 1019].

Between Ibstock and Osbaston, the Kilburn is about 0.75 m thick, but it is thickest around Ibstock reaching 1.42 m in the Ibstock Grange No 1 Borehole [4107 0954]. It thins to the south around Barlestone, where it is only 0.18 m thick in both the Osbaston Hollow and Barlestone Lodge boreholes. Near Merry Lees, 0.61 m of coal was proved in the Desford No.10 Borehole [4636 0519]. In the Desford No.17 Underground Borehole [4605 0722], the Kilburn occurs as two leaves, 0.3 m and 0.23 m thick, separated by 0.09 m of mudstone. The Kilburn is probably the 0.36 m coal encountered 23.8 m south-west of the Thringstone Fault in the Merry Lees Drift [4630 0565] (Butterley and Mitchell, 1945, p.708, figs. 2 and 5). Farther south, the seam is 0.41 to 0.61 m thick, being thickest in the Newbold Verdon Borehole. In the Osbaston Field Farm Borehole the seam is rather shaly.

In boreholes south and east of Newbold Verdon, the Kilburn exceeds 0.6 m only in the Lindridge Hall Farm Borehole [4644 0485], where it is 0.74 m thick.

The roof of the Kilburn is characterised by a 0.5 m band of black, commonly pyritous and bioturbated mudstone with fish remains. This passes up into mudstone or laminated mudstone and siltstone, often with abundant worm trails and ironstone nodules, becoming increasingly sandy westwards so that, between Ibstock and Barlestone, these beds are typically interbedded, fine-grained micaceous, sandstone and siltstone overlain by the thin seatearth of the Clod. South of Ibstock, in boreholes at Ibstock Manor [4602 0968], Ibstock Grange No.1 and Ibstock Lodge [4103 0894], poorly preserved mussels including *Carbonicola sp.* were found in dark mudstones 0.25 m to 0.4 m below the Clod.

The Clod is persistent but poor in quality and is not worked. It thickens westwards from 0.3 m to 0.7 m around Ibstock to 0.7 m to 0.9 m around Heather and Measham Hall. At Heather, the Clod splits into two leaves; in Heather Colliery Winding Shaft [3997 1114], the section reads downwards: coal, 0.6 m; black carbonaceous mudstone, 0.18 m; cannel coal, 0.3 m. The Clod thickens southwards from Ibstock reaching 1 m in the Osbaston Gate and Sefton House boreholes at Barlestone, although it was locally absent in the Barlestone Lodge Borehole, where its horizon is marked by dark shales resting on a sandy seatearth. Just to the south-east, in Desford No.9 Borehole [4353 0505], it is only 0.41 m thick. It is 0.85 m to 0.9 m thick in the Desford Colliery and Merry Lees area. The Clod is probably the 0.86 m coal found 35.7 m south-west of the Thringstone Fault in the Merry Lees Drift, and also in the Etna Heading, 90 m to the north-west (Butterley and Mitchell, 1945, pp.708, 709, figs. 2, 5). These authors tentatively regarded this coal as being the Alton, proved 69.8 m below the Lower Main in the Ibstock No.2 Borehole, but in view of the coal's proximity to the 'porcellanous breccia' which probably marks the Lower Main horizon, this seems unlikely. The beds are nearly vertical here, and the intervals between the porcellanous breccia, the 0.86 m Clod, and the 0.36 m Kilburn closely compare with the same intervals in nearby boreholes. The Clod is 0.75 to 0.94 m thick in boreholes around Newbold Verdon.

The roof of the Clod, like that of the Kilburn, is characterised by a persistent, thin band of bioturbated black mudstone with fish remains; mussels are common in the overlying few metres of mudstone and include *Carbonicola martini*, *C.* cf. *obtusa*, *C. sp.* ex gr. *pseudorobusta*, *Curvirimula subovata*, *C. trapeziforma* and *Naiadites sp.*, together with *Spirorbis sp.* and fish debris, suggesting an horizon within the Pseudorobusta Belt, late Communis Chronozone.. The Clod and Lower Main seams are only 6 m apart in Desford No.2 Borehole, but generally a persistent sandstone up to 6 m thick occurs 10 to 12 m below the Lower Main. About 3 m above this sandstone, an unnamed coal a few centimetres thick usually occurs with fish remains just above. The black mudstone with fish remains persists, even when the seam is absent, and lies at the base of the Lower Main seatearth; it is at the same horizon as the Twelve Inch of South Derbyshire (Spink, 1965, fig. 1).

The Lower Main was formerly known as the Roaster, a name which persisted until recently on the plans of opencast workings. North-west of a line from Whitwick to Heather, the Lower Main splits into the Lower Main (Lower Leaf) and Lower Main (Upper Leaf) which become increasingly separated northwards. In the extreme north, a thin Lower Main (Middle Leaf) coal is developed between them and this is the Middle Roaster of Spink (1965). The split in the Lower Main lies to the north-west of Whitwick Colliery, just south-east of Snibston No.1 Shaft [4257 1460] and Snibston Colliery, and to the east of Ravenstone. The line of split then passes south between Ibstock No.11 Borehole [4027 1116] and Heather Colliery Winding Shaft. South-east of this line, the Lower Main is 2.2 m to 3.1 m thick, has been extensively mined, and is virtually exhausted. A narrow washout in the Lower Main extends 550 m north-west from a point [4127 1286] 350 m north-east of Berryhills Farm. A smaller, parallel, washout extends 200 m south-east from St Mary's Church [4115 1313]. Between Nailstone and Barlestone, thin dirt bands sometimes occur near the top or bottom of the seam.

The horizon of the Lower Main, together with that of the Yard, is probably represented by the 'porcellanous breccia' proved 40 m south-west of the Thringstone Fault in the Merry Lees Drift, and also in the Etna Heading just to the north-west (Butterly and Mitchell, 1945, pp.708, 709, and figs. 2 and 5). Buttlerley and Mitchell considered the rock to be a fault breccia, cemented by igneous material now highly altered, but Horton and Hains (1972, pp.75, 76) concluded that the breccia resulted from the collapse of the surrounding sediments following the complete oxidation of the coals.

The Lower Main is present in all boreholes around Newbold Verdon, varying in thickness from 1.75 m in the Osbaston Fields Farm Borhole to 2.49 m at Newbold Verdon. In the former borehole, only 1.17 m of grey shale separate it from the base of the Trias, so its subcrop must lie just a few metres to the west.

In Desford No.2 Borehole, the Lower Main is split into four layers, with seatearth and black shale partings. In a tract about 1.5 km wide to the east of this borehole, the Lower Main and Yard lie within a zone of pre-Mercia Mudstone oxidation and are missing (Horton and Hains, 1972, figs. 8 and 9). Farther east, in the Lindridge Hall Farm, Ibstock No.2 and Desford No.13 boreholes, the Lower Main is again present, averaging 2.3 m thick.

North-west of the line of split, the Lower Main (Lower Leaf) is generally 0.9 m to 1.5 m thick and it has been worked both underground and opencast. At the Demoniac/Alton Opencast Site, a 0.97 m seam was recorded as the Yard, but Mitchell and Stubblefield (1948, p.14) are almost certainly correct in concluding that the seam worked was the Lower Main (Lower Leaf). The Yard was worked opencast at The Altons [388 153] and its incrop there is too far east for it to reappear at Demoniac/Alton, without the intervention of unproven faulting. The thickness of the Yard at the Alton and Bulwell Barn sites is 0.76 m and 0.86 m respectively, which makes it thinner than the Lower Main (Lower Leaf) anywhere hereabouts. The uncertainty at Demoniac/Alton seems to have arisen from the 'remarkable resemblance' of the two seams here (Mitchell, 1948, p.503).

North-westwards, the two leaves of the Lower Main diverge as shown below:

	Lower Main (lower leaf) (m)	Interval (m)	Lower Main (upper leaf) (m)
Heather Colliery Winding Shaft	1.2	0.13	1.4
Heather Old Colliery [3945 1147]	1.5	0.7	1.1
Stephenson Shaft	1.1	3.1	1.2
Snibston No.3 Shaft [4203 1547]	0.9	13.0	0.6
Heather No.5 Borehole	1.4	16.5	0.3 and 0.15
The Altons No.2 Borehole [3949 1460]	1.0	18.0	0.05 and 0.13
Hoo Ash Borehole	0.9	19.0	0.06, 0.15 and 0.25

Where the two leaves are close together, the interval consists largely of seatearth but as they diverge, thin sandstones are introduced.

The Lower Main (Middle Leaf) is developed mainly in the north and is recorded at only three localities within the present district by Spink (1965, p.54). It was encounterd in the Farm Town No.3 Borehole [3946 1523] where the section is:

	Thickness m
Lower Main (Upper Leaf)	0.7
Measures	7.6
Lower Main (Middle Leaf)	0.23
Measures	7.0
Lower Main (Lower Leaf)	1.1

The Lower Main (Upper Leaf) has not been worked, except near Whitwick where it is thickest, close to the line of split. To the west, it becomes thinner and is itself split in places.

North of Ibstock, the interval between the Lower Main and Yard is very variable in thickness and at the Coalfield Farm Opencast Site [394 113], the thickness of these beds is clearly fault-controlled, since the interval suddenly increases from 1.5 m to 4 m north-westwards across a small fault. The measures between the Lower Main and Yard include the seatearth-mudstone of the Yard and, in the south-west, a coal up to 0.07 m thick which occurs sporadically between the seams.

The Yard persists throughout the coalfield and has been worked extensively underground and opencast. It is generally 0.6 m to 1.3 m thick, reaching its thickest in the Ibstock Lodge Borehole. It thins northwards but is generally at least 0.9 m thick. It was thinner in opencast workings at Jubilee [399 135] (0.7 m) and The Altons (0.8 m). The Yard also thins to the east and south-east of Ibstock, being only 0.9 m in the Desford No.3 Borehole. In the Ibstock No.9 Borehole, the split coal at 72.5 m may represent an unusual development of the Yard. It occurs as a lower leaf of 0.71 m, and an upper of 1.32 m, separated by 0.36 m of dark grey fireclay. The lower leaf has previously been taken as the Lower Main (Greig and Mitchell, 1955, p.57), but the level of the Lower Main in workings to the east suggests that this coal is a lower leaf of the Yard, and that the Lower Main is cut out in the borehole by a fault at 76.0 m. In South Leicester Colliery No.2 Shaft, a 0.13 m seam is developed 0.7 m below the Yard, but Spink's assertion (1965, p.55) that this leaf is generally developed north of Heather and Ibstock is not borne out by the borehole and shaft sections. There is a local washout in the Yard at South Leicester Colliery (Mitchell and Stubblefield, 1948, p.10), and in the Lindridge area the coal is thin or absent due to post-Coal Measures oxidation (Horton and Hains, 1972).

The measures above the Yard are variable in thickness and characterised by very impersistent sandstones.

The Nether Lount is a thick seam which has been extensively worked underground and opencast, but which varies in thickness and tends to split. Thickness variations and the development of partings show no clear pattern. The seam attains a maximum of 3.2 m between Heather and Ibstock and in the Heather Brickyard Shaft, the section reads: coal, 0.76 m, parting, 0.08 m, coal, 0.76 m, parting, 0.03 m, coal, 1.37 m. East of Heather, the Nether Lount thins rapidly, apparently by the failing of the upper two leaves.

By contrast, farther south the Nether Lount consists of two leaves, the upper being the more persistent. The lower leaf is very variable in thickness, averaging about 0.65 m. It reaches its maximum of 1.45 m in the Ibstock No.9 Borehole, but in the Park Farm Borehole [4131 0895], only 0.5 km to the south-east, it is only 0.05 m. In the Barlestone Lodge Borehole, the lower leaf is absent, its horizon being marked by a seatearth 5 m below the upper leaf and it is also absent in the Desford No.3 Borehole. In Ibstock Lodge Borehole, the Nether Lount is overlain by a mussel band dominated by *Curvirimula subovata*; this is the youngest Communis Chronozone fauna recovered from the district.

The interval between the leaves of the Nether Lount varies in thickness from 0.3 m in Nailstone Colliery Shaft to about 3.0 m near Bagworth, south of Ibstock and around Barlestone. The interval consists of seatearth-mudstone, passing down into silty mudstone where the measures are thickest. The upper leaf averages

0.85 m in thickness and reaches its maximum of 1.55 m in the Nailstone Rectory Borehole [4210 0740]. In this and the Barlestone Lodge and Garland Lane boreholes, a thin dirt band (0.08 m) occurs near the top of the coal.

The subcrop of the Nether Lount below the Trias must lie to the east of the Heath Farm Borehole, as the seam was not encountered in it, and a few metres south of the Newbold Borehole, where only 1.0 m of measures separate the coal from the base of the Trias. The seam is 0.81 m and 0.61 m thick respectively in the Newbold Verdon and The Fields No.2 boreholes. In The Fields No.1 Borehole, the seam is split by a 1.3 m-thick seatearth parting into an upper leaf of 0.53 m and a lower leaf of 0.08 m.

The beds overlying the Nether Lount generally consist of a few metres of seatearth but where the sequence is thickest, south of Ibstock, mudstones yielding mussels and ostracods underlie the Middle Lount seatearth.

While not as thick as the Nether Lount, the Middle Lount is less variable and less ready to split. It has been widely worked underground and opencast. Over most of the area it is a single seam, 1.3 to 1.9 m thick. Its maximum thickness of 2.13 m was proved in the Ibstock No.9 Borehole. In the Ibstock Manor Borehole it is only 0.86 m thick, and the lower 0.25 m is canneloid. A thin parting appears near the base of the seam as it is traced northwards. For example, in The Altons No.3 Borehole, a 0.15 m parting occurs 0.16 m below the principal 1.5 m seam.

A thin ironstone band 0.13 m below the top of the seam in many sections in the north (Mitchell and Stubblefield, 1948, pp.10, 13) may be the tonstein 0.25 m below the top of the seam at Whitwick Colliery (Eden and others, 1963, p.52) and 0.2 m below the top of the seam in The Altons No.3 Borehole. Several washouts occur in the seam east of Coalville, trending WSW to ENE, and one was proved in the Ibstock Grange No.2 Borehole [4155 0975]. Around Nailstone, Barlestone and Newbold Verdon, the Middle Lount is split, generally with a thick lower leaf 1.0 to 1.5 m thick, and one or two thin upper leaves. A 0.76 m seam occurs 7.4 m above the Middle Lount in the Churchyard Farm Borehole [4293 0603], and is probably a local split. The upper leaves of the Middle Lount attain greater importance to the south-west; in the Sefton House Borehole for example (Figure 12).

The Middle Lount generally has three leaves around Newbold Verdon. The lowest leaf was not recorded in The Field No.2 Borehole, but core was lost from this level. Both the Nether Lount and Middle Lount are missing due to sub-Triassic oxidation near Desford. They are probably represented by the thin coals, in partially oxidised measures, 30.5 m to 42.1 m above the Clod in the Desford No.14 Borehole.

The 3 m of mudstone above the Middle Lount usually yield abundant well preserved, often solid, mussels including *Anthracosphaerium sp*, *Carbonicola crista-galli*, *C. rhomboidalis*, *C. venusta*, together with *Spirorbis sp*. and *Geisina arcuata*; the fauna correlates with the Crista-galli Belt, early Modiolaris Chronozone. Further mudstones overlie the band in the north, but to the south the mussel band is overlain by sandstones and siltstones, commonly rich in plant debris.

The Upper Lount is thin, variable and poor in quality. It has not been worked underground, but small tonnages were produced at the Ross Knob Plantation and Spring Lane opencast sites. One or two partings are usual and its uppermost leaf is generally the thickest. Excluding partings, the seam is 0.6 m to 1.8 m thick in the north, decreasing to 0.3 m in the south. In the Churchyard Farm and Sefton House boreholes, the Upper Lount is absent, but it may be represented by a thin seatearth overlain by black mudstone with mussels and ostracods, just below the seatearth of the Smoile. Alternatively, this horizon may correlate with the thin split from the Middle Lount encountered in the Park Farm and Nailstone Grange boreholes. A tonstein lies 0.8 m below the top of the Upper Lount at Ellistown Colliery (Eden and others, 1963, p.53).

A thin coal occurs locally 5 to 10 m above the Upper Lount in the Ibstock and South Leicestershire collieries. Mussels and fish remains occur in the roof of the Upper Lount in the south. Mussels were also found about 1.5 m below the overlying Smoile seam in the Pool House [4433 0505] and Craigmore Farm [4485 0527] boreholes, but it is not clear if these two bands are the same, since the boreholes did not penetrate far enough to prove the Upper Lount. Mussels curated with Leicester City Museum are probably from the roof of this seam ('12–21ft below Main') in Bagworth Colliery, (Mitchell and Stubblefield, 1948, p.42). The fauna is from the Regularis Belt within the early Modiolaris Chronozone.

The Smoile approaches workable thickness over much of the northern part of the coalfield but it has only been exploited at Whitwick. It is thickest, at 1.1 m, in the north and in the small area of incrop proved north of Heather. Traced southwards, it thins rapidly from 0.8 m in the Ravenstone Road Borehole [4107 1439] to 0.7 m in the Leicester Road, and 0.23 m in the Blackberry Farm, boreholes. South of Ibstock, it rarely exceeds 0.3 m in thickness, and is often split into two thin leaves.

The Upper Main lies just above the Smoile and is within 0.25 m in the South Leicester Colliery shafts and 0.5 m in Nailstone Colliery Shaft. The seams, however, are not known to merge. The Upper Main was formerly an important seam in the coalfield, but is now largely worked out. It is generally 0.8 m to 2.0 m thick. In the Noah's Ark Borehole [4356 0551], a 0.20 m rider occurs 0.36 m above the main 1.37 m seam. In Desford No 11 Borehole, the Upper Main is only 0.15 m thick probably due to oxidation beneath the sub-Triassic unconformity which lies 8.7 m above. The roof of the Upper Main is a sandstone in the north, which reaches 15 m thick in Whitwick Colliery No.6 Shaft [4313 1450] but dies out rapidly southwards.

The High Main is nearly everywhere split into upper and lower leaves; the lower is the more important and it has been extensively worked in recent years. In the Ellistown shafts, the seam is not split and totals 1.7 m. A 0.5 m band of impure coal 1 m below the High Main in this section is a local development and not an attenuated Lower Leaf. In the Upper Grange [4411 1181] and Bardon Hill Station [4389 1256] boreholes, the seam is again not significantly split, and in the latter the section is coal, 0.27 m; parting, 0.13 m; coal, 0.7 m. Elsewhere, the parting between the two leaves is thicker.

The High Main (Lower Leaf) reaches 1.3 m in The Hill Borehole [4390 0606], but farther north a thin parting is developed within it. In the Church Lane Borehole, for example, the section is: coal, 0.5 m; parting, 0.03 m; coal, 0.8 m.

The High Main (Upper Leaf) is a thin seam, 0.2 m to 0.7 m thick, and not worked.

MIDDLE COAL MEASURES (Westphalian B)

The mudstones of the Vanderbeckei Marine Band are rich in pyrite which frequently preserves macro- and trace-fossils. Clay ironstone is abundant as nodules and in bands, which in the Bardon Hill Station Borehole reached 12 cm in thickness. *Lingula mytilloides* is by far the commonest fossil in the Band, but arenaceous foraminifera, *Orbiculoidea sp.*, *Productus carbonarius*, *Dunbarella sp.*, *Myalina compressa*, *Posidonia sp*, turreted gastropods, *Retispira sp.*, *Belinurus sp.* and fish have also been recorded. So far, no goniatites have been found. The marine band consists of 1 to 7 m (average 3 m) of fissile, medium to dark grey mudstone, and is generally overlain by a mussel band dominated by *Naiadites quadratus*, marking a return to non-marine conditions (Mitchell and Stubblefield (1948, p.12)).

About 8 m of mudstone with *Anthracosia sp.* and *Naiadites quadratus* overly the Marine Band up to the horizon of the Cannel Seam. The Cannel includes much cannel coal and although it is generally about 1 m thick and is laterally persistent, it has not been worked. The seam gradually thickens southwards and, in places, a thin split develops; for example, in the Upper Grange Borehole, the sequence is: coal, 0.4 m; parting, 0.1 m; cannel, 0.5 m.

In the Neville Arms Borehole [4355 0872], sandstone rests directly on the Cannel, and in the Bagworth Borehole [4408 0791], the coal is washed out altogether. It is also absent from the Desford Colliery Shaft, although here the measures consist only of mudstone. A muddy 'transitional' tonstein in the Cannel was recorded in the Spring Farm Borehole [4391 1362] by Strauss (1971, p.1521).

Mussels were found in the roof of the Cannel in the Heath Road Villas Borehole [4541 0698], the Desford Colliery workings and the Upper Grange Farm Borehole [4411 1181]; in the last section, poorly preserved *Anthracosia* cf. *phrygiana* are recorded from a band about 10 m above the Cannel, suggesting a possible Phrygiana Belt, in the late Modiolaris Chronozone.

The Swannington Yard seam has not been worked anywhere in the district. In the north, it is normally split into two leaves and a typical section reads: coal, 0.4 m to 0.8 m; parting, 0.1 m to 0.3 m; coal, 0.1 m to 0.25 m. In the Ravenstone Road and Bardon Hill Station boreholes, the lower leaf is itself split. In the south, the Swannington Yard is a single 0.5 m seam, except north of Bagworth where it is split and dirty, and occurs as a thick upper leaf and several thin, poor quality lower leaves. Between Ibstock and Coalville, the seam is locally washed-out, as in the Hugglescote Grange Borehole [4324 1270] and in the South Leicester Colliery shafts. The roof of the coal consists of dark mudstones containing mussels, which passes up into mudstone interbedded with thin siltstones and sandstones. A thin seatearth usually occurs 6 to 9 m above the Swannington Yard and in the Bagworth Moats Borehole a 0.10 m coal occurs at this horizon. These beds are overlain by mudstones with mussels and then by the seatearth of the New Main. A fauna including *Anthracosia aquilina*, *A. lateralis*, *A. phrygiana*, *A. retrotracta* and *Anthracosphaerium sp.* was recovered from the mudstones in Upper Grange Farm and Wigg Farm boreholes, suggesting correlation with the Phrygiana Belt, late Modiolaris Chronozone.

The New Main is thick and persistent and has been extensively worked. It is usually 1.0 m to 1.7 m thick; exceptionally, as in the Spring Farm Borehole, it reaches 2.0 m. In the Hugglescote Grange Borehole, it is reduced to two leaves, each of 0.3 m. A distinct upper leaf is developed in the north of the district; for example in the Spring Lane Borehole [4161 1579] the section is: coal, with a thin parting, 0.3 m; parting, 0.9 m; coal, 1.3 m. West and south of Desford Colliery, a leaf of coal 0.05 to 0.15 m thick lies just below the main seam. A tonstein is widely developed in the New Main (Strauss, 1971, p.1526).

There is a mussel band in the roof of the New Main in places, for example, in the Wigg Farm Borehole, which contains poorly preserved *Anthracosia spp.* including a possible *A.* cf. *caledonica* suggesting an horizon close to the boundary between the Modiolaris and Lower Similis-Pulchra Chronozones. Elsewhere, thin sandstones overlie the coal. South-east of this area, the New Main Rider approaches the New Main very closely, almost joining it in the Desford Baths Borehole [4586 0656] where the separation is only 0.1 m.

The New Main Rider is a thin but persistent seam lying up to 10.25 m above the New Main and forming a useful marker horizon, with a mussel band in its roof. The coal is generally about 0.2 m thick but an exceptional 0.7 m was encountered in Snibston No.1 Shaft [4257 1460]. The seam thickens again around Desford, reaching 0.6 m in the Desford Baths Borehole, where it lies only 0.1 m above the New Main.

Above the New Main Rider are dark mudstones with several bands of abundant mussels and occasional fish fragments. The mussel band in the roof of the New Main Rider is particularly prominent and is dominated by *Anthracosia* cf. *beaniana*, *A.* cf. *phrygiana* and *A.* cf. *retrotracta*. The Threequarters, 0.1 m to 0.6 m

thick, has not been worked within the district. In the Bagworth Moats and Little Bagworth boreholes, a thin leaf of coal (0.05 m to 0.09 m) occurs just below the main seam. A mussel band is usually developed in the roof of the Threequarters and in the Bagworth Station, Bagworth Moats, Heath Road Villas and Desford Tip [4593 0722] boreholes, a thin coal lies 5 to 10 m higher in the sequence. This horizon is overlain by mudstones, in places containing abundant mussels. This fauna, from the interval between the Threequarters and Splent coals, includes *Anthracosia* cf. *beaniana*, *A.* cf. *retrotracta* and *Naiadites* cf. *quadratus*.

The Splent ranges in thickness from 0.5 m to 1.8 m; it has been worked separately at Whitwick, but farther south it closely approaches the overlying Five Feet and the two are worked together south of Ellistown.

In the south of the district, the parting between them averages 0.3 m. A mussel band occurs in the roof of the Splent in the more northerly boreholes such as Thornborough Lane, and at Whitwick No.3 Pit it yielded *Anthraconaia pulchra* and ?*A. ex gr. subcentralis*. (Mitchell and Stubblefield, 1948, p.42).

The Five Feet has been worked at Whitwick and Ellistown collieries and at the Limby Hall opencast site [414 165]. It is generally 1.6 m thick, ranging from 1.0 m to 1.8 m, and has been extensively worked jointly with the Splent in the south of the coalfield. Mussels and sometimes fish occur in the roof of the Five Feet, and a persistent mussel band is found 2 m to 4 m above, which includes *Anthracosia atra*, *A.* cf *concinna*, *A. fulva*, *A. nitida* and cf. *Anthracosphaerium sp.*, characteristic of the Atra Belt, Lower Similis-Pulchra Chronozone.

The Minge is a thick and economically important seam, especially south of Hugglescote where it reaches 2.0 m to 2.4 m in thickness, but its small area of incrop limits its reserves. Near Hugglescote the seam splits, so that the section in the Hugglescote Grange Borehole is: coal, 0.9 m; parting 0.4 m; coal, 1.3 m. Farther north it is a single seam, 1.1 m to 1.5 m thick, and perhaps the sudden thinning is due to the failure of the Upper Leaf.

Above the Minge, one or two thin coals are developed in the north of the coalfield and one of these was worked opencast during the 1926 General Strike in a small valley at Swannington. The workings extended for about 150 m south-eastwards from the former Methodist Chapel [4147 1598], but have been back-filled. In places, a poorly developed mussel band lies in the roof of the Minge.

Although the Excelsior has been known since the 19th century, its name was first published by Spink (1965, p.68). It was first worked in 1965 at Ellistown Colliery. The Excelsior is persistent but variable, and becomes increasingly split northwards. In the Bagworth Station Borehole, the Excelsior is a single 1.09 m seam. At Ellistown Colliery, the section is: coal, 1.5 m; seatearth, 0.3 m; coal, 0.8; at South Leicester Colliery the upper leaf has split and the section is: coal, 0.7 m; parting, 0.1 m; coal, 0.8 m; parting, 0.3 m; coal, 0.5 m. In the Church Lane Borehole, the seam is very much split, and in the Thornton Lane Borehole, 0.04 m of mudstone, resembling a tonstein in appearance, occurs near the top of the upper leaf.

A mussel band occurs in places in the roof of the Excelsior, and in the Bungalow Farm Borehole [4454 0944], mussels were found in a mudstone band within sandstone, 9.6 m above the Excelsior.

In the Bardon Hill Station Borehole, a group of thin coals with a mussel band above, which includes poorly preserved *Anthraconaia sp.*, occurs about 11 m above the Excelsior, but in the nearby Upper Grange Borehole, there is only a single 0.03 m coal in this part of the sequence. The succession above the Excelsior is generally difficult to correlate in detail, since there are several thin and impersistent coals. The problem is exacerbated locally by the intrusion of sills of the Whitwick Dolerite into this part of the sequence. A marine band, a few centimetres thick, occurs 24.7 m above the Excelsior (excluding a dolerite sill) in the Battleflat Borehole [4356

1105] (Horton, 1963, p.39), 23.8 m above the Excelsior in the Hugglescote Grange Borehole and 32.3 m above the Excelsior in the Bardon Hill Station Borehole [4389 1256]. It has not been confirmed in other boreholes through the same part of the succession. At Battleflat, the only marine fossil identified is *Orbiculoidea cincta* and at Hugglescote Grange only poorly preserved *Lingula* occur, whereas at Bardon Hill Station Borehole the band is represented by the marine ichnofossil *Tomaculum sp.* At this latter locality, the marine mudstones have been modified into a seatearth. It seems likely that the Haughton Marine Band was deposited over much of the district, but has subsequently been lost, or rendered difficult to identify, by pedogenesis. In the other boreholes the marine mudstones overlie seatearth. This marine band is so impersistent, and its fauna so sparse, that its correlation is uncertain, but the occurrence of *Orbiculoidea* suggests that it is the Haughton (Calver, 1968) rather than the Maltby Marine Band. RAO,MGS,BCW.

CHAPTER 6

Triassic

Rocks of Triassic age underlie the greater part of the district and are represented by the Sherwood Sandstone Group and the overlying Mercia Mudstone Group. At the base of the Sherwood Sandstone Group are locally-derived breccias, some part of which may be Permian in age. These are the Hopwas Breccia in the south-west fringing the Warwickshire Coalfield, and the Moira Breccia cropping out in the north, on the edge of the Leicestershire and South Derbyshire coalfields. In the west of the district, both breccias are overlain by sandstones and pebble beds, formerly known as the Bunter Pebble Beds but now locally named the Polesworth Formation. Overlying this in the west, but overstepping eastwards successively onto the Moira Breccia and older rocks is a sequence of interbedded sandstones and red mudstones ('marls'), formerly known as the Lower Keuper or Keuper Sandstone, and now called the Bromsgrove Sandstone Formation. The overlying Mercia Mudstone Group overlaps the Bromsgrove Sandstone, onto the ancient rocks of Charnwood Forest in the north-east, and onto the buried hills of diorite and associated igneous rocks in the south-east (Figure 27). It comprises all the beds formerly known as the Keuper Marl. Within the Mercia Mudstone, a number of formations, first described from the Nottingham area by Elliott (1961), can be distinguished.

Table 3 relates the lithostratigraphical subdivisions of the Triassic rocks of the present district to those of the central and western Midlands. The age of the Polesworth Formation is uncertain; nor is it known if there are non-sequences below and above it, matching those below the Kidderminster Formation and above the Wildmoor Sandstone (formerly the Bunter Pebble Beds and Upper Mottled Sandstone respectively) of the West Midlands. The Bromsgrove Sandstone, whose type area is south-west of Birmingham, is currently assigned largely to the Anisian stage, although its base is possibly Scythian in age and its top, Ladinian (Warrington and others, 1980, table 4). Palynological studies of the Mercia Mudstone Group in the district (p.67) indicate a late Anisian age for the Radcliffe and Carlton formations and a Ladinian to possible early Carnian age for the Cotgrave Skerry at the base of the Edwalton Formation.

Except for some of the sandstones, the local Triassic rocks are generally red in colour, reflecting their deposition in a semi-arid to arid environment. Sedimentary structures in the Polesworth and Bromsgrove Sandstone formations indicate fluvial conditions of deposition. There are fragmentary plant remains in the Bromsgrove Sandstone, and indeterminate molluscs and fish in micaceous sandstones of the Waterstones lithofacies at the top of this formation (p.63). The Waterstones fauna provides local evidence of a marine incursion, more strongly represented to the north-east by the Muschelkalk sea, then covering north Germany and the North Sea Basin. From Ladinian to Norian times, the climate seems to have been more intensely arid than previously. Gypsum, as primary nodules and as secondary veins of satin-spar, though present in the Bromsgrove Sandstone, is particularly characteristic of the Mercia Mudstone. A saline environment is also indicated by scattered pseudomorphs after halite crystals in the latter, though the present district lay beyond the limits of deposition of the Stafford Halite, the nearest bedded salt formation.

SHERWOOD SANDSTONE GROUP

Hopwas and Moira breccias

Both the Hopwas Breccia and the Moira Breccia are locally derived, with sandstones and subordinate beds of mudstone. They vary rapidly in thickness, being about 30 m generally and reaching a maximum of 65.5 m in the Caldwell No.22 Borehole just north of the present district. Accumulation began on the uneven land-surface produced by subaerial erosion, following the late Carboniferous or early Permian folding and uplift of the area. The breccias thicken into the Hinckley Basin, between the South Derbyshire and Warwickshire coalfields, which seems to indicate that deposition was also influenced by tectonic subsidence.

The type-area for the Hopwas Breccia is near the village of Hopwas [180 050] in the Lichfield (sheet 154) district where the breccias include a high proportion of Carboniferous Limestone pebbles, but elsewhere in the Lichfield district, these are replaced by clasts of quartzite, decomposed igneous rocks and Silurian sandstone (Barrow and others, 1919).

The Moira Breccia is normally a dark purplish brown, unconsolidated, sandy gravel. In places, it includes massive, consolidated beds with a calcareous cement. Subangular quartzite pebbles, pitted apparently by pressure solution when buried at depth, are common. In the Packington-Measham Hall area, the commonest pebbles are slightly rounded, tabular or blade-shaped and are most commonly clasts of Cambrian mudstone; Carboniferous sandstone forms larger and more spherical pebbles and other clasts are of Charnian tuff, quartz and granite, together with highly polished ironstone. The outcrop of the Moira Breccia is readily recognisable from the abundant tabular pebbles in the soil.

The name 'Breccia and Marl of Moira' was used on 1-inch sheet 155 (Atherstone) published in 1899, although in the accompanying memoirs (Fox-Strangways, 1900, 1907) the rocks were referred to as 'Permian breccias'. G. H. Mitchell, on the published 6-inch map Derbyshire 63 NE (1945), used the term Moira Breccia. Mitchell and Stubblefield (1948) equated these rocks with the Hopwas Breccia, using the term 'Moira or Hopwas Breccia'. Although the beds do not actually occur at Moira, their arcuate outcrop surrounds the locality. Recent practice, using Moira Breccia around the South Derbyshire and Leicestershire coalfields and Hopwas Breccia in the Warwickshire Coalfield area (Hains and Horton, 1969) is followed in the present account, since it

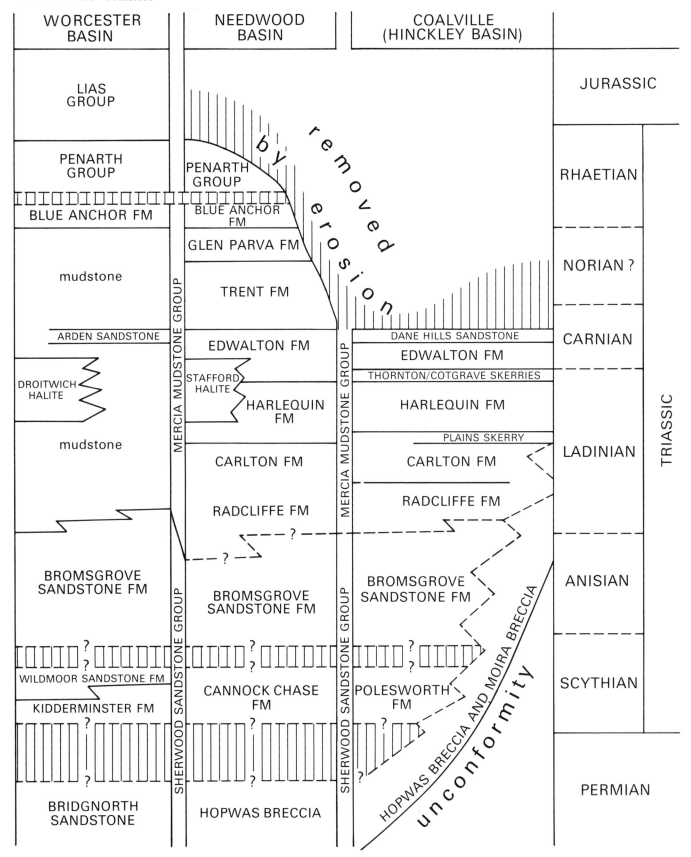

Table 3 Triassic formations and stages of the Midlands (based upon Warrington and others, 1980)

distinguishes between deposits with different constituents and provenance, and possibly of slightly different ages.

Thicknesses of Moira and Hopwas breccias proved in boreholes are shown in Figure 14. The Moira Breccia is over 30 m thick in a north-west-trending belt along the south-western side of the South Derbyshire Coalfield. Some of the thickness variations may result from the filling of valleys in an uneven topography. For example, there seems to have been a valley aligned south-east through Willesley along the Boothorpe Fault. It seems to join a north–south valley through Packington. Farther to the south-west, there appears to have been a south-west-trending ridge along which the Moira Breccia is relatively thin (e.g. 5.6 m at the Minorca Colliery [3552 1141] and 7 m in the Turnover Borehole [3423 0995].

Borehole provings of the Hopwas Breccia are meagre. The Lyndon Lodge No.2 Borehole [2738 0283] proved 12.2 m

and No.1 [2775 0315] proved 10.5 m. Down-dip from the northern end of the outcrop east of Polesworth, no Breccia was recorded in three boreholes. It was also absent from the Shuttington Fields Borehole [2642 0610]. No cores were taken, however, of this part of the succession in these holes, and it may be that the Hopwas Breccia was present but was logged as Polesworth Formation.

Brown (1889) described a section at Swadlincote [313 196] in the Loughborough (sheet 141) district, with an angular unconformity between 'Permian breccias' dipping at 23°, and the overlying, more gently dipping, 'Bunter' pebbly conglomerate. He considered this proof that the breccias are pre-Triassic in age. In the South Derbyshire Coalfield, he found (1889, p.28) that the breccias contain more angular clasts in the south and tend to die out northwards, being replaced by marls. He thought this indicated derivation from the south. He described suites of rock types in the breccias

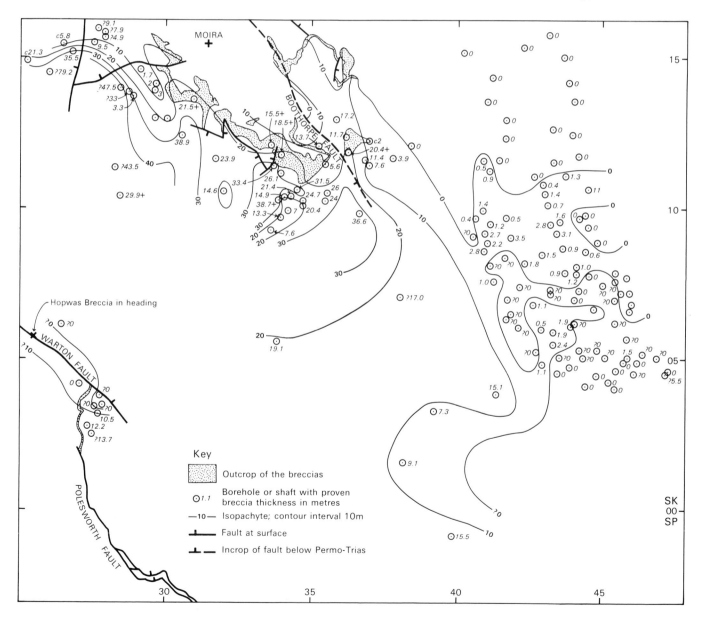

Figure 14 Isopachytes of the Hopwas and Moira breccias

from five localities in the South Derbyshire Coalfield, including Measham and Packington in the present district, and from a road section 0.5 km east of Polesworth, in what would now be termed Hopwas Breccia. The most abundant lithology was a quartzite resembling that from Hartshill, but with slight metamorphism indicated by strain-shadows of the quartz grains. He thought it likely that it originated from an outcrop of Cambrian quartzite now concealed beneath Triassic deposits, possibly lying near Market Bosworth. The sample from the Hopwas Breccia differed from the Moira Breccia samples in containing nearly 10 per cent of Carboniferous Limestone fragments, a rock-type rare in the South Derbyshire breccias.

The Hopwas Breccia was correlated by Smith and others (1974, p.27) with the Quartzite Breccia and the Barr Beacon Beds of the Birmingham area. They also suggested (p.17) that the Moira Breccia and Hopwas Breccia represent outlying patches of the Lower Permian breccias of north Nottinghamshire, and were possibly broadly contemporaneous with the Bridgnorth Sandstone.

Neither the Hopwas nor the Moira Breccia have been dated by fossils. At outcrop near Measham, the Polesworth Formation seems to pass laterally eastwards into the upper part of the Moira Breccia, which is likely, therefore, in its upper part at least, to be Triassic in age. Farther east, the Moira Breccia crops out beneath the Bromsgrove Sandstone at Packington and Measham Hall; it thins out rapidly eastwards and is absent around Heather and Normanton le Heath.

A thin basal breccia which has been correlated with the Moira Breccia (Hains and Horton, 1969, p.62) is found beneath the Mercia Mudstone in boreholes east of Ibstock and Coalville. In the Battleflat Borehole [4356 1105], the basal breccia is 1.3 m thick; exceptionally, in the nearby Wood Farm Borehole [4451 1057], it exceeds 11 m near the base of the Mercia Mudstone, many of its pebbles originating from the Charnian and its associated intrusives. The breccias have some of the characteristics of the Moira Breccia, particularly their lithology and constituent pebbles, but they also compare with the local breccias in Charnwood, since they are thin and appear to fill shallow channels and hollows in the pre-Triassic land surface. It is not clear therefore that they should be considered as part of the Moira Breccia. In particular, it is difficult to demonstrate their lateral continuity with the outcrops of the latter to the west. On balance, it seems better to regard them separately as local basal breccias.

DETAILS

Atherstone – Shuttington

At Grendon, two small outliers of Hopwas Breccia [280 993], in which probably only about 1 m of the deposit remains, cap slight rises of ground on the Lower Coal Measures outcrop. Ploughed-up in the soil are subrounded, quartzite pebbles, fragments of purplish red, calcareous sandstone, and pieces of a calcareous conglomerate of flattish sandstone pebbles. In a road cutting east of Dordon, at the south end of a small outlier [273 008] close to the coalfield boundary fault, the Breccia is seen resting with near-horizontal junction on pale grey Coal Measures mudstone. It consists of 3 m of purple breccia, containing shale chips 5 to 10 mm across and subangular

quartzite pebbles, many of them pitted and 150 to 200 mm in diameter. About 900 m to the north, 10.7 m of Breccia were proved in boreholes at the St Helena opencast site [2715 0169]. The basal Breccia in each of the cored boreholes lay unconformably upon the Coal Measures. Although the unconformity was steeply dipping at 35°, the contact was unfaulted and fragments in the Breccia were aligned parallel to it. In the Lyndon Lodge No.1 Borehole [2775 0315], the Breccia was 10.5 m thick.

North of the River Anker, the Hopwas Breccia forms a bold scarp. A roadside section on Stripers Hill, 850 m ENE of Polesworth Church, shows 6 to 7.5 m of Breccia, overlain by red pebbly sand of the Polesworth Formation. Brown (1889, p.24) listed the constituents of the Breccia from this locality as quartzite 84%, Carboniferous Limestone 9.4%, igneous rocks, more or less decomposed, 4% and grey, flinty slate 2.6%. Two small hilltop outliers of Breccia occur east of Bramcote Hall [273 043]. North of Shuttington, a heading in the Bench Coal at OD passed through the Warton Fault to encounter Hopwas Breccia [2552 0572] on the downthrow side.   BCW, KT

Overseal – Measham – Willesley

At Linton, just north of the present district, in a roadside section [2737 1680] described by Hull (1860, p.50, fig. 11) and still visible in 1975, the Moira Breccia is thrown against Bromsgrove Sandstone by the Netherseal Fault. The Breccia here is at least 15 m thick.

West of Overseal, the Grange Wood Borehole proved Polesworth Formation down to 297.2 m and Moira Breccia from 304.2 m to the Carboniferous at 307.5 m. In the Netherseal Borehole, closer to the Netherseal Fault, specimens of Moira Breccia occurred as high as 32.8 m above the top of the Carboniferous (G. Barrow, 1919, in MS). In the cores of the Caldwell No.2 Borehole [2568 1672], 400 m north of the present district, T. Eastwood recorded 65.5 m of Moira Breccia. The log of the downcast shaft of Netherseal Colliery [2757 1543] (Fox-Strangways, 1907, p.201, 'Boring No.4') shows the Breccia is 9.5 m thick, comprising 5.1 m of 'conglomerate (large pebbles and boulders)' overlying 4.4 m of 'red sandstone full of pebbles'.

In the Chilcote Borehole [2828 1145] (Fox-Strangways, 1907, p.331), the Moira Breccia is 43.5 m thick, comprising mainly conglomerate with beds of sandstone, red marl and 'mottled marl'. The Chilcote Waterworks Trial Borehole [2849 1039] proved 29.9 m of Moira Breccia, consisting mainly of calcareous breccia, without reaching its base.

North of Overseal Church, the Moira Breccia at outcrop is overstepped northwards by the Polesworth Formation. Near Short Heath [305 150] and on either side of the Hooborough Brook valley, the outcrop of the Breccia is marked by sandy soils with small, subangular pebbles about 25 mm in diameter and red-stained, subrounded, quartzite pebbles 50 to 75 mm across. An excavation [303 155] for a settlement pond for the Acresford Gravel Pit exposed 1.5 m of hard purplish sandstone with angular pebbles, in beds 0.3 m thick. The Acresford No.6 Borehole [3096 1358] started below the top of the Moira Breccia and entered Coal Measures at 21.5 m, suggesting that its full thickness here is about 25 m.

At Oakthorpe the retaining wall along part of the disused canal [317 130] is built of blocks of Breccia (Brown, 1889, p.11), possibly dug from an overgrown pit nearby [319 128].

G. H. Mitchell recorded (in MS) that, in 1950, on the north face [329 127] of the Oakthorpe Meer opencast site, the lowest 2.5 m of the Breccia was a coarse, bouldery bed with subangular and rudely rounded blocks up to 0.5 m across. Pebbles in a dull red, clayey matrix included igneous (syenitic) rocks, green tuffs, white porcellanite resembling that from Merry Lees, Desford (Horton and Hains, 1972), dolomite, clay ironstone replaced by hematite, sandstones and grits, all with an outer, hematitic coating.

Brown (1889, p.11) recorded unevenly bedded, highly consolidated Breccia seen to 4 m in a disused quarry beside the canal, 275 m west of Measham Church. This is one of the localities from which he described pebbles (1889, pp.24–28) as detailed above (p.54).

A bed of red clay is traceable eastwards for about 3 km from a fault [324 123] at 0.5 km south-east of Oakthorpe. For about 1.5 km the bed is underlain by Breccia, which crops out to the north of it, and is overlain by pebbly sand of the Polesworth Formation. Farther east, it lies within the Breccia and is, therefore, considered part of the Moira Breccia. The Breccia above it, east of Measham, may be stratigraphically equivalent to the Polesworth Formation beds farther to the west or may be overstepped by them. The Breccia below the red clay is about 17 m thick hereabouts, since the Chapel Street No.2 Borehole [3276 1240], which started in Breccia a little to the north of the red clay outcrop, entered Coal Measures at 15.2 m. The Pegg's Close Borehole [3357 1205], starting just above the base of the red clay, entered the Coal Measures at 15.47 m, having penetrated brown sandstone with well rounded pebbles of quartz and Coal Measures mudstone between 5.49 and 7.77 m. The strata above 5.49 m are unrecorded. Below 7.77 m, dull reddish brown and green, mottled Breccia, with angular to subrounded pebbles of Coal Measures and Charnian rocks rests on Coal Measures. The Gallows Lane Borehole [3390 1177] proved mainly sandstone with thin red mudstone bands between 6.1 and 8.9 m, on Breccia to 18.5 m, resting on Coal Measures. The clay was proved in the Measham No.10 Borehole [3386 1120] to be 14.5 m thick and to consist of chocolate-coloured mudstone with bands up to 0.3 m thick of coarse pebbly grit with a marly matrix, including micaceous partings, and passing down into micaceous siltstone with mudstone partings on its lowest 1.5 m.

The Snarestone Waterworks Borehole [3389 1011] proved Moira Breccia more than 38.7 m thick, down to the bottom of the hole at 121.9 m. The following percentages of pebbles at various depths were recorded in 1944 by Professor F. W. Shotton in a report in the BGS files.

Depth (m)	No of pebbles in samples	Carboniferous %	?Silurian %	Cambrian %	Precambrian %
85.9	31	10	6	36	35
92.0	78	6	14	23	48
98.4	82	17	11	20	25
103.9	122	8	2	45	34
110.3	88	2	4	20	56
121.0	50	—	—	10	86
Mean	—	7	6	27	47

The pebbles ascribed to the Carboniferous were of limestone and chert, those to ?Silurian of grey sandstones and grey shale. Silurian rocks are not known to occur in the district, even at depth, but Shotton (in litt., 1980) suggested derivation from buried Silurian rocks east of Coventry (cf. Shotton, 1927). The preponderance of Cambrian and Precambrian clasts (igneous rocks, tuffs and green and purple, baked shale) is noteworthy, as is the upward diminution in quantity of the Precambrian fragments. The outcrop of Coal Measures that trends ESE from Measham Colliery Shaft [350 120] is a topographical ridge in the pre-Permian land surface against which the Moira Breccia was banked. South of it, the Breccia includes the band of red clay described above. To the north the succession is thinner. The outcrop widens [3436 1243] west of Measham House but dies out 0.5 km to the north, possibly against another former ridge. The Moira Breccia crops out for 1 km or so along the line of the Boothorpe Fault north-west of Willesley Church, where it seems to have filled a valley in the pre-Permian land surface. Owing to the Breccias's resistance to erosion, it now forms a ridge. Trial boreholes here proved at least 5 m of Breccia.

Its outcrop extends for about 1.5 km north-eastwards from Willesley Church, where it is about 15 m thick, but the Breccia thins northwards along the face of an escarpment capped by basal Bromsgrove Sandstone. BCW

Packington to Measham Hall

The Moira Breccia varies rapidly in thickness across the line of a pre-Triassic valley running southwards between Packington and Measham Hall. It is thickest in the deepest part of the pre-Triassic valley south of Measham Hall (Figure 14). Nearby, 26 m were recorded in the Measham Colliery No.9 Borehole [3549 1049] and 24 m in the Valley Bungalow Borehole [3544 1026]. Only 5.6 m of Breccia were found in the nearby Minorca Colliery Shaft [3579 1138]. Eastwards, 3.9 m of Breccia were proved in the Opencast Executive Borehole 2005 [3773 1179] and it was absent in the Normaton No.6 Borehole [3836 1202].

Moira Breccia up to 3 m thick fills a hollow in Carboniferous sandstone in a quarry at Packington [3650 1487]. The unconsolidated Breccia is crudely bedded, with angular and rounded pebbles in a dark red, clayey matrix. Most of the pebbles are less than 6 cm across and are markedly tabular. Similar Breccia was encountered in trenches up to 1.5 m deep, 300 to 400 m E 10°N of Packington Church.

Brown (1889, p.24) gave the percentage composition of the Breccia clasts from Packington as: quartzite 50.8, 'gritty slates' 34.6, igneous rocks, more or less decomposed, 7.2, Coal Measures sandstone and ironstone, 5.8, 'grey flinty slate' 0.8 and volcanic ash, 0.8.

The Moira Breccia was encountered during opencast coal workings at the Stonehouse Farm site [366 135]. G. H. Mitchell observed 4.6 m of purple Breccia with a sandy marl matrix, strongly cemented in places, at the centre of the workings but this had thinned to only 0.6 m at the southern end, and it is absent both to the south of the Stonehouse Farm site and in opencast workings at Red Burrow, 200 m to the north-west. RAO

Appleby Magna – Snarestone – Market Bosworth

The Moira Breccia was proved in four boreholes in the Appleby Magna-Snarestone area, below the Bromsgrove Sandstone and resting unconformably on Coal Measures. The Measham Colliery No.1 Borehole [3670 0978] penetrated 36.6 m of Breccia with its base at 99.4 m depth, but in the Turnover Borehole [3423 0995], it was only 7 m thick down to 80.2 m. In both boreholes, the Breccia was predominantly of Charnian rock fragments with thin bands and lenses of green and red-brown mudstone and sandstone, up to 0.3 m thick. In the latter borehole, its junction with the overlying Bromsgrove Sandstone was sharp. About 76.2 m of combined Polesworth Formation and Moira Breccia occurred in the Appleby Road Borehole [3365 0931]; the basal 16.5 m of Triassic in the Shackerstone Borehole [3802 0704] may be Moira Breccia (cf. Richardson, 1931, p.75).

In the Twycross Borehole, Moira Breccia was cored from 500.03 to 503.30 m, where it rested sharply on a micro-diorite sill. The breccia was a compact rock with crystalline calcite showing prominently in the interstices between angular to subangular clasts of dark, fine-grained rocks including quartzite. The larger clasts, 3 to 5 cm in diameter, seemed to occur in bands within a matrix of small angular clasts, 2 to 5 mm in diameter. Geophysical logs suggest that rocks similar to these cores, but perhaps less compactly cemented, extend down from 484.2 m, with intercalated mudstone beds above 493 m. Above 484.2 m, the breccia is sharply overlain by the Polesworth Formation pebble beds and sandstones seen in the core taken between 470 and 473 m.

The Dadlington Borehole was not cored at the appropriate level but the geophysical logs suggest 15.5 m of Moira Breccia, comprising 8.5 m of compact rock, resting on Cambrian at 315.5 m, and overlain by similar Breccia with many intercalated mudstones. BCW

The Moira Breccia was 9.1 m thick in the Kingshill Spinney Borehole [3814 0154], 3.1 m at Bosworth Wharf [3916 0323] and 15.1 m in the Cowpastures Borehole [4124 0383]. Lithological details are given by Brown (1889).   KA,JB.

Bagworth – Desford

Boreholes in the southern part of the Leicestershire Coalfield prove 0.5 to 3 m of Breccia at the base of the Trias. In contrast, the rest of the local Trias is almost entirely pebble-free. The matrix of the Breccia is usually red-brown, gritty mudstone or poorly sorted sandstone, and contains angular to subrounded fragments of Charnian rocks, Triassic and Coal Measure mudstones, and a few Coal Measures ironstones. Fragments of shale, possibly Cambrian, also occur. Rounded pebbles of liver-coloured quartzite and white vein-quartz, (hereafter, 'Bunter' pebbles) are invariably present in subordinate quantities and are usually small, only occasionally reaching 50 mm.   MGS

In the Leicester Forest East Borehole, a few small subangular pebbles including purple ?quartzite and green-grey siltstone up to 10 mm in diameter, occur in the basal mudstone of the Bromsgrove Sandstone. They lie between 0.2 and 0.4 m above the base of the Trias, and there is no Moira Breccia in the section.   BCW

Polesworth Formation

The type area for the Polesworth Formation is the outcrop between Polesworth and Warton on the eastern side of the Warwickshire Coalfield, where the formation fills the lower part of the Hinckley Basin. It also crops out to the south-west of the South Derbyshire Coalfield, from Measham northwards (Figure 15).

The Polesworth Formation comprises poorly cemented, reddish brown sandstones, with thin beds of red mudstone and conglomerate. The pebbles are well rolled, up to 6 or 8 cm across, and include abundant 'Bunter' pebbles.

In some borehole records, it is difficult to separate the Polesworth Formation from the underlying Hopwas Breccia or Moira Breccia, since both may be recorded as 'conglomerate'. There is, however, sufficient unambiguous borehole evidence to establish its main variations in thickness (Figure 15).

The outcrops of the Formation are separated from those of its correlative in the west Midlands, the Kidderminster Formation (Warrington and others, 1980). The local sequence generally lacks the fine-grained sandstone of the Wildmoor Sandstone Formation (Upper Mottled Sandstone) of the west Midlands, which, there, overlies the Kidderminster Formation but underlies the Bromsgrove Sandstone Formation (Table 3). However, a thin, cyclic sequence of fine-grained, dark reddish brown ('foxy-red') sandstones was encountered just below the top of the Polesworth Formation in the Twycross Borehole (Figure 17) and these beds may correlate with the Wildmoor Sandstone.

The detailed sequence in the Polesworth Formation is difficult to establish. Its outcrops tend to be thickly strewn with pebbles, and quarries for gravel are generally in the more pebbly parts of the succession. Surface evidence, therefore, suggests that the Formation consists largely of pebble beds. On closer examination, however, finer-grained beds can be traced in many parts of the outcrop, though rarely for more

than a few hundred metres.

In the Twycross Borehole, cores were taken at four levels (Figure 18). Of the 20 m of core recovered, only 3 m were pebble beds. In this borehole, the sequence consists of fining-upwards alluvial cycles, with pebble-conglomerates in the lower parts of the cycles and overlain by beds of purplish grey and green, mottled mudstone. Cyclic pebble beds are commonly taken to indicate deposition by braided streams of low sinuosity. The mudstones of the Polesworth Formation presumably represent quiet-water deposits of limited lateral extent that settled in local hollows on the braid-plains.

The borehole sections in Figure 17 indicate a well defined junction between the Polesworth and Bromsgrove Sandstone formations. Samples from the Polesworth Formation, between 268.0 and 281.4 m in the Twycross Borehole, and from 155.75 m in the Appleby Parva Borehole, were examined for palynomorphs but proved barren; the time interval between the deposition of the Polesworth and the Bromsgrove Sandstone formations is not known.

It is possible that the Netherseal Fault was a growth-fault during deposition of the Polesworth Formation, since the beds thin rapidly from about 180 m on its western, downthrow, side to only about 50 m on its upthrow side. Undetected growth-faulting may trend north-westwards between Measham and Market Bosworth since the Formation is thin or absent to the north-east of this line.

Just west of the district, in the Statfold [2347 0699], Syerscote Barn [2266 0665] and Amington Hall [2279 0584] boreholes (Taylor and Rushton, 1971), the Polesworth Formation is absent and the Bromsgrove Sandstone rests directly on Coal Measures. In contrast, the Polesworth Formation is 200 m thick close to the Polesworth Fault. This suggests that the structures bounding the Hinckley Basin on its south-west side were also growth-faults, throwing down to the north-east during Polesworth deposition.

Though the Warwickshire Coalfield area was not affected by this subsidence, it may not have formed a barrier to deposition between the Polesworth area and the main sedimentary basin in the west Midlands. Indeed, a depositional connection across the Warwickshire Coalfield area would provide a readier explanation of the source of the Polesworth Formation pebbles than does the palaeogeographical reconstruction by Audley-Charles (1970, pl.7) based on Wills (1948, 1951, 1956), which shows the Formation laid down as a local delta-fan fed only by the 'Polesworth River' flowing from the south-east.

Regional studies of the origin of the pebbles in the 'Bunter Pebble Beds' of the Midlands largely point to their derivation from northern France and south-west England. They are thought to have been deposited by a large river that flowed along a hollow northwards through the Worcester Basin into the northern part of the Irish Sea, although also draining, at times, north-eastwards via Nottinghamshire into the North Sea (Audley-Charles, 1970, p.52, pl.7; Wills, 1970, p.246).

Pebbles from a disused quarry [2768 0382] 600 m south-east of Bramcote Hall, Warton (MR34820), and the Acresford Gravel Pit [3010 1327], 1.3 km at 075° from Netherseal Church (MR34819), have been examined in detail by G. E. Strong. He reports that pebbles of white vein-quartz, pale grey, pale brownish grey and liver-coloured,

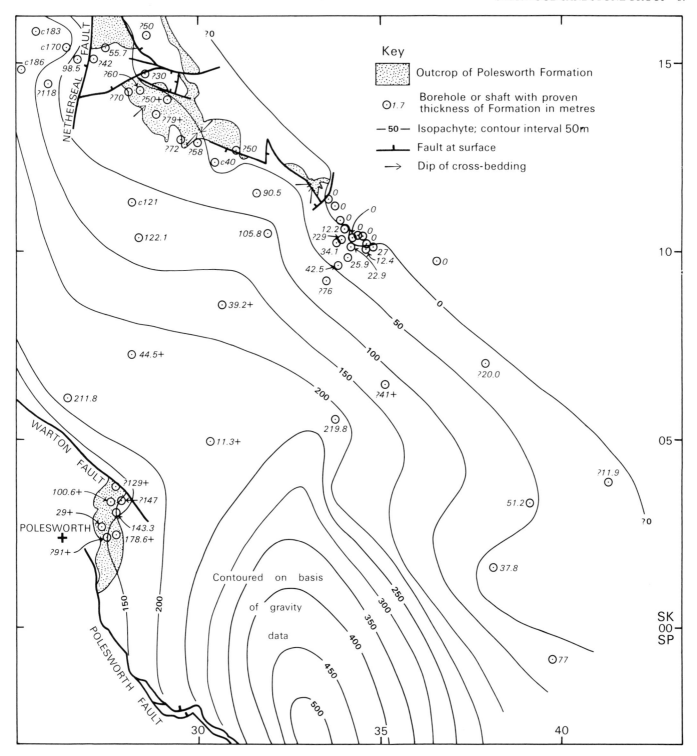

Figure 15 Isopachytes of the Polesworth Formation

fine-grained orthoquartzite, feldspathic in parts, and black aphanocrystalline chert are common to both localities. The Warton locality also yielded an olive-grey, fine- to medium-grained, feldspathic and chloritic subgreywacke, probably of Lower Palaeozoic derivation, and a black quartzite, composed in thin section (E53284) of inequigranular quartz with sutured grain boundaries, and containing abundant, scattered, irregular grains and aggregates of pyrite with minor goethite, possibly after primary carbonaceous material. A primary flaser texture is preserved. There are minor flakes of white mica forming microlaminae. The rock is probably Upper Carboniferous in age. The absence of distinctive lithologies amongst the pebbles precludes a precise definition of provenance. However, the proportion of quartzite and vein-quartz pebbles is comparable with that in pebble beds in the west Midlands.

Warton – Polesworth – Atherstone

A disused quarry [2678 0382] 600 m south-east of Bramcote Hall, Warton shows about 10 m of interbedded pebble beds and sandstone, dipping 20° to the south-east. Dips of 25° are shown by gravel and pebbly sandstone in a disused pit next to Lyndon Lodge [2739 0307]. In a field [279 036] 300 m west of Warton Church, slabs of coarse to gritty pale grey sandstone, 75 mm thick, and fragments of calcareous conglomerate including 50 mm quartz pebbles were ploughed up. These beds lie at, or just below, the top of the Polesworth Formation. Starting at this horizon, the Lyndon Lodge No.3 Borehole [2790 0347] proved red marl, silty marl and soft sandstone to 167.6 or 169.2 m (in two versions of the log) resting on grey Coal Measures mudstone to 170.1 m. The beds below 147.8 m do not resemble lithologically either the basal Polesworth Formation or the Hopwas Breccia, and may be red-stained Coal Measures. If this is the case, the total thickness of the Polesworth Formation (and Hopwas Breccia, if any) is about 150 m. The log of the Lyndon Lodge No.4 Borehole [2790 0376] also records about 12 m of red marl at the base of the Trias and again these beds may be stained Coal Measures. The Lyndon Lodge No.1 Borehole, which the mapping shows to have started near the top of the Polesworth Formation, proved 143.3 m of beds resting on Hopwas Breccia.

Augering, and the distribution of ploughed debris, near the cross roads [276 030], proved that pale yellow, coarse sand, about 7.5 m thick, is overlain by about 3 m of red sandy mudstone and silty sand which is capped by 2 to 3 m of sand with abundant pebbles. The red sandy mudstone presumably marks the top of a fluvial cycle and the overlying pebbly sand is the base of the succeeding cycle. About 250 m to the south-east, in the disused Kisses' Barn Quarry [2775 0260], the following beds dip at 9° to the south-south-east:

	Thickness m
Clay, sandy, reddish brown, with greenish grey mottling	to 0.5
Sandstone, reddish brown, medium-grained, fairly soft, with irregular, low-angle cross-bedding and lenticular inclusions of calcareous grit containing 1 to 2 mm diameter quartz grains and scattered 10 to 20 mm quartz pebbles	about 0.5
Limestone pellet-rock (intraformational conglomerate) including coarse quartz grains and 10 to 20 mm diameter quartz pebbles, channelling down 50 to 100 mm into underlying sandstone	about 0.2
Sandstone, massive, greyish yellow, with low-angle cross-bedding in sets 0.3 m to 0.6 m thick; includes ovoid carbonate concretions near the top and about 1 m below the top	to 2.0

South of the River Anker, similar beds are exposed in a small quarry [2761 0169] east of the railway cutting, where 2 m of massive, pink to yellowish brown sandstone with irregular curved bedding planes, possibly due to slumping of unconsolidated sand, is overlain by 0.4 m of conglomerate of quartz and quartzite pebbles in a pale grey, calcareous sandstone matrix, which passes up into yellowish to reddish brown, cross-bedded, soft sandstones with scattered quartz and quartzite pebbles. The northern part of the cutting, 1.2 km ESE of Polesworth Church, shows cross-bedded, yellow sandstone with pebbly bands. A little to the south [272 017], the cutting is overgrown, but there are traces of dark red, fine-grained, micaceous sand. Farther south [276 015], the cutting

exposes 6 m of friable, coarse, cross-bedded, yellow sandstone.

Boreholes in the Hinckley Basin

The top of the Polesworth Formation in the Twycross Borehole [3387 0564] is shown in Figure 17. The lithological log of this borehole based on cores does not correspond exactly with the geophysical logs. probably because the cores were only partially recovered, so the depths assigned to them were imprecise. A summary log of the borehole has been published (Institute of Geological Sciences, 1980a). At the top of the Polesworth Formation, a 9 cm bed of purple mudstone with green patches is the fine-grained member of an alluvial cycle and there are several other cycles in the sequence. In Figure 17, parts of the gamma ray and caliper logs of the Dadlington and Twycross boreholes are compared. In the Dadlington Borehole, which was drilled by rockbit, the Polesworth Formation is about 77 m thick and its top is taken at 223 m. As in the Twycross borehole, the upper part of the succession appears to consist mainly of sandstones, with two or three well marked mudstones. The geophysical logs of both boreholes pick out the contrast between the smoother gamma profile of the Polesworth Formation and the more indented profile of the succeeding Bromsgrove Sandstone sequence.

The upper part of the Polesworth Formation proved in the Appleby Parva Borehole [3067 0858] is also shown in Figure 17. The base of the overlying Bromsgrove Sandstone is taken at the base of a conglomeratic bed. It rests on 0.76 m of red-brown mudstone that passes down into coarse to pebbly sandstone. The borehole proved 39.17 m of the Polesworth Formation without reaching its base.

In the Newton Regis Borehole [2822 0728], core recovery was poor near the top of the Polesworth Formation, which is taken arbitrarily at 106.98 m. Coarse, loose, pebbly sands with a few bands of conglomerate and thin, red mudstone were recorded to the bottom of the hole at 151.49 m. The Shuttington Fields Borehole [2642 0610] proved a full thickness of 211.8 m of the Polesworth Formation resting on Cambrian shales, with no Hopwas Breccia. BCW

The Austrey House Borehole [3030 0488] was cored to 141.12 m. This depth was taken arbitrarily as the base of the Bromsgrove Sandstone. Below it, the log records 11.3 m of sandstone and conglomerate assigned to the Polesworth Formation.

The Gopsall Hall Borehole [3526 0649] penetrated 33.2 m of 'sandstone with pebbles' below 172.5 m and this is probably the Polesworth Formation. The Appleby No.6 Borehole [3100 0775], drilled to 185.8 m, may also have entered this Formation.

Strata assigned to the Polesworth Formation occur in the Bosworth Wharf [3916 0323] and Kingshill Spinney [3814 0154] borings (Fox-Strangways, 1907, pp.342–344), where 51.2 and 37.8 m respectively, of conglomerate with underlying sandstone, are present. In the Cowpastures Borehole [4124 0383] (Fox-Strangways, 1907, p.344), 11.9 m of 'Bunter' sandstone, overlying 15.1 m of breccia, may be part of the Polesworth Formation.

The Chilcote Waterworks Trial Borehole [2844 1039] proved 122.07 m of Polesworth Formation down to its base at 215.49 m. The sections in this and two production boreholes in the Pumping Station, which did not reach the base of the formation, show a succession similar to that in the Twycross Borehole. In the Chilcote Borehole [2828 1145] (Fox-Strangways, 1907, p.331), the Formation is 120.8 m thick and its base is at 171.3 m. The correlation of the upper part of the sequence with adjacent boreholes is shown in Figure 17.

In the Appleby No.1 [3205 1058] and No.2 [3176 1166] boreholes (Fox-Strangways, 1907, pp.333–334), the Formation is 105 m and 90.5 m thick respectively. Along a line between Snarestone and Measham, numerous boreholes prove that the Polesworth Formation dies out over a distance of about 3 km. South-westwards from this line, however, the Formation thickens rapidly.

Overseal – Netherseal – Measham

On the Polesworth Formation outcrop north of Netherseal Colliery [2755 1544], low features show that the beds dip gently to the south. The ridges are underlain by pebbly beds and the intervening slacks or hollows provide occasional exposures of fine-grained sand or red clay. The several small pits or quarries hereabouts are now mostly filled but all were excavated in the pebbly sandstones. Fox-Strangways (in MS) noted 'white false-bedded sandstone with a very few pebbles' in one pit [2772 1612], and 'massive false-bedded sandstone with small pebbles' dipping 3°S, in another [2767 1566]. He thought (1900, p.33; 1907, p.64) that the ridges were relics of Triassic topography exhumed from beneath an unconformable cover of Keuper Sandstone but this is not the case.

The Netherseal Colliery Shaft [2752 1545] proved a total of 55.7 m of Polesworth Formation beneath the Bromsgrove Sandstone and resting on Moira Breccia. The narrowness of the outcrop of the Polesworth Formation through Overseal, especially just south of the village, is largely due to steeper dips, although the Formation here is less than 50 m thick (Figure 15).

At Netherseal, a disused pit [2872 1384], 850 m ESE of Grangewood House, showed:

	Thickness m
4 Sand, yellowish brown, indurated to massive sandstone, seen for	1.2
3 Sand, red-brown clayey, and micaceous sandy clay, in one place showing 1 cm thick, cross-bedded layers dipping NE	about 0.6
2 Sandstone, massive, coarse, buff, with lines of pebbles along cross-bedding, dipping NE	0.9 to 1.2
1 Pebble-gravel, sandy, of closely-packed, mainly quartz and quartzite pebbles, mostly 1 to 5 cm in diameter but with some up to 15 cm; some 10 to 15 cm seams of sand	seen for 1.8

Beds 1 to 3 represent part of a fining-upwards cycle.

The Acresford Gravel Pit [303 133] closed about 1977. Sections exposed during the previous decade were mostly in up to 10 m of coarse-grained, yellow, cross-bedded sand and soft sandstone with sparsely scattered pebbles. In places, the sands include lenticular sandy pebble beds up to 1.8 m thick, and a few seams of red clay up to 0.3 m thick. No consistent pattern of cross-bedding was noted.

About 400 m to the south-east, the outcrop of a lenticular bed of red clay is marked by a slack about 500 m long. In Stanleigh Plantation, the clay has been dug, possibly for use as 'marl' on the surrounding sandy fields. At the southern edge of the wood [3066 1300], 1 m of massive, coarse, reddish to yellowish brown sandstone overlies 1.5 m of red clay.

East of Saltersford Brook [314 126], the outcrop of the Polesworth Formation is faulted out for about 1 km. Cuttings and small quarries at Measham Station [3225 1190] show up to 2.4 m of fine to coarse and gritty, soft sandstone, with cross-bedding dipping to the north and south (Plate 8).   BCW

Plate 8 Cross-bedded sandstone in the Polesworth Formation, Measham Station. A 11080

Bromsgrove Sandstone Formation

The outcrops and isopachytes of the Bromsgrove Sandstone Formation are shown on Figure 16; it dies out in the north-east of the district beneath the Mercia Mudstone which overlaps it on the margins of Charnwood Forest.

The Bromsgrove Sandstone consists of alternating sandstones and mudstones, each up to about 10 m thick. Detailed sections from the Twycross and Leicester Forest East boreholes (Figure 18) show that the sandstones and mudstones are generally the coarser and finer grained parts respectively of fining-upwards, alluvial cycles (Allen, 1965; Bridge and Leeder, 1979). The lower part of each cycle consists of grey to brown or red-brown sandstone, coarse at the base and fining upwards, and deposited in a river channel-bed, or by lateral accretion in a point-bar extending downstream from the apex of a meander. In the upper part of each cycle, red-brown mudstones, commonly with thin beds of fine-grained sandstone, represent overbank deposits. These were laid down by vertical accretion in quieter waters in overbank areas, flanking or separating distributary channels on the floodplain. The sandstone at the base of each cycle typically rests with a sharp break on the underlying mudstone, cuts down into it and commonly includes angular or rounded mudstone fragments. Because rivers meander randomly over their floodplains, the deposits of individual cycles do not continue very far laterally and individual sandstones have not been traced farther than about 5 km. Near its type locality south-west of Birmingham, the sandstones in the Bromsgrove Sandstone cyclic succession are thicker than the mudstones throughout (Wills, 1970). In the present district, the ratio of sandstone:mudstone is about 1:2. Perhaps the main stream-channels traversed this marginal part of the basin of deposition less often than the more central tract near Birmingham.

The sandstones of the Bromsgrove Sandstone sequence give good features at outcrop, and the mudstones form slacks or hollows. Exposures are generally few, but some show that a single feature may include deposits of more than one cycle, particularly where the intervening mudstone is less than 0.5 m thick. South of the River Anker, sandstones and mudstones cannot be separately mapped, since the features are obscured by a thin wash of sand, with pebbles derived from the Polesworth Formation outcrop up-slope.

In the Twycross Borehole (Figure 17), the Bromsgrove Sandstone is 136.45 m thick and divisible into three parts, which correspond in upward sequence broadly with the 'Conglomerate', 'Building Stones', and 'Waterstones' subdivisions of the former 'Keuper' Series of the central

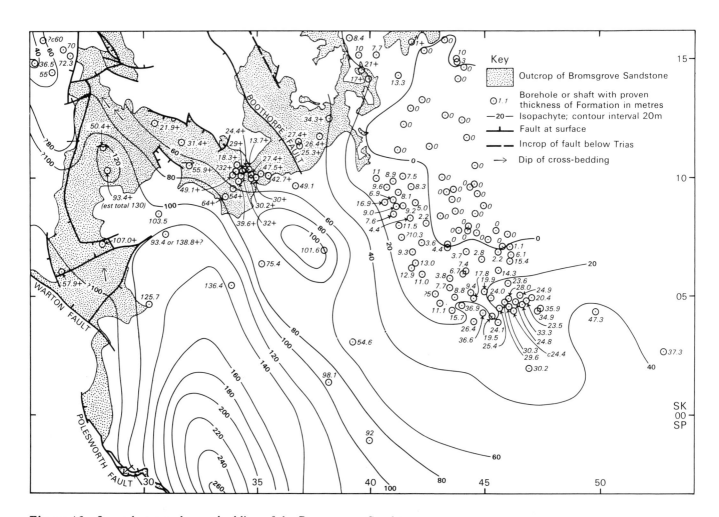

Figure 16 Isopachytes and cross-bedding of the Bromsgrove Sandstone

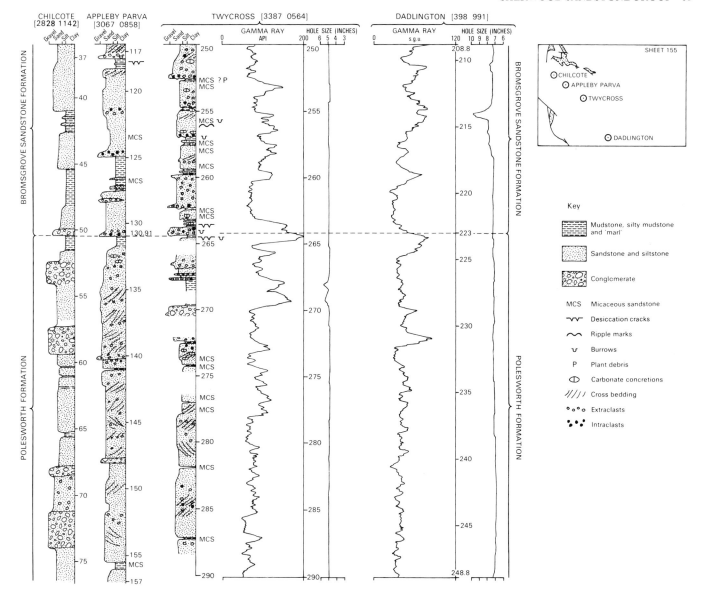

Figure 17 Boreholes showing the junction of the Polesworth and Bromsgrove Sandstone formations

Midlands (Warrington, 1970, pp.204–205, 210–212). In the lowest part, which is about 65 m thick, the sandstones are coarse grained and some near the base contain pebbles which are smaller, up to 5 cm across, and less well rounded than those in the Polesworth Formation. It has been suggested that these beds were deposited by rivers of low sinuosity, perhaps braided (Warrington, 1970, p.204). The base of the sequence is taken at a pebble bed containing mudstone clasts and vertebrate bone fragments. The bed correlates with similar pebble beds at the base of the Bromsgrove Sandstone in boreholes in the western part of the district (Figure 17). In the lowest part of the Bromsgrove Sandstone in the Twycross borehole, carbonate concretions occur in both sandstones and mudstones and, like the cornstones of the Old Red Sandstone of the Welsh Borderland, are probably pedogenic calcretes produced by the evaporation of carbonate waters on the alluvial flood-plain. Burrows and plant debris occur sporadically in this part of the Formation.

In the middle part of the Formation more arid conditions are indicated by gypsum nodules, possibly after primary anhydrite, by the absence of plant debris, and by occasional layers with flowage-bedding (Figure 18) suggesting saline muds. Pebbles are absent from the sandstones of this part of the formation, and it seems likely that the beds were laid down by widely meandering, sinuous rivers. Some beds in the Bromsgrove Sandstone of the present district, for instance a bed at about 140 m depth in the Twycross Borehole, lithologically resemble building-stones worked near Bromsgrove but have not been quarried to any extent.

The gradual incoming of micaceous sandstone in the upper part of the Bromsgrove Sandstone produces beds traditionally known as the Waterstones, characterised by mica-

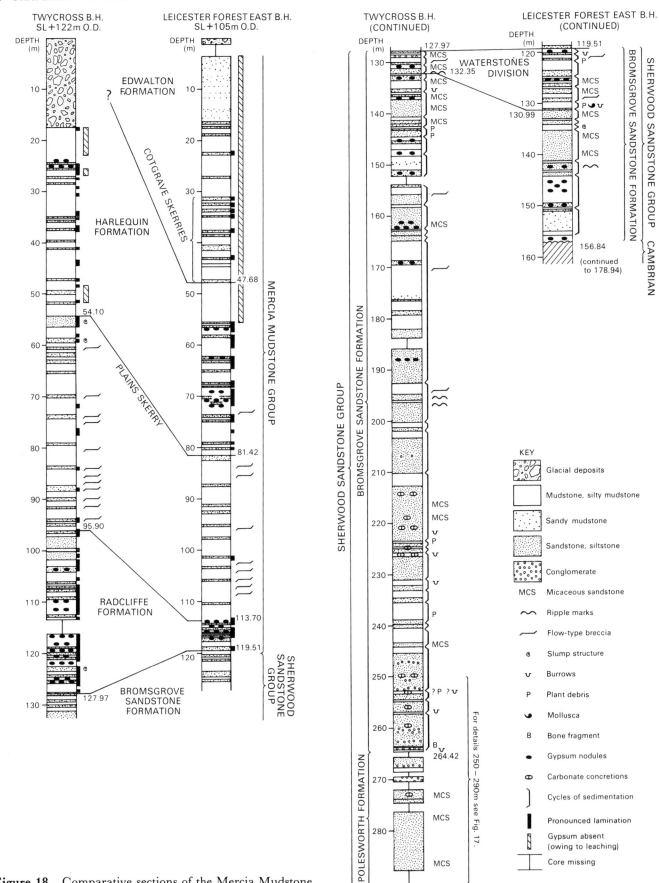

Figure 18   Comparative sections of the Mercia Mudstone and Sherwood Sandstone groups in the Twycross and Leicester Forest East boreholes

covered bedding planes in both sandstones and mudstones, and with the appearance of watered silk. These beds, 4.38 m thick at Twycross, are made of fining-upwards cycles. Some of the mudstones are of Mercia Mudstone-type, and to that extent, these beds are a passage between the Sherwood Sandstone and Mercia Mudstone groups. In the Twycross and Leicester Forest East boreholes, the predominance of fine-grained sandstone in the fining-upwards cycles makes the succession more comparable to the Sherwood Sandstone Group, but near Nottingham its correlative, the Colwick Formation, is included in the Mercia Mudstone Group (Warrington and others, 1980).

In the Leicester Forest East Borehole there are anhydrite nodules but no pebbles in the lowest beds of the Bromsgrove Sandstone, which rest directly on the Cambrian rocks. The 'Waterstones' here are 11.48 m thick, and bivalves, fish remains, burrows and plant stems occur in the interbedded mudstones.

Although the 'Waterstones' is present in the two boreholes, it has not been identified with certainty at the surface. The laminated, silty beds near the base of the Mercia Mudstone Group in places, for example near Austrey [295 063], may include some of these beds.

Palynology samples from the Formation at Measham, and from the Twycross, Leicester Forest East and Appleby Parva borehole cores, have all proved barren.

Details

Overseal – Netherseal – Chilcote

North of Overseal, the Bromsgrove Sandstone dips gently south-westwards. The sandstone beds do not make strong features hereabouts and their outcrops are largely conjectural. A low ridge at the foot of this slope, 800 m west of Overseal Church, is formed by north-east-dipping beds between two north-west-trending faults.

The lowest sandstone of the Bromsgrove Sandstone sequence is exposed at the crossroads [2876 1289] in Netherseal village as 1.5 to 3 m of massive, coarse-grained, buff sandstone, cross-bedded in sets about 0.6 m thick, with foresets dipping to the north-west. Pebbles are sparse, subrounded and 3 mm to 12 mm long; a few are well rounded. They include white quartz and red-brown to dark brown, fine-grained siliceous rocks. Sandstone cropping out in the River Mease [2865 1275] is probably part of the same bed. Just overlying it, in a cliff on the north bank, 1.8 m of thinly bedded red-brown sandstone is succeeded by 1.2 m of red mudstone. These beds probably comprise one sedimentary cycle.

A higher sandstone, making a bold feature on the hillside south of Grangewood Hall, is over 10 m thick but may include more than one cycle, as shown in the following quarry section [2818 1342], 900 m south-east of Grangewood Hall. Beds 1 and 2 represent the top of one cycle and beds 3 and 4 the lower part of an overlying one:

	Thickness m
4 Mudstone, dark red, weathered	0.6 to 1.0
3 Sandstone, massive, coarse-grained, reddish brown, cross-bedded, with quartz grains up to 2 to 3 mm across, Foresets inclined to the NW. Partings of red-brown and greenish grey, laminated mudstone, 25 to 50 mm thick, die out laterally; with 25 to 75 mm pebbles of greenish grey mudstone	about 2.4
2 Sandstone, red-brown, thinly bedded, clayey	0.6
1 Mudstone, red-brown, sandy	seen for 0.3

	Thickness m
Obscured section but probably sandstone	about 3.0

Another locality where a single, mapped sandstone includes the deposits of two sedimentary cycles is a disused digging [2720 1165], 300 m ENE of Seal Fields Farm:

	Thickness m
Mudstone, red, passing by intercalation into	0.60
Sandstone, greenish grey laminated, passing by intercalation into	0.15
Sandstone, reddish brown, coarse, with cavities (3 to 25 mm), in beds up to 0.15 m thick, channelled into the underlying bed	0.45
Mudstone, red	about 0.45
Sandstone, reddish brown, soft, massive, medium-grained; cross-bedding dipping E	to 1.50

Greig and Mitchell (1955) recorded the Bromsgrove Sandstone in the Netherseal Borehole [2678 1521] as 103.8 m thick, but it now seems more likely to be only 72.3 m thick, beds of 'sand marl', previously included in the Formation, now being assigned to the Mercia Mudstone. The Grange Wood Borehole [2652 1547] proved a Bromsgrove Sandstone thickness of 70 m. A little to the south, the Coalpit Lane and Church Flats boreholes apparently proved only 36.5 m and 55 m respectvely, but the Triassic succession was not cored in these holes and part of the strata assigned to the Polesworth Formation may be Bromsgrove Sandstone.

In the Chilcote Borehole [2828 1145] (Fox-Strangways, 1907, p.331), the base of the Bromsgrove Sandstone is taken below 0.45 m of coarse, grey sandstone with pebbles (Figure 17). At Chilcote Waterworks, a trial borehole [2844 1039] and two production boreholes 1 and 2 [2840 1040] proved fining-upwards alluvial cycles in the upper part of the Polesworth Formation, as well as throughout the Bromsgrove Sandstone, so the base of the latter is difficult to fix. Since elsewhere the Bromsgrove Sandstone has a pebbly base, the boundary in the Waterworks trial borehole is taken at 93.42 m at the base of a 'coarse sandstone with pebbles'.

A small quarry [2849 1141] by Chilcote Church showed:

	Thickness m
Mudstone, dark reddish brown (10R3/4)	to 1.8
Sandstone, laminated, with cross-bedding dipping E, and channelling into bed below	0.6
Sandstone, soft, massive, greyish orange, with small cavities of 5 to 50 mm in diameter	to 1.8

Above the strong sandstone features between Chilcote and Stretton en le Field, is a wide outcrop of reddish-brown, silty mudstone around Park Farm [298 111].

Measham – Willesley

A disused digging (Quarry Plantation) [3210 1145] on Birds Hill is said locally to have provided the stone for Appleby Magna Church.

South of Measham, the sections in quarries of the Redbank Brick and Terra Cotta Works are shown in Figure 19. The northern quarry [335 112] exposed 4 m of mudstone, overlain by massive sandstone. The latter is the bed flooring the southern quarry [336 112], which worked 10.5 m of dark reddish brown, unbedded, sandy mudstone. The upper mudstone had formerly been worked in quarries adjacent to the works [3355 1080] and east of the railway [3385 1070 and 3390 1055].

At Willesley [340 147], the basal sandstone of the Bromsgrove Sandstone thickens from about 1.5 to 3 m on the south-west side of the Boothorpe Fault, to at least 10 m on its north-eastern side. Small exposures show coarse sand with small quartz pebbles. The increased thickness is maintained for about 1.5 km to the north-west, the sandstone thinning near Rotherwood, across another north-west-trending fault in the Carboniferous rocks. The underlying Moira Breccia also thickens along this tract; both beds are filling a south-east-trending valley shown on the isopachyte map (Figure 14).   BCW

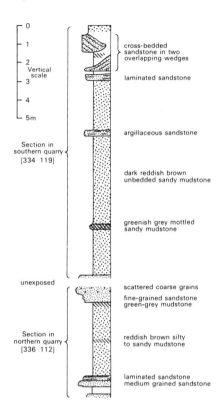

Figure 19   Sections in Bromsgrove Sandstone from quarries at the Red Bank Brick and Terra Cotta Works, Measham

Normanton le Heath – Swannington

The base of the Bromsgrove Sandstone was exposed in the Red Burrow opencast coal site [3699 1345] where G. H. Mitchell recorded;

	Thickness m
Sandstone, green	0.8
Mudstone, red	1.8
Sandstone, yellow, passing to green at depth, pebbly near base	3.7
Mudstone, red, sandy with pebbles	0.6
Sandstone, grey with pebbly bands, resting unconformably on Coal Measures	0.3

The Bromsgrove Sandstone was formerly worked for building stone in small quarries north of Alton Grange [3905 1484], where 1.0 m of pale brown, flaggy to massive sandstone overlies 0.6 m of thinly bedded, red-brown sandstone and mudstone.

Beds near the base of the Bromsgrove Sandstone are poorly exposed in a disused railway cutting at Foan Hill, Swannington [4193 1604]:

	Thickness m
Mudstone, red	2.0
Sandstone, grey, coarse, micaceous	0.8
Mudstone, red and grey	2.0
Sandstone, brown, coarse	0.2
Mudstone, red	1.5
Sandstone, brown, coarse, micaceous	1.1
Mudstone, red	to 1.0 (seen)

RAO

Appleby Parva – Snarestone

The Appleby Parva Borehole proved 103.5 m of Bromsgrove Sandstone. The lowest beds are shown in Figure 17. The full succession contains 13 fining-upwards cycle with an average thickness of 7 m. At the top, the 'Waterstones' consist of 3 m of thinly interbedded sandstone and red-brown, micaceous mudstone.

The uppermost 20 m of the Bromsgrove Sandstone crop out near Snarestone, where fine-grained sandstones and siltstones, 1 to 2 m thick, have been mapped as features. A composite section, in a brick pit [384 095] south-east of Snarestone, shows:

	Thickness m
Soil and gravel wash	1.0
BOULDER CLAY	
Clay, red-brown, sandy, pebbly	3.0
MERCIA MUDSTONE	
Clay, red-brown, silty	3.3
BROMSGROVE SANDSTONE	
Sandstone, grey-green, medium grained	0.8
Mudstone, red-brown, silty	1.8
Sandstone, green-grey, medium-grained, passing laterally into thin sandstones and red-brown, silty mudstones	2.2
Sandstone, green-grey, medium-grained	to 0.8

Weathered sandstone blocks from the section show ripple marks and cross-bedding.   KA

Newton Regis – Austrey – Warton

North of Newton Regis, the Bromsgrove Sandstone is disposed about a north-trending, faulted anticline whose axis swings south-eastwards near the village. There are sections displaying up to 1.5 m of massive, cross-bedded sandstone in a disused quarry [2753 0820], in roadside sections in Newton Regis and Seckington, and in a disused quarry [2624 0755] at Seckington.

In a small quarry [2826 0642] west of Austrey, a 1 m face shows massive sandstone with cross-bedding inclined to the north. Around the quarry, the southward-projecting lobe of the outcrop is apparently a channel filled with sandstone. Along the western side of the lobe, the sandstone base curves sharply northwards, as it cuts down into the underlying red mudstone. Similar channelling seems to occur, locally, 1 km WSW of Newton Regis Church [270 070] and 0.5 km east of Seckington Church. If the channels are generally aligned north – south, then this might explain why the sandstones around Newton Regis are more persistent in this direction than east – west.

The slope [285 045] north of Warton is underlain by an alternating sequence of sandstones and red mudstones, though the sandstone features are not prominent. About 9 m of this succession were proved in a nearby well [2838 0472].

In a disused quarry [2840 0350] in Warton, beds near the top of the Bromsgrove Sandstone are exposed:

	Thickness m
Marl, red-brown, silty with laminae of deep red-brown mudstone and buff and greenish grey, cross-bedded sandstone	to 0.6
Marl, silty, and fine sandstone, interbedded	0.76
Marl, red-brown, silty with laminae of chocolate, smooth, micaceous mudstone and fine-grained, greenish grey, cross-bedded sandstone; also thicker bands of sandstone	3.96
Mudstone, deep red-brown, silty, well laminated, micaceous	1.22
Sandstone, fine, greenish	to 0.6

In a nearby quarry [2853 0318], red-brown, micaceous mudstones with thinly cross-bedded, buff sandstones were seen for 2.1 m.

Polesworth – Atherstone

South of the River Anker, it has proved impracticable to map separate sandstones and mudstones within the Bromsgrove Sandstone. The banks of a sunken lane [2740 0067] expose about 3 m of interbedded fine-grained red-brown sandstone and red-brown, micaceous, silty mudstone in 0.5 m layers. In the Dordon Farm opencast site [2746 0004], massive, yellow Bromsgrove Sandstone was faulted against Coal Measures. In a stream section [2856 9864] north-east of Woodlands Farm, Waste Hill, 2 m of interbedded, grey, fine sandstone and red-brown and greenish grey mudstone and micaceous, silty mudstone, lie at the top of the Formation.

Along the outcrop of the Bromsgrove Sandstone east from Merevale Abbey, sandstone debris is plentiful. A disused quarry [2910 9773] exposes 1.5 m of massive, yellow-brown sandstone, with cross-bedding dipping south-westwards.

A small quarry in Sandhole Spinney [3008 9755] shows, close against the faulted contact with the Cambrian, small exposures of orange-brown sand and sandstone, some beds being coarse to gritty, with small (1 to 2 cm) quartz and quartzite pebbles. At one point, the pebbly sandstone is overlain by fine-grained greenish grey clayey sandstone with highly micaceous bedding planes, 0.5 m thick, followed by 1 m of yellow-brown, medium-grained sand; at another, it is overlain by dark red mudstone. The beds are probably near the base of the Bromsgrove Sandstone, but the section may be complicated by undetected faulting. The pebbly sands are reminiscent of the Polesworth Formation, as Fox-Strangways (1900, p.34) noted, but they lack the large, well rounded, quartzite pebbles typical of the Polesworth sequence.   BCW

In the Austrey House No.1 Borehole [3030 0488], the Bromsgrove Sandstone is 125.7 m thick. Other water bores around Orton and Sheepy Magna penetrate the top of the formation.

The Bromsgrove Sandstone in the southern part of the Leicestershire Coalfield has been proved, in numerous borings for coal, to consist mainly of sandstone with subordinate beds of sandy mudstone. The sandstones are generally less than 1 m thick, usually greenish grey although sometimes red, fine- to medium-grained, poorly sorted, silty, sometimes cross-bedded, commonly micaceous and cemented by gypsum. The mudstones are generally red-brown, silty, commonly sandy, gritty and micaceous, with occasional gypsum crystals and nodules.

The top of the Bromsgrove Sandstone in these sections is taken at the top of the highest thick sandstone, above which the succession is dominantly argillaceous.   BCW

MERCIA MUDSTONE GROUP

The Mercia Mudstone outcrop covers the greater part of the district, overlapping the Sherwood Sandstone around the coalfields, and overstepping onto the Precambrian rocks of Charnwood Forest and the diorites of the south-east.

The Mercia Mudstone Group consists of red-brown, silty mudstones with subordinate greenish grey bands and patches. At intervals are thin beds, mostly less than 0.3 m thick, of pale greenish grey, quartzose siltstone, or, in the higher part of the Group, sandstone, traditionally termed 'skerries'. Dunham (in Stevenson and Mitchell, 1955, p.62) gave the following average analysis of 'Keuper Marl' for the Midlands, calculated from 39 analyses in Geological Survey records: SiO_2 53%, Al_2O_3 14.3%, Fe_2O_3 3.7%, FeO 1.4%, MgO 4.8%, CaO 5.4%, Na_2O 0.5%, K_2O 3.5%, TiO_2 0.7%, remainder, including CO_2, H_2O, FeS_2 and organic matter, 12.7%. The mudstones of the Group give fertile, red, clayey soils.

Because they are more resistant to erosion, the skerries tend to form distinct topographical features. In the west, a group of siltstones give rise to a steep, west-facing multiple escarpment west of Twycross. East of Market Bosworth, where the topography is a dissected plateau capped by glacial deposits, the skerries are mostly seen as shelf-like features on the valley sides.

The highest beds of the Mercia Mudstone have been removed by erosion in the present district. The full thickness beneath the Penarth Group near Leicester is about 180 m. It is likely that the highest hills of Charnwood Forest were formerly covered by Mercia Mudstone, which must then have been present in much greater thicknesses than is now preserved in Charnwood. In the west of the district, an even greater thickness of sediment accumulated in the subsiding Hinckley Basin. Gravity data suggest that, in the centre of the basin near Fenny Drayton [350 970], as much as 300 m of Mercia Mudstone are presently preserved, and as these beds are only the lower half of the full sequence, then the Group may originally have been 600 m thick, hereabouts.

In borehole cores, gypsum occurs as small nodules, up to 10 cm across at certain horizons, and as secondary veins of satin-spar at all levels. Gypsum is readily leached by circulating groundwater, however, and has not been seen in exposures or near-surface excavations in the present district. The depth of leaching proved to be 56 m at Leicester Forest East and 23 m at Twycross. Occasional cubic pseudomorphs after halite, with sides up to 1 cm long, occur in the siltstones, and mud cracks are common.

The red colour, mud cracks and evaporites in the Mercia Mudstone are consistent with deposition in an arid climate. The area of deposition was probably subdued topographically, resembling the present-day Ranns of Kutch, an arid coastal flat in north-west India, which is subject to annual marine inundation as well as to local flooding by seasonal rivers. Throughout the Mercia Mudstone, there is a rhythmic alternation between beds of finely laminated mudstone and siltstone, and beds of massive, unbedded mudstone, from a few centimetres to more than 5 m thick. At different horizons, laminated or massive mudstones may predominate. The laminated facies probably results from deposition in ephemeral bodies of shallow water, whilst the massive mudstones are likely to result from the accumulation of wind-blown sediment (Arthurton, 1980, p.54).

The nodular gypsum occurs most commonly in the laminated mudstones and may originally have formed as anhydrite nodules, at about 0.5 m below the contemporary sediment surface under sabkha conditions. The satin-spar veins are secondary and formed after lithification. Their ubiquitous occurrence supports the view (Burgess and Holliday, 1974, pp.22–23) that the source of the sulphate in the veins was connate brines, rather than (Shearman and others, 1972) that the satin-spar results from an increase in volume of calcium sulphate in the system consequent upon the alteration of nodular anhydrite to gypsum.

A marginal facies of coarse breccias and sandstones, described by Bosworth (1912) and Watts (1947), is developed around the Precambrian inliers of Charnwood Forest. It is too local to be mapped separately, passing laterally within a few metres into normal Mercia Mudstone. The breccias fill depressions in the Triassic land-surface. The sandstones are interbedded with red-brown mudstones and commonly contain clasts of local origin. Although these deposits have been described as 'littoral' by Taylor (in Sylvester-Bradley and Ford, 1968, p.161), they are not typical beach or intertidal deposits and there is no evidence of prolonged wave action having been involved in their formation. Small-scale base metal mineralisation of red-beds type is associated with the marginal Mercia Mudstone deposits, as described elsewhere (p.130).

Elliott (1961) subdivided the Mercia Mudstone around Nottingham into six formations, each characterised by a different assemblage of small-scale sedimentary structures. Though these structures are clear in borehole cores, they are not usually preserved in weathered exposures. However, by relating the boundaries between formations to certain skerry bands, Elliott (1961) was able to map them in his type area. The Leicester Forest East and Twycross boreholes have proved that these formations can be recognised in the present district (Table 3).

The Plains Skerry, which lies about 3 m below the base of the Harlequin Formation, and the Cotgrave Skerry, at the base of the Edwalton Formation, can both be identified in the boreholes. A group of closely associated skerries can be traced at outcrop from Ellistown [430 110] to near Thurlaston [500 990]. The Cotgrave Skerry lies within this succession but cannot be individually identified. A particularly well developed bed among them is here given the local name Thornton Skerry. The Plains Skerry cannot be distinguished at outcrop, though it is likely to be among the siltstones cropping out near Twycross. In the absence of recognisable marker bands, it is difficult to assign local sections to the formations described in Nottinghamshire by Elliott. To aid the classification of surface exposures in terms of these formations, therefore, the main features of each formation as displayed (Figure 18) in the Twycross and Leicester Forest East boreholes, are described below.

Radcliffe Formation

The formation is characterised by beds about 0.5 m thick of finely inter-laminated, dark red-brown mudstone and pale brown siltstone, which alternate with layers of less distinctly bedded mudstone and siltstone. Not far above the lowest appearance of laminated silty mudstone the fining-upwards alluvial cycles that occur throughout the underlying Bromsgrove Sandstone die out and mud cracks become common. Some beds in the lower part of the sequence are micaceous, so that the Radcliffe Formation and the underlying Waterstones are not readily distinguishable in the field. In practice, the base of the Radcliffe Formation has generally been taken at the top of the highest well defined sandstone bed in the Bromsgrove Sandstone because this provides the nearest practical boundary recognisable in the field. In the upper part of the Radcliffe Formation are layers of 'flow-type breccia' described below. The formation is 6 m thick at Leicester Forest East and 32 m thick at Twycross.

Carlton Formation

The formation, 32 m thick in the Leicester Forest East Borehole and 41.8 m at Twycross, is characterised by a lack of well defined lamination and many beds, 0.5 m to 1.5 m thick, show the structure described by Elliott (1961) as flow-type breccia. In the present district, flowage is more conspicuous than brecciation. The breccia seems to have developed in layers, within which the original sediment was disrupted by desiccation cracks. Subsequently, while confined beneath later sediments, the layers were brought to a near-fluid state by an excess of pore-water, probably saline; flow was then induced by the weight of the overlying sediment. The less-disturbed layers include angular mudstone fragments that appear to have become detached from blocks of laminated sediment between desiccation cracks, and the more disturbed layers consist of cumuliform wisps of silt suspended in a fine-grained mudstone matrix.

At Twycross, the Plains Skerry, 2.28 m thick, is a hard, pale greenish grey siltstone with unbedded layers up to 0.3 m thick alternating with thinner, laminated to finely cross-bedded layers with dark greenish grey mudstone partings. Slump structures (Elliott, 1961, fig. 3), occur in a 10 cm layer, 0.74 m above the base of the bed and in a 6 cm layer 1.52 m above the base. In the Leicester Forest East borehole, however, the Plains Skerry, 1.03 m thick, lacks such slump structures.

Harlequin Formation

The formation is 34 m thick at Leicester Forest East, and more than 36.7 m at Twycross. It is characterised by thinly laminated silty mudstones similar to those of the Radcliffe Formation, interbedded with unlaminated mudstones.

At Leicester Forest East, as near Nottingham (Elliott, 1961, p.219), laminated beds, each about 0.5 m thick, are dominant towards the base of the formation and unlaminated beds are common towards the top. Nodular gypsum is well developed about 10 m above the base of the formation at Leicester Forest East and is also present, at 30 m above the base at Twycross. 'Vein-type breccia' (Elliott, 1961) was noted at two levels in the Harlequin Formation at Twycross. The brecciation may have originated as desiccation cracking. The rock is traversed by cracks filled with mudstone, which follow a crude rectangular pattern.

Edwalton Formation

This is the highest Mercia Mudstone formation preserved in the district. At its top near Nottingham is a sandstone called the Hollygate Skerry; it occurs in two outliers south of Kirby Muxloe but its main outcrop lies farther east in the Leicester district (Fox-Strangways, 1903). There it is called the Dane Hills Sandstone and is correlated (Warrington and others, 1980) with the Arden Sandstone of Warwickshire. The Leicester Forest East Borehole, beginning just below the Hollygate Skerry, reached the base of the Edwalton Formation at 47.68 m, so the full thickness of the formation is probably about 50 m. Down to about 40 m below ground surface, the beds were broken and weathered, the gypsum was completely leached and most of the bedding was destroyed. Greenish grey mottling of red mudstone and distinct bands of greenish grey mudstone are a feature of this formation (Elliott, 1961). Many of the mudstones in the uppermost 17 m are sandy. Several thin beds of fine-grained sandstone between 32 and 47 m in the borehole probably include the Cotgrave Skerry. The base of the formation is taken at the lowest of these, which is 0.21 m thick. The bed is coarse and gritty in its lower part, and includes rounded quartz pebbles up to 2 mm across.   BCW,RAO

Mercia Mudstone palynology

Palynology samples from the lower part of the Mercia Mudstone in a pipeline trench [296 054 to 302 058] south of Austrey yielded miospore assemblages (Table 4); these include *Tsugaepollenites oriens* and *Stellapollenites thiergartii* and indicate a late Anisian (Middle Triassic) age. Similar assemblages were recovered from the Radcliffe Formation at 109.00 to 109.15 m and from the Carlton Formation at 64.77 to 77.10 m in the Twycross Borehole (Table 4).

Miospores of Early to Middle Triassic age were recovered from the Harlequin Formation at 21.75 m in the Twycross Borehole (Table 4).

Samples from the Edwalton Formation in the Leicester Forest East Borehole were mostly barren, but one, from near the base of the Formation at 44.7 m (Table 4), yielded *Echinitosporites iliacoides* and *Retisulcites perforatus*, an association indicating a Ladinian (late Middle Triassic) or possibly early Carnian (early Late Triassic) age.

The biostratigraphical evidence from the Mercia Mudstone of the present district indicates that the Cotgrave Skerry correlates with part of the Lettenkohle and Lower Gipskeuper succession in the Trias of Germany, and that the Carlton and Radcliffe formations correlate with beds in the higher part of the Muschelkalk in Germany.   GW

Petrography of the Skerries

Thin sections were made of skerries from several localities. The E numbers are those of the slides in the BGS Sliced Rock Collection.

National Grid Reference	Approximate height of sample above base of Group (m)	
1 E47895 [371 984]	unbedded siltstone	110
1 E47896 [3115 0826]	finely bedded, dolomitic siltstone	15

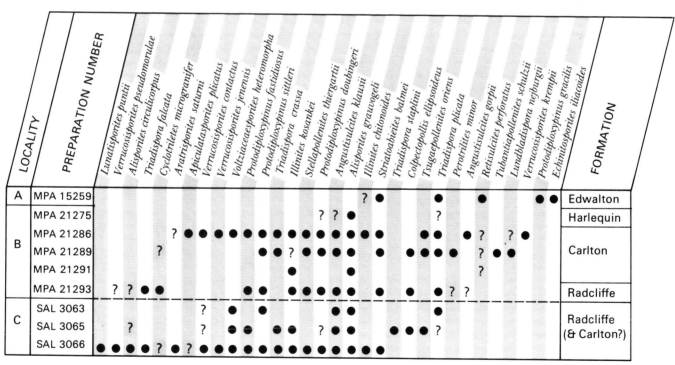

Table 4 Principal miospores from the Mercia Mudstone Group. A—Leicester Forest East Borehole, B—Twycross Borehole, C—trench section near Austrey (preparations are held in the palynology collections at the British Geological Survey, Keyworth, and are registered in the SAL and MPA series)

1	E47897	[3784 0582]	dolomitic siltstone with slumped bedding	30
1	E47899	[3862 0695]	finely cross-bedded, dolomitic siltstone	40
1	E47900	[3041 0520]	finely bedded, dolomitic siltstone	15
1	E47901	[3213 0528]	finely cross-bedded, dolomitic siltstone	55
1	E47902	[3088 0877]	coarse, dolomitic siltstone with hematite staining	8
1	E47903	[3284 0531]	finely cross-bedded, dolomitic siltstone	70
1	E47904	[3215 0509]	finely bedded and slumped, dolomitic siltstone	52
1	E47905	[3895 0466]	finely cross-bedded, dolomitic siltstone, with slumped and graded bedding	55
2	E48071	[418 107]	siltstone	30
3	E48162	[4881 0385]	irregularly laminated, silty dolomicrite with pseudomorphs after halite	80
3	E48163	[4698 0366]	parallel-laminated, fine-grained sandstone with goethite- or limonite-rich layers	80
3	E48164	[5039 0258]	quartzose siltstone with limonitic weathered surface	80
3	E48165	[5248 0433]	quartzose feldspathic siltstone	100
3	E48166	[5170 0037]	fine-grained, calcareous sandstone (Arden Sandstone)	120

1 examination of thin sections by K. Ambrose and R. R. Harding
2 examination of thin sections by R. R. Harding
3 examination of thin sections by G.E. Strong

In thin section, the skerries are moderately well sorted siltstones with thin, alternating carbonate- and quartz-rich layers. Most of the clastic components are angular to sub-rounded grains. Occasional euhedral dolomite rhombs are present and dolomite/ankerite appears to form most of the carbonate cement. A calcite cement with 'ghost' ooliths or pellets is the only carbonate in E 48166 and there are no carbonates in E 48163–5.

Clasts in dolomite-rich layers are finer grained than those in quartz-rich layers, except in E 47900. Quartz grains in each layer usually form 60 to 80 per cent of the clasts. Dolomite varies from 40 per cent in E 47905 to 95 per cent in E 47896. Micas commonly form less than 10 per cent (15 per cent in E 47900); muscovite is dominant, with biotite wholly or partially altered to chlorite, for example in E 47900.

Of the accessory grains, opaque minerals and plagioclase are ubiquitous. Where determinable, the plagioclase is in the albite-oligoclase range, with rare grains of more calcic plagioclase in E 47897. Other accessory minerals include apatite, zircon and microcline. Authigenic albite occurs in E 48165.

The matrix is mainly composed of chlorite and clay minerals, stained red by hematite in E 47899 and E 47902. Minute dolomite granules up to 20μ are also common in the matrix in the quartz-rich layers. Evidence of pressure solution of quartz is seen in E 47895, E 47901, E 48162 and E 48164.

The origin of the dolomite is uncertain. Most grains are anhedral but the presence of some rhombs implies authigenic growth. Cavities were observed in some skerries in the field and may be due to solution of carbonates or gypsum. KA

DETAILS

Clifton Campville

North of Botany Bay [2595 1545], red-brown micaceous silty clay and silt marks the outcrop of beds in the Mercia Mudstone sequence. At Botany Bay, a fine-grained sandstone dips gently southwards; a temporary excavation 60 m north-east of the nearby road junction exposed red mudstone interbedded with 4 cm layers of ripple-marked, micaceous, silty sandstone and some blocky, soft, red-brown, silty sandstone. A similar bed of sandstone crops out on a hill [2695 1345] west of West View Farm, but its outcrop ends abruptly to the south and the west against faults.

A siltstone lying 15 m above the base of the Mercia Mudstone south-east of Clifton Campville, changes along its outcrop from greenish grey siltstone with pseudomorphs after halite [264 103], through coarse to gritty, greenish grey sandstone in layers up to 50 mm thick [2598 0937], to red-brown, fine-grained sandstone and pale grey, fine- to medium-grained sandstone [258 079] south of Thorpe Constantine church. In a disused marl pit [2586 0983], a skerry about 15 m higher in the sequence gives fragments of coarse to gritty, greenish grey sandstone with cavities 2 mm across.

Temporary excavations, 200 to 500 m west of Thorpe Constantine church, showed pale red-brown, silty mudstone and dark red-brown, laminated mudstone within the Radcliffe Formation. The skerry west of Seckington consists of siltstone and fine sandstone; it probably lies within the Radcliffe Formation but its exact position in the succession is uncertain.

Austrey

Between Newton Field [282 094] and No Mans's Heath, small exposures of red-brown, thinly laminated, silty clay and micaceous siltstone lie within the Radcliffe Formation. Thin, apparently lenticular, beds of laminated siltstone or silty sandstone make slight, shelf-like features east of Newton Gorse [284 078] and south of Austrey, and dip gently to the south-east. Tufa [296 053] on the edge of the alluvial tract appears to have been deposited by springs from these beds.

On the crest of the low scarp at Orton Gorse [2975 0429], a bed containing grey to reddish brown, laminated, silty sandstone and interbedded, laminated, red-brown, silty mudstone and green-grey siltstone were seen in temporary sections. The bed continues south of Warton and also crops out around the northern margin of the alluvial tract 1 km SSW of Orton Gorse where it is folded into a gentle anticline. Ploughed fragments [297 028] from this outcrop show slump structures. The rock (E 48161) is a pale yellowish brown, micaceous, dolomitic, coarse siltstone to very fine sandstone. The dominant constituent is quartz, with grain-size ranging from 10μ to 64μ approximately. The grains are corroded, with recrystallisation of silica as chert. Interstitial carbonate occurs as subhedral and euhedral rhombs of dolomite, with minor calcite. Minor components are albite, goethite, hematite, zircon and opaque minerals.

A disused clay pit [2965 0337], on the south side of the Warton road, shows 2 m of unbedded, red-brown, silty mudstone overlying 1 m of thinly interbedded, red-brown and green-grey mudstone and greenish grey to pale brown silty sandstone, in layers up to 100 mm thick. Some of the beds are micaceous, and some silty sandstone layers show slump structures. The lithologies are typical of the Radcliffe Formation.

Grendon House Farm [2970 0185] is situated on a low ridge

formed by an outlier of siltstone, and capped by red-brown mudstone. The main outcrop of the siltstone forms a low escarpment 0.5 km east of the farm, and is traceable southwards for about 3 km, marked by green clay and fragments of greenish grey, blocky, non-laminated siltstone in the soil. The bed is particularly prominent where it crosses the railway line [290 992].

Appleby Parva – Shackerstone – Carlton

At Appleby Parva, a disused quarry [3087 0875] exposes 0.6 m of red-brown, silty, micaceous mudstone on 0.6 m of red to yellowish grey, fine-grained, flaggy sandstone with ripple marks and solution cavities. This is the middle of three skerries which die out northwards towards Appleby Magna. To the west, the highest and lowest skerries come together [298 091] and the bed can be traced southwards through Austrey. It is exposed in a stream [304 051] east of Austrey House Farm as 6 m of soft, red, flaggy, fine-grained sandstone passing upwards into harder, green siltstone.

West of Norton-juxta-Twycross is a sequence of mudstones with two interbedded siltstones. The base of the lower is exposed [3195 0710] as 1 m of grey-green siltstone resting on red-brown mudstone, and the higher bed is seen in a nearby ditch [3137 0720].

In the stream running south past Lea Grange Farm [322 054], 1 m of green-grey, massive siltstone with small-scale cross-bedding, pseudomorphs after halite and solution cavities, is exposed. Northwest of the farm, this bed underlies red-brown mudstone. At a locality [3127 0582] 120 m south-west of Orton House Farm, traces of malachite occur in cavities in the skerry. The overlying beds crop out in a ditch 250 m south of Lea Grange Farm, as red-brown mudstone 1.5 m, overlain by grey-green, fine-grained, thinly bedded siltstone with ripple marks, cross-bedding and pseudomorphs after halite 0.1 m, red-brown mudstone 1.5 m, and green-grey, massive siltstone. Higher beds, dipping NNE at 5°, are exposed in a north-east-trending ditch [3274 0516] as red-brown mudstone with grey-green mudstone lenses 3 m, overlain by grey-green siltstone and mudstone 0.15 m, soft, pale brown, fine-grained sandstone 0.75 m, grey-green siltstone 0.1 m, and red-brown mudstone 3 m. The siltstones are thinly bedded, with pseudomorphs after halite, ripple marks and cross-bedding.

Between Snarestone and Ibstock, no skerries were mapped in the basal 20 m of the Mercia Mudstone, but excavations in these beds at Beanfield Farm [354 085] show finely laminated, hard, red-brown, micaceous mudstone and paler, fine-grained sandstone with ripple marks, small-scale cross-bedding and pseudomorphs after halite.

Other exposures of the lowest beds, in ditches [378 093] northeast of Newton Burgoland, show red-brown, silty mudstone with beds of soft, red sand, some over 1 m thick. Augering proved the sand beds to be lenticular.

South-east of Shackerstone, siltstones are more common. A composite section in the banks of the track 500 m north-east of Tivey's Farm [376 075] reads:

	Thickness m
Topsoil and weathered red-brown mudstone	0.6
Mudstone, red-brown	0.4
Mudstone, red-brown and green-grey, with green-grey siltstone	0.3
Mudstone, red-brown	0.65
Siltstone and mudstone, grey-green	0.4
Mudstone, red-brown	0.1
Unexposed	2 to 3
Mudstone, red-brown	2.0
Mudstone, grey-green	0.3
Mudstone, red-brown	1.0

The skerries east of Shackerstone pass laterally into mudstone to the north-east, and southwards apparently merge into one bed.

An exposure [3899 0540], 700 m north-west of Carlton Post Office, shows about 1 m of grey-green siltstone and fine-grained sandstone. A building stone of similar appearance, used to construct several old buildings in Carlton and known locally as 'Carlton Stone', is reputed to have been quarried [389 054] northwest of the village, although no trace of the former workings can be seen.   KA,GES

Orton-on-the-Hill – Sibson

The siltstone which forms the crest of the prominent escarpment at Orton-on-the-Hill reaches 2 m in thickness. A temporary excavation [3046 0390] for a sewer pipe proved 1.5 m of siltstone on red marl. The Orton to Twycross road traverses an extensive dip-slope inclined to the south-east, which carries sandy soils with siltstone debris.

At Glebe Farm [3188 0344], small-scale, but well developed, dip-and-scarp topography is produced by successive siltstone beds with a gentle ESE dip. Two of the siltstones merge south of Hill Farm [3242 0304]. South of Twycross, a group of laterally impersistent siltstones, all less than 1 m thick, dip gently to the south-east, but their relation to the beds near Hill Farm is uncertain.

Siltstones cropping out east of Sheepy Parva [332 014] are probably higher in the sequence than those at Orton. In a roadside ditch west of Cool Hill Farm [3434 0116], less than 0.5 m of greenish grey, micaceous, weathered siltstone is visible.

The siltstone which crosses the Upton-Sibson road at [3626 9993], shows lateral variation in grain size. The bed is here a greenish grey siltstone less than 1.0 m thick, but to the north-east it thickens to about 1.5 m of green, fine- to medium-grained, micaceous sandstone. Beyond, the bed thins rapidly and where it dies out [3709 0046], only thinly bedded siltstone is found.

The Bosworth Wharf Borehole [3916 0323] encountered 106.7 m of Mercia Mudstone below glacial deposits (Fox-Strangways, 1907, p.342). Gypsum bands, mostly 0.3 m thick but up to 0.9 m thick, were recorded throughout but only one skerry is recorded near the base.   JB

Ratcliffe Culey

East of Ratcliffe Culey, the beds dip generally to the east or south-east at 5° to 10° but, in places, there is gentle monoclinal folding and elsewhere the beds are flat-lying.

A 1 m bed of green siltstone and sandstone, within red-brown mudstone, is exposed [3237 9965] in the south bank of the River Sence. Its sandy base appears erosional. The bed consists of laminated siltstones and sandstones, 10 to 100 mm thick, interbedded with red-brown mudstones, and comprising a fining-upwards sequence.   MRH

The skerries near Upton are siltstone and very fine-grained sandstone beds. The lowest is exposed in a ditch [3746 9960] as 0.5 m of grey-green, fine-grained, cross-bedded sandstone. The same skerry, more than 1 m thick, crops out in the banks of a stream [3690 9828].

In a pond bank [3745 9833], the overlying skerry, mainly of green siltstone, is exposed below 1.5 m of red mudstone, and can also be seen in the river [3826 9775], 400 m NNE of Foxcovert Farm.   KA

Coalville – Ellistown – Heather

The lower beds of the Mercia Mudstone are exposed in the south face of a disused pit [4295 1527] at the former Hermitage Brick and Terra Cotta Works, Whitwick, as about 6 m of red-brown, thin-bedded, sometimes blocky, silty, micaceous mudstone with thin sandstone bands. The latter are either grey, coarse and friable or red-brown, silty and less friable. A prominent sandstone in the sequence consists of 0.32 m of soft, brown, very silty, fine-grained sandstone overlain by 0.20 m of grey-brown, coarse, well sorted sandstone. This bed also occurs at the top of a disused pit [4248 1482] at the former Mansfield Brickworks, Whitwick, where the section was described by Bosworth (1912, p.123).

Thin-bedded, laminated and wispy-bedded, red and green mudstones, siltstones and sandstones are exposed in 4 m high faces at the southern end of the Coalville Brick Co. pit [4165 1485]. The beds show small-scale ripples and cross lamination; pseudomorphs after halite occur in the sandstones. The pit at Ellistown Pipeworks Ltd [4357 1042] is now largely filled, but up to 5 m of thin-bedded, red, silty mudstones and siltstones, with minor green siltstones, and sandstones, are still exposed.

Large exposures in the Mercia Mudstone are to be found at Ibstock, in the two pits worked by Ibstock Building Products Ltd. The main diggings [4196 1080] are 400 m south-east of Ibstock Brickworks where the 26 m section beneath the boulder clay reads:

	Thickness m
Mudstone, red	4
Sandstone, grey-green, soft, silty, micaceous, with muddy wisps and partings	2
Mudstone and siltstone, red-brown and dark brown, interbedded and interlaminated, with minor grey-green laminae; minute, incipient load casts of siltstone into mudstone; ripple marks oriented 290°, and occasional pseudomorphs after halite. Thin sandstone bands up to 0.1 m	2.5
Sandstone, grey-green	0.5
Siltstone and mudstone, red-brown, thin bedded, with thin, grey-green sandstones up to 0.1 m; vugs occur in all lithologies	9

The lower part of this section is also seen in the other Ibstock pit [4115 1095], 150 m north-west of the brickworks:

	Thickness m
Mudstone and siltstone, thinly bedded, red-brown, silty, with subordinate sandstone bands. Much water coming from 0.1 m green sandstone, 0.3 m from base	3
Sandstone, grey-green, with false bedding dipping west. Channels up to 0.2 m into siltstones beneath	0.7–0.9
Mudstone and siltstone, red-brown, with thin, red-brown sandstone bands	3.5

The water-bearing sandstone in this section also occurs in the previous Ibstock pit, just above the 0.7 m sandstone.

The base of the Mercia Mudstone is exposed in the former Heather Brickyard [4014 1075], now part of a gravel pit worked by Heather Minerals Ltd. The section was recorded by Fox-Strangways (1900, p.34) and, as it is not well exposed now, his section, with some modification, is given below:

	Thickness m
Boulder Clay	2
Sand and gravel	7.5
Red mudstone	3
Sandstone	0.3–0.9
Red, silty, micaceous mudstone	3.0–3.4
Yellow-brown, soft sandstone, dark brown in basal 0.5 m; thin mudstone partings and frequent scattered mudstone pebbles	1.0

Hard sandstone	0.05
Red mudstone	0.9
Yellow sandstone	0.3
Red mudstone	1.8
Conglomerate, very hard in centre of pit, soft and decomposed to east	0.3
Sandstone on Coal Measures	0.9

Coal Measures were reportedly met with during recent excavations at the western end of the pit [3990 1080] but are now obscured by tailings ponds. Fox-Strangways placed all the section in the 'Lower Keuper Sandstone', (that is, the Bromsgrove Sandstone), but during the present survey, the yellow-brown sandstone has been taken as the highest bed of the Bromsgrove Sandstone and the overlying beds are therefore assigned to the Mercia Mudstone.　　RAO

Ibstock – Bagworth – Thornton

The mudstones cropping out south and south-east of Ibstock lie near the base of the Group and are more micaceous than those higher in the sequence. There are no mappable skerries hereabouts, but debris from small excavations along the valley south of Ibstock Grange [4101 0927 to 4203 0938] includes fragments of green siltstone and soft, fine, white sandstone in addition to red clay.

About 1 km east of Ibstock Grange, two skerries, several metres apart and each about 1 m thick, form strong features. They lie about 70 m above the base of the Trias. The outcrop of the upper one, the Thornton Skerry, is marked by many fragments of fine-grained sandstone in ploughed fields; the lower is exposed in deep ditches [4274 0994 and 4203 0938]. Along the south side of the valley south of Ibstock Grange, both skerries pass beneath the boulder clay capping the hill. The lower skerry is probably indicated by large blocks of massive fine-grained, white sandstone around a disused pit [4222 0942].

The two skerries can be traced northwards from Ibstock Grange as far as Ellistown. The Thornton Skerry is probably that described by Bosworth (1912, p.78, fig. 42) from the brickyard [433 117], now filled in, at South Leicester Colliery. He recorded 0.5 m of grey, hard, dolomitic sandstone 'crowded with small cavities which appear to be hollow casts of small molluscs'. It seems more likely that the cavities are due to the solution of gypsum. The lower skerry does not appear in Bosworth's measured section and presumably lies below it.

The Thornton Skerry crops out west of Bagworth Colliery, as a low feature associated with patchy, siltstone debris and green clay. Bosworth (1912, p.77) described a section at Bagworth Brickworks pit [probably 440 096], now filled in. It showed 7 m of mudstone with skerries at 2.7 m, 3.8 m and 5.7 m above the floor of the pit. The two lower bands, 0.7 m and 0.8 m thick respectively, probably represent the Thornton Skerry.

The two skerries probably crop out in the drift-free area 1 km south-west of Bagworth, where patches of siltstone brash occur, but they could not be mapped here. North-east of Bagworth Colliery, the Thornton Skerry is thin along the southern slope of the valley above Bagworth Park and seems to be absent on the north side. South-eastwards, however, it forms a consistent feature marked by patches of siltstone flakes and green clay, and caps the prominent 20 m ridge east of Bagworth Park, although it is still very thin hereabouts. Farther south, it is thicker and passes into fine-grained sandstone forming a strong feature on the flanks of the ridge at Thornton. At a building site [4638 0786], a section in the lower part of the skerry exposed 0.6 m of massive, medium-grained, pink sandstone resting on red marl. At Thornton, the skerry is probably at least 2 m thick, but it thins towards Charnwood Forest, and cannot be traced up the valley towards Markfield.

The Thornton Skerry, and another skerry 7 to 8 m higher in the sequence form prominent features on the ridge just south-east of

Great Fox Covert [464 055]. Both these beds are probably exposed in Desford Colliery Brickworks pit [461 065], north-west of Heath Farm, but they cannot be identified with certainty. The section shows 20 m of mudstone alternating between silty, massive, unbedded mudstone and laminated beds, and including several skerries. These lie generally 2 m to 3 m apart, are mostly 0.5 m to 1 m thick and consist of thin green siltstones and fine-grained sandstones a few centimetres thick, separated by shaly red marl. Both the bases and the tops of the skerries are gradational.

On the east slopes of the valley south of Thornton Reservoir, the Thornton Skerry crops out just below the base of the boulder clay, and forms prominent spurs, commonly covered with abundant sandstone debris. Farther south, its broad outcrop caps the ridge north-west of Botcheston and forms prominent spurs above the village. In roadside exposures [4828 0542 and 4840 4520], its lower part consists of thin beds of fine-grained sandstone with mudstone partings. South of Ratby Burroughs and Bury Camp, its outcrop is largely obscured by hillwash, but in the valley north-east of Ratby Burroughs, it forms a strong feature, associated with much green siltstone and fine sandstone debris. In a road cutting [5085 0540], the skerry is 2 m of interbedded green-grey, soft sandstone and red mudstone.   MGS

Desford – Kirby Muxloe

The Thornton Skerry and overlying skerry are both well developed at Desford, especially around the spur about 1 km north of the village. The Thornton Skerry produces a marked feature, though only red clay with green-grey patches, or green-grey clay, is ploughed, so the bed is probably a hard, thin siltstone. Its outcrop widens east of Desford and the skerry changes south-eastwards from siltstone [4818 0411], through silty sandstone [4832 0373] to fine-grained sandstone. Below this bed, east of Desford, a slight feature on the lower slopes of the valley is formed by a thin siltstone accompanied in places by a bed of greenish grey mudstone. The skerry at Desford above the Thornton consists of thinly planar-bedded, fine-grained sandstone weathering into slabs up to 40 mm thick. Two disused diggings [4715 0377 and 4797 0367] were worked for building or walling-stone.

The Thornton Skerry crops out as a sandstone along the south side of the valley south-west of Kirby Muxloe, and is well developed south of Oaks Farm [511 031]. The outcrop northwards from the disused railway line [520 036] is uncertain, but the greenish grey flaggy sandstone in the stream [5220 0417] is probably the same bed.

North-east of Kirby Muxloe, a bed of hard, greenish grey siltstone emerges from beneath boulder clay [5242 0413], to form a prominent scarp northwards to beyond the M1. A temporary section [5272 0476] on the southern bank of Kirby Lane showed 3 m of red mudstone with thin green silty bands, overlain by the siltstone, 50 mm thick, and overlain in turn by 2.4 m of red mudstone. In the motorway cutting, 90 m to the north-east, the siltstone is 100 mm thick, and about 600 m north-east of the motorway, the feature is more subdued because the lithology here is softer, fine-grained sandstone [538 060].

Earl Shilton – Enderby

Near Folly Farm, [4676 9970], north of Earl Shilton, the outcrop of a thin, fine-grained sandstone, commonly with pin-hole cavities, forms a slight feature on the lower slopes of the valley and is probably the Thornton Skerry. It can be followed to near Hill Farm, Thurlaston [494 994], and has apparently been worked in places [4807 9977 and 4943 0021]. Probably the same sandstone emerges, from beneath glacial gravel [469 983] near Earl Shilton and extends as far as Barrow Hill [489 971], where it banks against the upstanding mass of diorite with no apparent change in lithology. West of

Normanton Turville, slabs of coarse-grained grey sandstone with cavities abound in ploughed fields [488 983] along the outcrop. Two skerries lower in the sequence can be traced for short distances near Knoll Farm [4905 0020], though only one of these is shown on the 1:50 000 map. It is likely that all the skerries cropping out north of Earl Shilton correlate with the Cotgrave Skerries in the Leicester Forest East Borehole (Figure 18).

Around Huncote Grange [5175 0085] and Old Warren Farm [5310 0230] are two outliers of Dane Hills or Arden Sandstone. The bed is probably only about 1 m thick and includes bands of greenish grey mudstone. At 800 m north of Huncote Grange, an 18 cm bed of pale grey sandstone crops out [5168 0167] in a pond bank. Green mudstone overlies the sandstone hereabouts. Around Old Warren Farm, the sandstone forms a broad dip-slope inclined to the east and is again overlain by green mudstone.   BCW

The eastern end of the diorite quarry [5405 9998] at Enderby shows good exposures of basal Mercia Mudstone with breccias:

	Thickness
	m
Boulder clay, red, mudstone-rich, with flints, 'Bunter' quartzite pebbles and large diorite boulders	15
Mudstone, green, sandy, with thin brown and red sandstone bands containing many large wind-polished grains	1.2
Mudstone, reddish purple, thinly bedded and shaly, with black iron-rich veins and patches	1.2
Mudstone, reddish brown, rubbly, hard and sandy, with green nodules and thin bands resting with uneven unconformity upon diorite	2.1

A basal breccia up to 1 m thick, consisting of greenish grey, well cemented, hard, calcareous mudstone, with abundant pebbles and boulders of locally-derived diorite, occurs in hollows in the diorite surface. In Narborough Quarry [5240 9755], 6 m of red mudstone with thin green bands, dip westwards at 15° at the west end of the quarry and are practically horizontal over the crags of diorite elsewhere in the excavation.   EGP

Bardon Hill – Cliffe Hill – Spring Hill

At Bardon Hill Quarry [459 131], the pre-Triassic topography is very irregular. Hollows are filled with thin-bedded mudstones and sandstones with catenary bedding (Watts, 1947) or with coarse angular breccias. On the north side of the quarry [457 133], a depression is filled with 4 m of an unbedded breccia of angular dacite blocks, which resembles a recent head deposit except for its red mudstone matrix. It passes upwards into 10 to 15 m of blocky mudstones, laid down across the filled depression, and including a 0.5 to 1 m bed of hard, grey sandstone near the base which is packed with angular Charnian fragments, 2 cm to 5 cm across. The largest clasts occur where the sandstone is thickest. The top 4 m of mudstone is massively false-bedded with foresets dipping eastwards. On the opposite side of the quarry, a channel, 6 to 8 m, deep is filled with a very coarse, crudely bedded, buff breccia with angular Triassic and Charnian clasts, overlain by 1 m of mainly greenish grey mudstones with red-brown mottling, and crowded with angular Charnian and vein-quartz clasts.

Similar beds are seen elsewhere in the quarry and the Trias base climbs to within 30 m vertically of the summit of Bardon Hill. It is notable, in these and similar exposures, that the Charnian debris in the mudstones is very localised. Red mudstones commonly persist to within a few metres of the unconformity. Where they abut against older rocks, the Triassic beds show westerly depositional dips up to 20°, subparallel to the unconformity but less steep. The dip rapidly decreases away from the unconformity and in most exposures, the Trias is almost flat-lying.

The sub-Trias unconformity is well exposed along the north side of Cliffe Hill Quarry [474 108] (Plate 9). The lithologies resemble those at Bardon Hill Quarry except that blocks of diorite occur, in addition to Charnian clasts.

Many of the diorite blocks have been deeply weathered in situ (Ford and King *in* Sylvester-Bradley and Ford, 1968, p.330). At the north-west end of the quarry [4738 1080], where Triassic rocks are banked against an east-facing slope of diorite, the section is:

	Thickness m
Mudstone, dark red-brown, packed with small clasts of diorite and Charnian rocks and larger boulders of rotten diorite up to 10 cm in diameter; cut by narrow sandstone dykes and sills in top 0.5 m	0.3
Sandstone, red or red-brown, soft, muddy, thin-bedded, with many small diorite and Charnian clasts up to 3 cm across. Some green-tinged beds in basal 30 cm	1.8
Mudstone, red-brown, thickening eastwards and becoming blocky and gritty	0.3
Sandstone, muddy, brown and grey-green, thin-bedded, with blocks of diorite at base and passing down into breccia-filled hollows in diorite	2

The prominent 1.8 m sandstone can be traced for at least 300 m along the quarry face before passing into a thinner, muddier sandstone.

Several minor folds about NNE–SSW axes are developed in the Trias and dips are not usually related to the topography of the underlying buried landscape, as they are in Bardon Hill Quarry. Many NNE–SSW, small, normal faults, throwing up to 0.7 m, cut the Triassic beds, producing 'horst and graben' structures.

A Trias-filled depression, 10 m deep and 20 to 30 m across, occurs on the north face of Peldar Tor Quarry [4470 1596] at Spring Hill. A very coarse, locally derived breccia is overlain and overstepped by thin-bedded, red and grey-green mudstones and thin siltstones. A similar depression, 5 m deep, occurs 300 m to the east.

RAO

Markfield – Newtown Linford – Groby

Sections of Mercia Mudstone have been described (Poole, 1968) from motorway cuttings north of Markfield, at Charley Knoll [487 158], Birch Hill [482 134] and The Hollies [478 116]. In the cutting [481 103] west of Markfield, 1.5 m of buff to brown-weathering mudstone rest on 2 m of reddish purple mudstone with green spots, veins and thin bands. The beds are finely micaceous and contain scattered, small, sand grains. They must lie in a deep pre-Triassic gully between the igneous outcrops of Cliffe Hill and Markfield.

At a bend of the stream in Bradgate Park [5318 0987], 9 m of red mudstone contain thin green bands. In the disused Bradgate Granite quarry [5125 0924], 1.5 to 3 m of red mudstone with thin green bands rest on the upper surface of the diorite. Exposures of Mercia Mudstone resting on diorite occur in the large quarry [5137 0916], 90 m farther east, where 3 to 4.5 m of red mudstone are con-

Plate 9 Mercia Mudstone filling a hollow in the pre-Triassic land surface, eroded in steeply dipping Charnian tuffs. Cliffe Hill Quarry. A 12341

tinuously exposed along the south-west face, dipping to the south-west at 12°. In the north of the quarry [5157 0912], a 3 to 4.5 m boulder bed consists of large, locally-derived, diorite boulders up to 1 m across, set in red mudstone. The boulders are deeply weathered and concentric sphaeroidal shells have flaked off to be incorporated in the mudstone. The deep weathering took place in Triassic times and the underlying diorite is deeply reddened along all its joints. In the western face of the next quarry to the south-west [511 089], a 15 m high buried hill of diorite is revealed, with red mudstones and basal breccias dipping away to the north and south. Breccias of locally derived diorite are up to 6 m thick on the north side of the hill. At the northern entrance of the quarry [5131 0893], about 6 m of red mudstone with ripple-marked, green, silty bands rest on 1.5 m of rubbly, red mudstone with large, angular, diorite boulders. In the south of the quarry [5139 0868], 12 m of mudstone with 0.15 to 0.6 m green, silty, flaggy bands are slumped close to the diorite surface and include boulders from it.

In the disused quarry west of Groby [520 073], 6 m of red and green, banded mudstone with ripple-marked, silty beds rest upon 1.5 m of well cemented, basal breccia derived from the underlying diorite. The dips in these beds are largely depositional, being northwards on the north side of the diorite ridge and southwards on the south side. Farther east, about 10 m of red mudstone with breccias at the base are exposed in a disused railway cutting [5224 0760].

In a disused quarry [5255 0787] just north of Groby, 3 m of mudstone rests on 6 m of diorite. In the upper part of the section, the mudstone is predominantly red with thin, green, fine-grained cross-bedded sandstone bands. The lowest 1.5 m contains an increasing number of small, angular, diorite fragments with the basal 0.6 m forming a well cemented, hard, massive conglomerate. The uppermost 0.15 m of the diorite itself is deeply weathered and indurated, with Triassic reddening penetrating deeply into all joints and cracks. In the main Groby quarry [525 083], pre-Triassic red staining extends at least 60 m below the former land surface, down all major joints as far as the lowest levels of the quarry.   EGP

74

CHAPTER 7

Intrusive igneous rocks

The older igneous intrusive rocks of the district are not generally well dated and may represent more than one intrusive episode, but they are all pre-Carboniferous in age. In contrast, the Whitwick Dolerite dates from post-Carboniferous times.

PRE-CARBONIFEROUS INTRUSIONS

The acid-intermediate intrusions of Charnwood and South Leicestershire were classified as syenites by the early surveyors (Hill and Bonney, 1878; Watts, 1947), but are more correctly termed diorites, granodiorites and tonalites (Eastwood and others, 1923; Snowball, 1952; Le Bas, 1968). They fall into three groups. Two crop out in Charnwood and form the 'northern' and 'southern' varieties of Hill and Bonney; they are here referred to as the North Charnwood Diorites and the South Charnwood Diorites. The third group occurs farther south and was formerly called the 'South Leicestershire Syenites' but is here termed the South Leicestershire Diorites. Each of these groups is distinctive petrographically, structurally and geochemically. Their terminology is summarised in Table 5, their geographical distribution is shown in Figure 1, and their modal and chemical compositions are listed in Tables 6 and 7.

Although the diorites of Charnwood are undoubtedly post-Charnian in age, the younger limit of the intrusive episode is difficult to determine. In the Nuneaton inlier, Wills and Shotton (1934) found that the diorite (markfieldite), which they considered identical to the South Charnwood variety, was unconformably overlain by Lower Cambrian quartzite. If the Nuneaton and South Charnwood intrusions are indeed contemporaneous, then the South Charnwood Diorites are late-Precambrian in age, and an

Rb-Sr isochron age of 552 ± 58 Ma obtained from them (Cribb, 1975), supports this view. In contrast, the North Charnwood Diorites yield an Rb-Sr isochron age of 311 ± 92 Ma (Cribb, 1975), but this may relate to Hercynian mineralisation (Le Bas, 1981, p.45). Spots produced by thermal metamorphism of the Charnian in contact with North Charnwood Diorite at Longcliffe Quarry [494 170], are deformed by the later cleavage, proving a pre-cleavage date for the intrusion (Boulter and Yates, 1987).

The only known intrusive contact between the South Leicestershire Diorites and their country rock was formerly exposed just east of the present district at Coal Pit Lane Quarry, Enderby [542 992] (Hill and Bonney, 1878, p.227; Bonney, 1895, pp.25–29), but there is no certainty as to whether the country rock was Charnian or Cambrian. Le Bas (1968, pp.41–42) favoured a Cambrian age and writes (personal communication, 1987) 'A sample of the poorly bedded sediments from Coal Pit Lane showed no trace of cleavage, which makes it unlikely that it is Charnian. It compares better with the baked Stockingford Shales exposed at Nuneaton.' An Rb-Sr isochron age for these intrusions of 546 ± 22 my (Cribb, 1975), however, is not significantly younger than that obtained for the South Charnwood Diorites, and suggests that the South Leicestershire intrusions are also late-Precambrian in age. Nevertheless, Le Bas (1972) has argued that they are Caledonian in age and a Zircon U-Pb age of 452 ± 5 my from Enderby (Pidgeon and Aftalion, 1978, p.205) apparently substantiates this view. Sills cutting the Tremadoc in the Merry Lees Drift, described as 'feldspar-phyric dolerites' or 'quartz porphyrites' by Phemister (in Butterley and Mitchell, 1945, p.707), were classified as 'carbonated and chloritised quartz-diorites' by Le Bas (1972, p.75) and equated with the South Leicestershire Diorites of Barrow Hill [487 972] and the Stocks House

Table 5 Intrusive rocks and their isotopic ages

Diorites	South Charnwood Diorites	South Leicestershire Diorites and Quartz-diorites		North Charnwood Diorites
Synonyms in previous literature	Southern Syenites Southern Diorites Markfieldite	South Leicestershire Syenites		Northern Syenites Northern Diorites
Country rock	Charnian	?Charnian or Cambrian		Charnian
Radiometric age	552 ± 50 m.y.(Rb:Sr)	546 ± 22 m.y.(Rb:Sr) 452 ± 5 m.y.(U:Pb)		311 ± 92 m.y.(Rb:Sr)
		DIORITE	Q.DIORITE	
Main localities	Bradgate Park Groby Markfield/Cliffe Hill	Enderby Yennards Quarry Barrow Hill	Narborough Huncote/Croft Stoney Stanton/ Sapcote	Bawdon Castle Hammercliffe Lodge Newhurst-Longcliffe

Table 6 Modal analyses of intermediate rocks

	SOUTH CHARNWOOD DIORITES	NORTH CHARNWOOD DIORITES	SOUTH LEICS DIORITES	SOUTH LEICS DIORITES	MOUNTSORREL GRANODIORITE
	Cliffe Hill, Markfield, Bradgate, Groby	Newhurst, Longcliffe, Ulverscroft	Enderby, Yennards	Croft, Stoney Cove, Narborough	Mountsorrel
Number of rock slices analysed	11	4	4	6	2
Average of counts per slice	1973	1966	1922	1990	2000
Range of content in % with means in brackets					
Plagioclase (P)	10–47	2–12	0–34	0–48	33–36
Alkali feldspar (A)	4–17 (11)	1–3 (2)	0–6 (2)	0–8 (3)	21–22 (21)
Quartz (Q)	15–27 (18)	7–12 (10)	7–12 (10)	15–22 (18)	29–32 (31)
Amphibole	0–7	0–15	0–14	0–5	0–2
Chlorite	5–23	15–27	9–16	8–15	2–4
Opaques	1–4 (3)	2–9 (6)	1–8 (5)	1–4 (3)	0–2 (1)
Apatite	0.4–1.1	0.4–0.5	0.2–0.6	0.2–0.5	0.1–0.2
Epidote (E)	6–26	28–42	0–28	1–16	1–2
Sericite (S)	0–28	1–24	0–41	2–46	1–3
Carbonate (C)	0–6	0–3	0–8	0–7	0–1
Brown alteration	4–8	5–9	4–9	5–7	0–1
Nomenclature Streckeisen (1976)	Granodiorite (9)	Diorite (4)	Diorite (4)	Quartz-diorite (5)	Granite (2)
P + A + Q = 100%	Quartz-diorite (1)			Granodiorite (1)	
P includes E,S,C	Monzodiorite (1)				

Borehole [4680 0212]. If this correlation is correct, then the South Leicestershire suite postdates the Tremadoc and the 546 Ma age determination for the Barrow Hill diorite is suspect. Pharaoh and others (1987b) show that the South Leicestershire Diorites are more enriched in certain incompatible trace elements than the Charnwood Diorites, and are more likely to be Caledonian in age.

South Charnwood Diorites

The South Charnwood Diorites crop out in an arc partly broken by faulting, between Bradgate Park and Stanton under Bardon. All the outcrops probably form part of a single large intrusion (Hill and Bonney, 1878, p.212), elongated parallel to the Charnian structure and considered by Watts (1947, p.72) to be laccolithic. At Cliffe Hill Quarry, cross-cutting intrusive relationships with the Char-

nian are observed and there is a large, foundered xenolith of Charnian within the intrusion. Smaller, enclosed and partly assimilated, Charnian xenoliths occur at Bradgate Quarries. Only one major phase of intrusion can be recognised, although aplite dykes occur near Groby and in Groby Quarry.

The South Charnwood Diorites are more leucocratic than the North Charnwood Diorites, with visible quartz, and they are generally less sheared. Plagioclase is the most common mineral, though it is rarely fresh. The crystals are equant, tabular and subhedral in shape, with an andesine core and variable, oscillatory and normal zoning to an oligoclase rim. Some plagioclase grains from a quarry near Groby [5178 0774] have cores of An_{70} (E 43848); in others, the cores are now of sericite and epidote. The latter minerals are the most common alteration products, occurring with lesser amounts of calcite, chlorite, actinolite, prehnite, albite and quartz. In

Table 7 Whole rock analyses of Leicestershire dioritic rocks

	S. Charnwood diorites							N. Charnwood diorites			S. Leicestershire diorites						
	I	II	III	IV	V	VI	VII	VIII	IX	X	XI	XII	XIII	XIV	XV	XVI	XVII
SiO_2	65.81	49.40	59.17	61.54	57.01	55.70	54.3	48.3	48.32	52.36	61.84	64.30	65.16	63.89	60.31	67.05	55.50
Al_2O_3	13.06	15.85	15.02	15.13	15.86	17.50	14.7	12.7	16.95	14.19	15.51	17.89	16.19	16.52	16.27	15.78	17.89
Fe_2O_3	9.81	5.44	3.18	2.23	4.41	5.20	5.5	8.7	5.15	7.96	1.90	4.75	0.29	1.33	1.89	1.48	2.42
FeO		5.29	4.91	4.71	4.14	4.70	6.3	6.8	6.11	5.11	2.84		3.88	2.78	3.18	2.44	3.88
MgO	1.01	3.89	2.49	2.12	3.11	2.90	4.6	6.8	3.98	4.09	3.22	1.12	1.72	2.18	3.20	1.53	4.63
CaO	4.55	8.70	4.31	3.60	6.04	6.70	5.7	5.2	8.76	7.49	5.03	3.98	3.85	3.90	5.50	3.47	7.20
Na_2O	2.34	2.33	3.38	3.33	2.81	2.40	2.9	2.0	2.12	2.00	4.12	3.84	4.49	4.28	3.63	3.92	4.22
K_2O	2.85	1.27	2.21	2.88	2.17	2.10	2.04	1.68	1.00	1.28	1.98	3.37	2.05	2.16	1.69	2.69	1.34
H_2O^+	0.54	3.60	3.06	2.68	2.54	2.30	3.4	4.4	4.07	3.10	1.97	1.60	1.31	1.76	2.77	0.78	0.33
H_2O^-		0.54	0.20	0.15	0.18				0.27	0.14	0.34		0.19	0.15	0.17	0.17	0.21
TiO_2	—	1.04	0.78	0.69	0.73		0.73	1.44	1.11	1.18	0.66		0.57	0.63	0.78	0.48	0.80
P_2O_5	—	0.18	0.33	0.27	0.26		0.21	0.17	0.18	0.22	0.21		0.19	0.19	0.20	0.21	0.24
MnO	—	0.18	0.17	0.16	0.14		0.17	0.21	0.19	0.20	0.08	tr	0.08	0.09	0.09	0.06	0.10
CO_2	0.03	2.08	0.63	0.29	0.23		0.8	1.0	1.55	0.67	0.52		0.14	0.03	0.04	0.30	1.13
Traces		0.22	0.25	0.23	0.27				0.21	0.27				0.21	0.22		
Total	100.0	100.01	100.09	100.02	99.90	99.50	99.05	98.90	99.98	100.26	100.34	100.85	100.11	100.10	99.94	100.36	99.89
Li		36	26	26	22				32	14				37	38		
F		280	400	410	340				260	310				410	410		
S		120	100	200	110				50	60	600			60	30		
V		330	100	60	140				290	330				65	80		
Cr			<10		<10		26	37									
Co		20	20	<10	20				30	20				<10	<10		
Ni			<10		<10		9	19									
Cu		170	70	70	95		98	103	160	110				30	40		
Zn		125	115	95	120		93	122	115	140				80	90		
Ga		<10	<10	<10	<10				<10	<10				<10	<10		
Rb		26	65	80	70		67	25	27	28				60	38		
Sr		320	270	190	340		281	187	325	400				520	640		
Yt		23	25	26	25		22	22	23	30				25	18		
Zr		30	50	90	100		89	67	40	100				160	160		
Ba		420	1040	1000	1000		857	633	490	850	180			520	440		
La							14	11									
Ce							31	22									
Pb		13	13	14	13		32	27	8	12				9	19		
Th		<19	<18	<14	<13		9	5	<15	<12				<15	<15		
U		<2	<2	<2	<2				<2	<1.5					3		

I Syenite from Groby, analyst H. Rogers, quoted from Watts (1947).

II Basic diorite (E 43847), Groby Old Quarry [5178 0774]. New analysis by J. I. Reed, S. C. M. Dwarka, W. A. McNally and J. P. Arnold, with spectrographic determinations of V, Cr, Co, Cu, Ga, Sr, Zr and Ba by R. G. Burns.

III Diorite (E 43849), Groby Quarry [5266 0823]. Analysts as for II.

IV Diorite (E 43850), Bluebell Quarry, Groby [5239 0857]. Analysts as for II.

V Diorite (E 43853), Markfield Quarry, east side [4860 1030]. Analysts as for II except for J. I. Reed.

VI Syenite form Markfield, analyst J. Hort Player, quoted from Watts (1947).

VII Diorite from Cliffe Hall Quarry [4710]. Mean of 10 analyses by Thorpe (1974, p.122).

VIII Diorite from Newhurst Quarry [487 179]. Mean of 10 analyses by Thorpe (1974, p.122).

IX Diorite (E 43857), Newhurst Quarry [4862 1796]. Analysts as for V.

X Diorite (E 43859), Longcliffe Quarry [4933 1697]. Analysts as for V.

XI Quartz diorite, Sapcote Quarry, analyst E. G. Radley, quoted from Eastwood and others (1923).

XII Syenite or hornblende tonalite from Huncote, Croft Hill, analyst E. E. Berry, quoted from Berry (1882, pp.197–199).

XIII Augite-biotite tonalite, Croft Quarry [512 964]. Quoted from Le Bas (1972).

XIV Quartz-diorite (E 43873), Croft Quarry [512 962]. Analysts as for V.

XV Diorite (E 43878), Warren Quarry Enderby [539 002]. Analysts as for V.

XVI Porphyritic microtonalite, Coal Pit Lane, Enderby [542 992]. Quoted from Le Bas (1972).

XVII Quartz-augite diorite, Barrow Hill Quarry, Earl Shilton [487 972]. Quoted from Le Bas (1972).

some rocks, such as E 44567 from Groby Quarry, highly altered plagioclase is found next to clear, zoned andesine; in others, alteration products have developed preferentially in some zones; this variation in the sericite–epidote ratio probably reflects original differences in the An content. Some of the diorites appear to be sericitised with little or no epidote, whereas others show extensive epidote and chlorite, with sericite virtually absent.

The plagioclase grains commonly form nuclei to patches of radiating graphic quartz and K-feldspar intergrowth, (e.g. E 44538 and E 47543, Plate 10) which is characteristic of Markfieldite as originally defined (Hatch, 1909). This intergrowth is commonly interspersed with more discrete, but irregular, growths of cracked or strained quartz, or with patches of brown, turbid K-feldspar. Mantling of plagioclase by K-feldspar occurs sporadically (E 44545), though quartz (or quartz with K-feldspar) is generally adjacent to the outermost zone of the plagioclase. These textural relations indicate that the period of quartz crystallisation overlapped that of plagioclase. There are sporadic occurrences of euhedral quartz adjacent to both amphibole (E 43847) and K-feldspar (E 43849), and quartz commonly occurs in clusters or as single grains, with marginal, irregular, crudely concentric (E 44561) cracks filled by K-feldspar or a white mica. Pyroxene, amphibole and opaque minerals have been altered, giving rise to chlorite, epidote, titanite and hematite. One rock, from Cliffe Hill Quarry, contains subhedral, partly altered, hypersthene, anhedral augite marginally altered to amphibole, subhedral elongate crystals of inverted pigeonite, and slightly zoned amphibole (E 43852). The hypersthene and inverted pigeonite indicate that Ca-poor pyroxene crystallised from a slowly cooling melt over a range of perhaps 50° or more (Brown, 1957). Microprobe analysis of an augite from the diorite at Groby Old Quarry [5178 0774] E 43847 indicates a composition of $Wo_{43} En_{40} Fs_{17}$. In most samples of the diorite, however, no fresh pyroxene remains; its former presence is inferred from the elongate masses of pale green chlorite that contain lamellae of chlorite or amphibole in a herringbone pattern.

Amphibole is much commoner than pyroxene, and occurs surrounding pyroxene cores and in discrete, euhedral to anhedral grains. It is generally homogenous, showing green to pale green to straw-yellow pleochroism, although some grains are zoned, with olive to brown cores (E 43847, E 43852), and others are brown only in those parts of the grains adjacent to opaque minerals. Inclusions of opaque grains, apatite and tiny grains causing different-coloured haloes in the host, are found sporadically. Marginal alteration to chlorite is common. These 'primary' amphiboles are easily distinguished from the much smaller acicular crystals of actinolite, that are common as alteration products associated with chlorite and titanite.

The opaque minerals are largely magnetite, ilmenite and hematite with minor amounts of iron and copper sulphides. The oxides are subhedral and commonly contain euhedral prisms of apatite. Many grains are extensively altered and now consist of a skeletal, opaque framework lying in a microcrystalline mass of titanite and chlorite.

Where the diorites are heavily altered, up to half the rock consists of chlorite. Generally, this is green to pale green, but there are some dark green and straw-yellow varieties. It occurs in a platy habit (with $2V_x$ of 40°) and as masses of tiny unoriented crystals, and in most rocks it contains tiny inclusions with pleochroic haloes (probably zircon, allanite and monazite). In some rocks, the chlorite shows moderate birefringence but generally this is low, and anomalous interference colours of blue, pale blue and lilac are quite common. Chlorite occupies the cores of amphibole grains (e.g. E 43851), where it may be pseudomorphous after pyroxene; elsewhere, chlorite patches are rimmed by actinolite (E 43853) or more commonly amphibole is fringed by chlorite.

The epidote is typically pale yellow-green and shows a slight pleochroism, with minor zoning in a few rocks, indicated by yellow-green cores and colourless rims (e.g. E 43849). More extreme zoning occurs in rocks from Cliffe Hill (E 43851, 44560), where brown allanite forms the cores of apple-green epidote (Plate 10d). It is uncommon for epidote to grow as well formed crystals at the expense of plagioclase or amphibole, and it is generally found in anhedral granules, mixed with titanite in places, and replacing other minerals indiscriminately. Epidote with quartz, chlorite, calcite, prehnite, pumpellyite, apatite or hematite is a common constituent of the veins which occur in varying abundance throughout the diorites.

Titanite, apatite, zircon and allanite, with minor sulphides and iron and titanium in various states of oxidation, are accessory minerals. Apatite shows the largest range in habit with broken, euhedral, stumpy prisms lying alongside needles with a length to breadth ratio of up to 200:1. It appears to have formed early, as inclusions in amphibole and opaque minerals, and continued to form during cooling, as inclusions in quartz and K-feldspar. Zircon and allanite form tiny, high relief inclusions in amphibole and chlorite, many surrounded by pleochroic haloes (E 44569, Plate 11a).   RRH

Details

Diorite is well exposed in several crags in Bradgate Park, such as those in Bowling Green Spinney [5303 1063]. Strong vertical slickensides occur on outcrops along the steep-sided valley between Newtown Linford and Cropston Reservoir. These are particularly well seen at the Wishing Stone [5262 1004] and near Bradgate House [5340 1017]; possibly the valley is eroded along a fault line.

RAO

The pre-Triassic Newtown Linford Fault (Figure 2) displaces the diorite southwards to Groby. In the western face of the main Groby Quarry [525 083], 37 m of diorite are exposed beneath the Mercia Mudstone. Red staining extends down the joint planes in the diorite to the floor of the quarry. The diorite is generally coarse- or medium-grained, and purple and green mottled. It is cut by strong jointing at 022° and 328°, with horizontal slickensiding along the joint faces. Thin quartz veins are also aligned in these directions and occur indiscriminately in some highly crushed areas [5267 0824]. Near the junction with the Mercia Mudstone [5246 0826], pyrite occurs in the quartz veins and secondary malachite on the joints.

Numerous shear zones, oriented between 315° and 350°, cut the diorite in the main quarry and in a subsidiary quarry to the north [524 085]; in these zones, the diorite is heavily altered and has a schistose texture. The shearing may be associated with the Newton Linford Fault. Parallel shear zones are common in the outcrop north of Groby Pool. A 2 m-wide, fine-grained, grey, aplite dyke

a

b

c

d

Plate 10a Diorite, Bradgate Quarry [510 087]. Crossed polarisers; width of field 2.9 m. E 47543

Quartz-K feldspar granophyric intergrowth of variable coarseness; some appears to radiate from the plagioclase tablet in the centre of the field

Plate 10b Diorite, Bradgate Quarry [510 087]. Width of field 2.9 mm. E 47543

Two cracks across altered plagioclase in the top of the field are filled with calcite and the same cracks lower down, across granophyric intergrowth, are filled with quartz and epidote

Plate 10c Diorite, Cliffe Hill Quarry [4720 1082]. Width of field 2.9 mm. E 44546

Altered plagioclase in the upper left field is adjacent to amphibole and relict twinned pyroxene in the upper right and lower left centre field, with inclusions of magnetite and apatite; quartz-K feldspar intergrowth is in the bottom right field

Plate 10d Diorite, Cliffe Hill Quarry [4756 1065]. Width of field 0.75 mm. E 44560

Very pale yellow crystal of clinozoisite with a core of brown allanite lying in altered plagioclase, adjacent to calcite and displaced by a hematite and calcite-filled fracture. Granophyric intergrowth is in the bottom left field

occurs in the north face of the main quarry [5250 0842], and a dark green aplite dyke, 1 m wide, strikes at 207° across the northern quarry [5243 0858].
EGP

An aplite dyke in the north-east face of Groby Quarry [5279 0833] consists of patches of sericitised plagioclase in a groundmass of granular quartz, sericite, chlorite, carbonate, opaque minerals, titanite, apatite, feldspar, and possibly clinozoisite. The dyke is cut by veins of quartz and carbonate. Chlorite occurs in platy or fibrous crystals with anomalous blue interference colours, and apatite is present as stumpy prisms of variable size. The margin of the dyke has the same mineralogy as its interior, but the groundmass is finer grained and contains quartz clusters and lozenge-shaped patches of chlorite and carbonate. Alteration is more extensive in the dyke than in much of the host rock, so the dyke may have formed a channel for the metasomatic fluids.
RRH

A fine-grained, greenish grey, aplite dyke, 3.8 m wide, occurs [5202 0763] 300 m west of Groby Church, and intrudes coarse, dark grey diorite to the south, and quartz-veined, coarse, pegmatitic, pink and green diorite to the north. The dyke trends at about 300°, dips north-eastwards at 65° and can be traced for about 200 m.

Up to 18 m of diorite are exposed in the faces of the most southerly [5087 0901] of the Bradgate Granite quarries, which lie east and south of Bradgate House. Reddening everywhere extends downwards along the joints from the overlying Triassic rocks. Near the entrance to the quarry [5120 0862], the diorite is well jointed, with the main east–west joints dipping south at 22°. A second set of strong joints trends at 354° and dips westwards at 75–80°. About 18 m along the northern face from the entrance of the quarry, the diorite is netted with quartz veins, up to 8 cm thick, which are associated with the northerly-trending joints. A few metres farther west, a large mass of fine-grained, slaty, Charnian country rock is caught up in a crushed and quartz-veined zone, and dips about 30° to the north. In the northernmost corner of the quarry [5107 0873], the diorite is crushed and sheared into elongated, lozenge-shaped masses enclosed within highly sheared bands; the quarry terminates here in a 25 m-wide shear-zone which trends at 292°. The diorite in the south-west of the quarry [5103 0863] is veined with quartz along joints at 009° and 101° some of which are bordered by a fine-grained, green zone, up to 15 cm wide, rich in pyrite. Small galena crystals occur in the quartz veins. The Bradgate Granite Quarry

300 m to the north-east [514 088] shows up to 25 m of diorite, with reddening penetrating deeply down the joints. The eastern face of the quarry is composed of coarse, dark purple and green diorite with a few bands of coarse, pink pegmatitic diorite interbanded with partly assimilated, fine-grained, green, slaty bands and fine-grained, dark grey quartzites which dip northwards at about 20° to 25°. The Bradgate Granite Quarry which lies a further 300 m to the north-east [515 092], has faces up to 30 m high in diorite. At its southern entrance, a coarse-grained, dark green, black and pink mottled diorite has strong joints, striking at 275° and dipping southwards at 70°. Chlorite and hematite follow these joints. Cross-joints trend at 095° with a dip of 50°; some of them are followed by white quartz veins, and on the eastern face, about 55 m north of the entrance, these veins contain small patches of malachite. About half way along the western face [5138 0921], the diorite is badly crushed and has quartz-veins aligned at 095°. South of this crush-zone, the rock is composed of very coarse, pegmatitic, green diorite but, to the north of the zone, it is more variable with purple, grey and pink varieties, some fine-grained.

An exposure about 140 m long of coarse, pink and grey pegmatitic diorite occurs in the western half of New Plantation [5010 0902], north of Groby Parks Farm. The contact with the Swithland Greywacke Formation to the east is gradational, the slates gradually passing into grey, schistose rock and then into coarse, pink pegmatite. The quartzite underlying the slates is particularly hard, probably due to thermal metamorphism, with thin, white, quartz veins following the bedding. Farther north-east, a small exposure [5080 0984] north of Bradgate House reveals 2.7 m of granitised tuffs, where the gradual passage from tuff to diorite is clearly seen.
EGP

Markfield Quarry [486 103] is the type locality for 'Markfieldite', as defined by Hatch (1909). The quarry is now partly flooded, but there is a 50 to 60 m-high face of massive, fresh diorite which strongly resembles that seen in Cliffe Hill Quarry [475 107].

At Cliffe Hill, the bulk of the diorite in the quarry is uniform, coarse, pink, and granitic in texture, with no great variation in primary amphibole content. No internal contacts are seen and there appears to have been only one phase of intrusion. There are several contacts with the Charnian tuffs. A clearly cross-cutting, irregular, intrusive contact dipping steeply north-eastwards is seen high on the north-east face of the quarry. The face is complicated by faulting, but the diorite is seen to be chilled and darker in colour for at least 10 m from the contact, whereas the tuffs seem hardly metamorphosed, possibly because of their highly siliceous composition. The contact is almost planar for about 600 m to the north-west and then swings sharply north-east to form a small projection into the wall-rock. The clay mineral, palygorskite, is abundant in joints near the diorite–Charnian contact, where this approaches the sub-Mercia Mudstone unconformity. A raft of tuffs, about 4 m thick and 30 to 40 m long, occurs in the middle of the quarry. The upper and lower contacts with the diorite are roughly concordant with the bedding. The granodiorite is markedly chilled and darker in colour for at least 15 m from the contact although, again, the Charnian appears to be hardly metamorphosed.

Diorite protrudes through the Trias to form the hill north-west of Stanton under Bardon [458 109], where a large area is strewn with diorite boulders.
RAO

North Charnwood Diorites

The North Charnwood Diorites include the main intrusions at Hammercliffe Lodge [491 126] east of Copt Oak, Bawdon Castle [495 144] and farther north, in the Loughborough district, at Newhurst [484 180] near Shepshed and at Longcliffe [494 170] near Nanpantan. There are also several smaller diorite intrusions associated with them. Most of the intrusions occur near the axis of the Charnwood Anticline and are confined to the outcrops of the Blackbrook and Beacon Hill Tuff formations. They form thin sheets intruded roughly parallel to the Charnian strike. Watts (1947, p.73) concluded that the diorites were associated with the NW–SE faults, but this does not appear to be the case from the field evidence. At Longcliffe Quarry [494 1700], for example, both margins of the intrusion are in unfaulted contact with hornfelsed and cleaved Charnian, and such faults as are exposed nearby are later than the diorites and are not parallel to their margins. The diorites here are coarse grained and dark grey.

The major minerals of these rocks are, in order of declining abundance, epidote, chlorite, sericite, plagioclase, opaques, titanite, quartz, calcite and prehnite. Apatite and zircon are accessories. Fresh plagioclase is rare throughout and most grains remain in outline only, their interiors having been altered to epidote, sericite, chlorite, carbonate, prehnite, pumpellyite or albite in varying proportions, as in Plate 11b (E 43859). Chlorite in platy or feathery form occurs throughout. It is generally pale green or yellow-green, with dark grey, blue or brown anomalous interference colours. Zircon is a rare inclusion in the chlorite (E 46055) and the pleochroic haloes, so well developed in the South Charnwood Diorites, are absent. Altered primary hornblende and needles of secondary actinolite occur sporadically and intimately mixed with chlorite and epidote. Epidote ranges from fine, granular to coarse, euhedral, in colours varying between pale green, yellow-green and pale brown (E 43858). Very fine granular titanite has largely replaced the original opaque oxides, and along with flecks of chlorite, it occupies the spaces between the remaining skeletal frameworks of an opaque mineral, probably titaniferous magnetite, which commonly forms triangular or rectangular trellises (E 43856). Anhedral crystals of quartz and quartz-K feldspar intergrowth lie between altered plagioclase grains and commonly contain grains of epidote and needles of apatite, with very small flakes of chlorite and brown prehnite (the 'Brown alteration' row of Modal Analysis Table 6). Calcite, in fine-grained disseminations and more coarsely crystalline patches, is present throughout.

Details

Small exposures of dark grey, medium- and coarse-grained, non-porphyritic diorite are found on Cattens Rough [495 147] and Bawdon Rough [495 144]. The intrusion appears to have an outcrop about 100 m across, which is offset by a fault passing between Cattens Rough and Bawdon Rough. No contacts with the Charnian are presently exposed, but Hill and Bonney (1878, p.217) found outcrops of Charnian less than 1 m from chilled diorite on Bawdon Rough. Hornblende schist found in the lane [4974 1405] west of Bawdon Castle suggests that there is a fault in the diorite, hereabouts.

Coarse, massive, hornblende-diorite is exposed in places along the ridge north-west from Hammercliffe Lodge [491 126]. The rock is predominantly dark grey, becoming pink and feldspathic in the most north-westerly outcrop. Numerous, north-trending, en-échelon veins of pink aplite cut the diorite [4898 1269] 200 m east of Copt Oak Farm. These are elongate, rhomboidal, up to 5 cm across and sometimes interconnected. A south-eastward extension of the intrusion, beneath the Trias for at least 400 m, is indicated by the logs of two wells. The first [4957 1212] penetrated 'green mottled stone',

probably weathered diorite, at a depth of 8.4 m to the bottom of the well at 21.0 m. The second [4959 1204] was drilled to a depth of 150 m; core lying close by and presumably from this borehole is of typical diorite. No evidence could be found for the small outcrop of diorite shown on the previous one-inch geological map at Ulverscroft [499 121].   RAO

South Leicestershire Diorites

Within the present district, the South Leicestershire Diorites include the diorite intrusions at Enderby, The Yennards and Barrow Hill and the quartz-diorite intrusions at Huncote, Narborough and Red Hill (Figure 1). The coarse varieties resemble the South Charnwood Diorites, but others are much finer grained. They are believed by Le Bas (1972, p.73) to form the north-western quadrant of a huge, Caledonian, intrusive complex, but in contrast to the diorites of Charnwood, their structural setting is virtually unknown.

Le Bas has extended the intrusive complex to include three diorite occurrences further north, in the Merry Lees Drift, and in boreholes at Barron Park Farm and Stocks House, as described below.

The intrusions lying within the Coventry and Market Harborough districts were described by Eastwood and others (1923) and Le Bas (1968, 1972). In his earlier paper, Le Bas proposed that the intrusive complex be divided into quartz-diorite, hornblende-tonalite and porphyritic microtonalite, and tentatively suggested that intrusion took place in that order. No intrusive contacts between the different lithologies are seen and neither the number of intrusive phases, nor their relative ages, can be demonstrated in the field. Chemical and modal analyses have enabled some of these rocks to be reclassified, notably in the case of the intrusion at Enderby, which is now regarded as a diorite.

Plate 11a Diorite, Groby Quarry [5425 0826]. Width of field 0.75 mm. E44569

Subhedral zircon (0.15 mm long) with conspicuous halo, lying in chlorite which also contains other haloed inclusions, titanite (centre field) and calcite (upper centre right field)

Plate 11b Diorite, Longcliffe Quarry [4933 1697]. Width of field 2.9 mm. E43859

Plagioclase, altered to epidote and chlorite, lies adjacent to a little quartz-K-feldspar intergrowth, and to quartz with an acicular apatite inclusion (left centre field). Subhedral, opaque grains are common and the rock is cut by an epidote vein (north-east trend in upper left of field) and by a calcite vein which also cuts the epidote vein (middle top field)

Plate 11c Diorite, Enderby Warren Quarry [539 002]. Width of field 0.75 mm. E43878

Epidotised plagioclase (left centre field), quartz (centre), and chlorite (lower centre field) with opaque grain, apatite and ?zircon (with dark halo) inclusions

Plate 11d Quartz-diorite, Croft Quarry [512 962]. Crossed polarisers; width of field 2.9 mm. E43876

Zoned and variably altered plagioclase (centre field), with groups of quartz grains showing coarse intergrowth with K-feldspar at some grain edges

a

b

c

d

The diorites and quartz-diorites of South Leicestershire contain the same minerals as the Charnwood Diorites, but in different proportions and with different textures. The commonest mineral is roughly equant, tabular plagioclase with oscillatory and normal zoning in the andesine to oligoclase range, which is variably altered to sericite, epidote or calcite. In some rocks, such as E 43880 from Enderby Quarry, altered cores are successively surrounded by clear oligoclase and turbid alkali feldspar zones, whereas in E 47546 from Narborough, the rims are the most altered part of the grain. The variable character of the alteration is also shown by the occurrence of relatively fresh plagioclase with chlorite patches, probably pseudomorphs after amphibole and biotite; in contrast is the association of fresh augite with plagioclase replaced by sericite, prehnite and epidote in two rocks, E 43873 and E 46053, from Croft Quarry [5125 9630], just south of the present district. In the last specimen, the augite is anhedral and associated with chlorite and opaque grains, whereas in the only other sample examined which contained pyroxene (E 43880), its habit is granular and its host is saussuritised plagioclase. The primary amphibole in this rock, subhedral prisms of a faintly pleochroic, pale green hornblende, is typical of the South Leicestershire intrusive rocks. Secondary amphibole, in the form of sheaves of pale green actinolite needles, is well developed in samples from Croft, Warren Quarry in Enderby and The Yennards. As shown in Plate 11c (E 43878), it forms at the boundaries of chlorite and quartz. Chlorite is ubiquitous, showing variations in its green colour, in the intensities of its pleochroism, and in its degrees of birefringence. It commonly contains inclusions, both of opaque granules and of tiny, high-relief minerals surrounded by pleochroic haloes. The latter could, in some instances, be zircon, larger grains of which are identified outside chlorite in about a third of the rock samples; allanite and monazite are also possibilities. Apatite commonly occurs as stumpy prisms associated with amphibole and opaque minerals, and it shows a gradation to the acicular habit (E 43878), which is more commonly found in alkali feldspar and quartz. In the former habit, it retains its crystal shape relatively well, where included in amphibole, but many larger grains in the body of the rock are zoned and corroded, and show variable, brown, dusty alteration (for example E 43871, E 43874, E 43881 and E 47546). Subhedral to anhedral grains of titanite, of similar size to the corroded apatites, occur in some examples (E 43871) from Stoney Cove [495 941] and Croft (E 43874), but in others this primary form is absent and the titanite occurs as secondary, fine-grained, granular masses replacing opaque grains. As with the other minerals, the alteration of opaque grains is variable, with titanite, hematite and chlorite being formed in different proportions at different places. Most opaque crystals are equant subhedral, but some rocks from The Yennards, such as E 43881 contain acicular prisms and needles which are probably ilmenite. Epidote is of patchy distribution and is commonest as subhedral to anhedral granules replacing plagioclase and as interstitial material associated with chlorite and prehnite. In places, it is associated with calcite, as in E 43872 from Stoney Cove, or as larger crystals adjacent to euhedral quartz, as in E 43880 from Enderby. Quartz is commonly interstitial and poikilitic, grading to graphic intergrowth with alkali feldspar

(E 43871), but it can also occur in rounded grains with septae of K-feldspar, as illustrated in Plate 11d (E 43876, E 47546), closely resembling the similar grains found in the South Charnwood Diorites. The alkali feldspar generally appears homogenous but in one specimen, from Croft (E 43875), string perthites are well developed. Also in Croft Quarry, analcime is quite common in cracks and cavities.

DETAILS

At Warren Quarry, Enderby [540 000], about 40 m of dark purplish grey diorite mottled with pink feldspar were formerly exposed. The rock has a coarse, even-grained, granite texture. Late-stage hydrothermal alteration along joints has produced a series of dark bands, about 15 cm thick and about 3 m apart, dipping 25° to 30° south-east. Similar bands characterise the other South Leicestershire diorites (Le Bas, 1968, p.45). Pre-Triassic reddening extends down the joints to the quarry floor.

A matted white fibrous mineral, palygorskite, occurs on joints in this quarry in the form known as 'Mountain Leather'. Under the microscope, it is a fibrous, matted aggregate of low to moderate birefringence with a mean refractive index of 1.520. It is associated with calcite (°W = 1.658) in rhombs up to 0.3 mm across. Details of the mineralogy of palygorskite are given by Tien (1973). The mineral, which is also found in the overlying Triassic breccias, is thought to have been deposited by meteoric waters seeping downwards in Triassic times (Evans and King, 1962, King, 1968). Similar diorite is exposed in the nearby flooded quarries at Enderby Hill [533 996] and Froane's Hill [534 000].

At Yennards Quarry [489 970] and Barrow Hill Quarry [487 972], both now flooded, the rock is a dark grey, porphyritic microdiorite with feldspar phenocrysts 1–2 mm long (Eastwood and others, 1923, p.18). RAO,EGP

The earliest published records (Hill and Bonney, 1878, p.225; Harrison, 1884, p.41) of the Barron Park Farm Borehole, [5098 0454] drilled in 1830, mention 'syenite' at a depth of 36 m. The detailed log by Fox-Strangways (1907, pp.351–352) ends with 0.35 m of 'red stone' seen to 36 m, and no specimens or analyses of the 'syenite' are now available. Nonetheless, it is assumed that this sinking penetrated diorite below the Mercia Mudstone.

In the Stocks House Borehole [4680 0212], 'whinstone or soft granite' is recorded (Fox-Strangways, 1907, pp.104, 347–348) from 152.2 m to the bottom of the hole at 163.7 m. Watts (in Fox-Strangways, 1907, p.104) compared this rock with the 'camptonites' (lamprophyres) intruded into the Cambrian of Nuneaton, but it is considered by Le Bas (1968, pp.43, 57) to be identical with the hornblende-quartz-diorite of Barrow Hill. BCW

Two quarries were formerly worked in porphyritic microquartz-diorite at Narborough (Bonney, 1895, pp.27–28). The New Narborough Quarry [524 975] is almost completely flooded; the only accessible exposure is a medium- to fine-grained rock, stained dark red-brown. Feldspar phenocrysts are 2 to 3 mm long and chloritised, hornblende phenocrysts reach 1 to 2 mm in length. The Red Hill Quarry [531 975] has been filled in.

The disused Huncote Quarry [5129 9650] exposes about 60 m of quartz-diorite. It is a coarse pink, and dark greenish brown, pegmatitic variety identical to that in Croft Quarry. The upper 3 m is weathered and rubbly, probably partly due to Triassic weathering, below which it becomes hard and massive. Three sets of well developed, high-angle joints strike at 81° to 85° (with associated crushed and sheared zones), 208° to 217° and 338° respectively. To the south, just into the Coventry district, the Croft Granite Quarry [512 963] works the same intrusion. Here about 45 m of pink quartz-diorite are exposed, which is fairly soft in the uppermost 15 m, possibly because of Triassic weathering, becoming very hard and grey below. Scattered, dark green, rounded, pyroxene-rich xenoliths occur. North-east-trending, late-stage, carbonate veins are found in places and analcime occurs in vugs. The rock is well jointed with high-angle joints at directions of 077° to 092° and 338°, and low-angle, southerly dipping joints at 8°. Some of the north-west and west-trending joints have associated sheared zones dipping northwards at 60°.   EGP

ANALYSES OF CHARNWOOD AND SOUTH LEICESTERSHIRE INTRUSIVES

Modal analyses

Modal analyses of 25 rocks from the Charnwood and South Leicestershire intrusions are summarised in Table 6, together with two from Mountsorrel, just to the east in the Leicester district, for comparative purposes. Plagioclase is variably altered to sericite, carbonate, prehnite and epidote, so these constituents should be considered in any assessment of the original total plagioclase content of the rock. Pyroxene is included in the amphibole totals. The figures in the row labelled 'Brown alteration' represent a mixture of titanite, epidote, prehnite and chlorite with iron staining, which is not easily resolvable by the optical microscope. With the range and extent of alteration, evident both from the petrographical descriptions and from the modal analyses, the rocks can only be tentatively named. For present purposes, the epidote, sericite and carbonate contents are assigned to plagioclase and the amended plagioclase totals are used with quartz and alkali feldspar to classify the rocks according to Streckeisen (1976). Eleven specimens of South Charnwood Diorites were analysed (Table 6). The ranges include measurements from both non-porphyritic and porphyritic rocks from the northern face of Cliffe Hill Quarry. In mineral content, the two types are not significantly different.

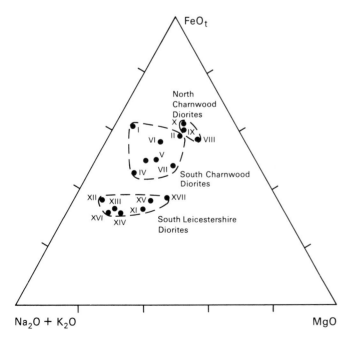

Figure 20   FeO/Na$_2$O + K$_2$O/MgO plot for Leicestershire diorites

Of the 11 specimens, 9 fall in the granodiorite field (where quartz is greater than 20% and plagioclase is 65–90% of the feldspar), one is quartz-diorite (plagioclase more than 90% of the feldspar), and one a monzodiorite (quartz less than 20%).

Analyses of four North Charnwood Diorites show that three of the rocks differ significantly from the South Charnwood Diorites in both quartz and K-feldspar content, with a less significant difference in their opaque mineral concentrations. Other differences or similarities may be present but are masked by the degree of alteration.

Ten specimens analysed from the South Leicestershire intrusions show that the samples from Enderby and The Yennards are diorites, and those from Narborough, Croft and Stoney Cove are predominantly quartz-diorite. This apparent division may not have great genetic significance, for textural features, such as quartz-feldspar and plagioclase-alkali feldspar relations, suggest a closer affinity between the diorites of the Enderby-Sapcote area than between the diorites of Enderby and Newhurst, Shepshed. This conclusion is supported by the grouping of the chemical analyses shown in Figure 20, where the South Leicestershire rocks fall in a cluster nearer the Na + K apex than the Charnwood rocks. Nevertheless, the mineralogical differences between the South Leicestershire rocks revealed by modal analysis do suggest that, given more closely spaced sampling than available exposure allows, a reliable structure and shape for the whole intrusion or intrusions could be worked out.

Whole rock analyses

The term 'syenite' was used for the Charnwood and South Leicestershire intrusions during the 19th century and was perpetuated by Watts (1947). Since then, the use of the term has been questioned and some geologists (Snowball, 1952; Evans, 1968) have proposed that diorite is more appropriate. In the compilative table of chemical analyses (Table 7), the older analyses are listed under the names that were originally assigned to them. Eight new analyses for 12 major, and 18 trace, elements have been made by the Laboratory of the Government Chemist and are presented in Table 7, together with the more reliable of the analyses of the Charnwood and South Leicestershire rocks published up to 1975. Four new analyses, marked by E numbers, of the South Charnwood Diorites are closely similar to previous major and minor element analyses (Thorpe, 1974), and provide new data for the elements Li, F, S, V, Co, Ga and U. These are all present in amounts characteristic of intermediate intrusive rocks. New analyses of the North Charnwood Diorites and South Leicestershire intrusions indicate, similarly, that their compositions are those of intermediate rocks, and are similar to analyses published by Le Bas (1972) and Thorpe (1974). Further analyses of the Charnwood rocks with additional trace element data will shortly be published (Pharaoh and others, 1987b).

The total iron, magnesium and alkali contents of the analyses have been recalculated to sum 100% and the resulting proportions plotted on a triangular diagram (Figure 20). Their distribution emphasises the chemical affinity of the North Charnwood Diorites and South Charnwood Diorites, which lie close together but separate from the

South Leicestershire Diorites which are relatively alkali-rich. The mafic to leucocratic distribution shown on the diagram is also indicated in detail by the average magnesium content in the three groups which decreases from 6.3% in the North Charnwood Diorites to 2.5% in the South Leicestershire intrusions, the average ferrous iron content (6.2% to 3.2% respectively), the titanium content (1.39% to 0.65% respectively), and the manganese content (0.21% to 0.08% respectively). For all these elements, the South Charnwood Diorites show intermediate values. Silica, alumina and soda are all more abundant in the South Leicestershire intrusions that in the Charnwood Forest rocks.

Mineral analyses

Electron microprobe analyses have been made of various minerals occurring in the Charnwood Forest rocks (Table 8). An analysis of augite from Groby is given in column I and from the same rock, analyses of two amphiboles indicate that they are magnesic- and ferro-hornblende (column II and III). Titanite (columns IV and V) generally occurs as an alteration product. In E 43850, the titanite mantles ilmenite and in E 43847, it mantles an opaque mineral of rather different character. Under incident light, the opaque core was seen to consist of inner and outer components both of irregular shape. The inner core has a composition similar to garnet but with too low a silica content, and the outer core is intermediate between this and the rim of titanite (which itself has rather low TiO_2). It would seem that the titanite is replacing either an igneous mineral of early formation, a xenocryst, or part of a xenolith. Other occurrences of this nature were sought but only a few grains, too small to analyse, were found. The slightly low totals of the titanite analysis probably mean that H_2O, F and possibly some rare earths are present. The analysis of chlorite (column VI) has been made up to 100% assuming H_2O of 11.1%, and the calculated Si, Fe and Mg proportions indicate that it is ripidolite (Deer and others, 1962, vol.3, p.137). A zircon inclusion in chlorite from Groby Quarry displayed a marked pleochroic halo and was found to contain about half a per cent each of Th and U (column VII). The analysis total is low and the mineral was scanned for other elements, but none showed above the limit of detectability (about 0.1 to 0.3% depending on the element). In analyses quoted by Deer and others (1962, vol.1, p.62), more than 3% H_2O is present in zircons with high contents of phosphorus and rare earths, but a maximum of 0.8% is quoted for purer zircons; it thus seems possible that some H_2O and minor amounts of perhaps Y and Hf and other rare earths could account for a significant part of the missing 2.3%.

The opaque oxides in both the Charnwood Forest rocks were found to be complex mixtures of titanomagnetite, ilmenite and titanite (columns VIII to XIV). The 'Magnetite' analysis (column VIII) with 18.8% SiO_2 is probably of a magnetite partially altered to titanite in an extremely homogeneous way, although the possibility that pure titanite lies within 5μ of the polished surface of the magnetite (giving a false analysis) cannot be discounted. In the North Charnwood Diorite, the magnetites analysed exhibit variable reflectivity and colours under incident light, which may be accounted for by the variable content of TiO_2; they are virtually free of MgO. The most notable feature of the ilmenites is their high and variable content of MnO, and this, together with the variability of magnetite composition, indicates extensive inhomogeneity in physical or chemical conditions, or both, during formation of the minerals. Both the epidote (column XV) and the chalcopyrite (column XVI) are of average composition.

Andesite dyke

A 1 m-wide andesite dyke [4599 1330], cuts the dacites of Bardon Hill Quarry 110 m north of Bardon Hill Trigonometric Point, and may have been intruded penecontemporaneously with them. The rock is dark grey with 2´ to 3 mm long phenocrysts of hornblende. The dyke strikes east–west and dips steeply northwards and is probably a continuation of that formerly seen in Bardon Hill Quarry, (King, 1968, p.113), which has now been quarried away.

Lamprophyres

Numerous ultrabasic sills intrude the Cambrian rocks of the Nuneaton Inlier, particularly the Outwoods Shales and Merevale Shales. The intrusions are up to 35 m thick and are approximately concordant with the bedding, except in the Merevale area where they show transgressive relationships locally. Their age is uncertain but they do not appear to cut the Upper Old Red Sandstone of the Nuneaton Inlier, so a Caledonian age is favoured (Le Bas, 1968, p.47; Hawkes in Taylor and Rushton, 1971, p.43). Allport (1879), Watts (in Lapworth, 1898), Thomas (in Eastwood and others, 1923) and Le Bas (1968) have described the rocks, which have been called both camptonites and diorites. This uncertainty is due to the extensive alteration affecting them.

From a study of surface and borehole material from the nearby Merevale area, Hawkes (in Taylor and Rushton, 1971, pp.38–43) concluded that the original constituent minerals of the rocks were probably oligoclase, brown hornblende and diopsidic pyroxene. Accessory constituents include apatite, ilmenite, titanite and pyrite. Hornblende and pyroxene commonly show extensive alteration to chlorite, carbonate and secondary iron oxide. The mineralogy is characteristic of spessartites, a type of lamprophyre often associated with diorites. The sills may be consanguineous, therefore, with one of the diorite suites in Charnwood or South Leicestershire.

In the Twycross Borehole [3387 0564], the Moira Breccia rests unconformably, at 503.30 m depth, on highly altered, fine-grained, igneous rock, probably a sill, which at 506.95 m showed a gently inclined, sharp but irregular, contact on purple-stained, laminated, silty mudstone and siltstone of presumed Cambrian age. Though originally described (Institute of Geological Sciences, 1980a) as a microdiorite, the rock compares more closely with the carbonated and chloritised spessartite lamprophyres of the Nuneaton Inlier. A thin section (E 52741) confirms the equigranular, holocrystalline texture apparent in hand specimens, and shows a mesh of plagioclase laths (averaging 0.6×0.2 mm), mainly randomly oriented or forming radiating clusters in places. A principal composition within

Table 8 Electron-microprobe analyses of minerals from Leicestershire diorites

	South Charnwood diorites									North Charnwood diorites							
	I	II	III	IV	V	VI	VII	VIII	IX	X	XI	XII	XIII	XIV	XV	XVI	
SiO_2	54.0	46.9	44.2	32.7	30.8	25.4	31.8	0.1	0.0	1.4	0.3	1.7	0.3	38.0		Si	0.1
TiO_2	0.1	1.3	1.6	27.2	32.8	0.0	0.1	52.8	12.2	3.8	13.1	51.5	51.9	0.2		Ti	0.1
Al_2O_3	0.8	6.4	5.8	5.9	3.7	17.9	0.7	0.0	0.0	0.1	0.1	0.0	0.1	22.4		Al	0.0
Fe_2O_3							2.2		55.2	62.6	55.5			14.8			
FeO	10.2	23.7	23.1	3.2	2.0	34.5		37.7	28.4	29.6	28.1	39.6	37.9			Fe	31.2
MnO	0.4	0.4	0.5	0.1	0.0	0.8	0.2	9.7	1.1	0.2	1.4	5.0	9.0	0.2		Mn	0.0
MgO	13.6	8.5	7.8	0.8	0.1	10.2	1.1	0.0	0.0	0.0	0.1	0.0	0.1	0.0		Mg	0.0
CaO	20.1	9.3	9.9	27.2	29.0	0.1	1.8	0.3	0.2	1.2	0.1	0.4	0.5	23.5		Ca	0.0
Na_2O	0.1	1.9	2.0	0.0	0.0	0.0											
K_2O		0.7	0.8	0.0	0.0	0.0											
H_2O																	
P_2O_5							0.6										
V_2O_3		0.0	0.1	0.4	0.0			0.1	0.8	0.8	0.8	0.2	0.0			V	0.0
Cr_2O_3									0.0			0.0	0.8				
CoO																	
ZrO_2							57.7										
ThO_2							0.6										
UO_2							0.5									Cu	33.3
Ce_2O_3							0.4									S	34.3
Total	99.3	99.1	95.8	97.5	98.4	88.9	97.7	100.7	97.9	99.7	99.5	98.4	100.6	99.1		99.0	

Augite E 43847
Based on 6 O atoms

Si 2.02
Al 0.03 ⎫
Ti 0.00 ⎪
Fe 0.32 ⎪
Mn 0.01 ⎬ 1.93
Mg 0.76 ⎪
Ca 0.81 ⎪
Na 0.00 ⎭

Wo 42.5
En 40.0
Fs 17.5

Magnesio-hornblende (II) E 43847

Si 6.91 ⎫ 8.00
Al 1.09 ⎭
Al 0.02 ⎫
Ti 0.14 ⎪
Fe′″ 1.17 ⎪
Fe″ 1.75 ⎬ 5.00
Mn 0.05 ⎪
Mg 1.87 ⎭
Ca 1.47 ⎫
Na 0.54 ⎬ 2.14
K 0.13 ⎭

Ferro-hornblende (III) E 43847

Si 6.87 ⎫ 8.00
Al 1.06 ⎪
Fe 0.07 ⎭
Ti 0.19 ⎫
Fe′″ 0.69 ⎪
Fe″ 2.24 ⎬ 5.00
Mn 0.07 ⎪
Mg 1.81 ⎭
Ca 1.65 ⎫
Na 0.60 ⎬ 2.41
K 0.16 ⎭

Titanite E 43850
Based on 20 (O,OH)

Si 4.09
Al 0.58 ⎫
Ti 3.28 ⎪
Fe 0.22 ⎬ 4.11
Mg 0.03 ⎪
V 0.00 ⎭
Mn 0.00 ⎫
Ca 4.13 ⎬ 4.13
Na 0.00 ⎪
K 0.00 ⎭

Chlorite E 43850
Based on 36 (O,OH)

Si 5.56 ⎫ 8.00
Al 2.44 ⎭
Al 2.18 ⎫
Fe 6.31 ⎪
Mn 0.14 ⎬ 11.99
Mg 3.34 ⎪
Ca 0.02 ⎭
OH 16.28

Epidote E 43859
Based on 13 (O,OH)

Si 2.95 ⎫ 3.00
Al 2.00 ⎭
Al 2.00 ⎫
Ti 0.01 ⎪ 2.88
Fe′″ 0.87 ⎭
Mg 0.00 ⎫
Mn 0.01 ⎬ 1.97
Ca 1.96 ⎭
OH 0.45

No entry means that the oxide was not looked for

0.0 means oxide <0.05%

Amphibole analyses recalculated to 23(O), 2(OH) using computer program of Rock and Leake (1984)

I–IV   Basic diorite (E 43847), Groby Old Quarry [5178 0774].
   Analyst R. R. Harding.
   I augite   II, III amphibole   IV titanite
V, VI   Diorite (E 43850), Bluebell Quarry [5239 0857].
   Analyst R. R. Harding.
   V titanite   VI chlorite
VII   Zircon; Diorite (E 43849), Groby Quarry [5266 0823].
   Analyst Mrs A. E. Tresham.

Total iron is quoted as FeO except where Fe″ is calculated by change balance, and epidote where total iron is given as Fe_2O_3

VIII   Ilmenite; as for V except that the analyst was Mrs A. E. Tresham.
IX–XIV   Quartz diorite (E 43859), Longcliffe Quarry [4933 1697]. Analyst Mrs A. E. Tresham.
   IX–XI magnetite   XII, XIII ilmenite   XIV epidote
XV   Chalcopyrite; Quartz diorite (E 43858), Longcliffe Quarry [4933 1697]. Analyst Mrs A. E. Tresham.

the oligoclase to andesine range is indicated. No primary ferromagnesian minerals remain unaltered, but pseudomorphs composed of chlorite, calcite, quartz and ferric oxide show six-sided and elongated, prismatic sections indicating original amphibole. Of the accessories, apatite, as clear, colourless euhedra, commonly elongated to 1×0.04 mm, is conspicuous. There is much interstitial carbonate and ferric oxide with a little secondary (or deuteric) quartz. This degree of alteration is commonly observed in lamprophyre specimens.   JRH

POST-CARBONIFEROUS INTRUSIONS

The Whitwick Dolerite is an olivine-dolerite intruded as a number of sills at horizons above the Excelsior seam in the Middle Coal Measures of the Leicestershire Coalfield. The sills lie in the core of the main syncline of the Leicestershire Coalfield and are commonly the youngest pre-Triassic rocks preserved (Figure 21). The dolerite is recorded in borehole and shaft sections from Whitwick to near Bagworth but does not crop out at the surface. It does not extend north of Whit-

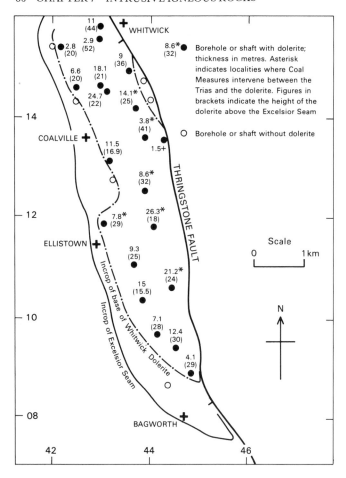

Figure 21   The thickness, stratigraphical horizon and subcrop beneath the Trias of the Whitwick Dolerite

wick and either it dies out hereabouts, or its incrop turns sharply east on the dip to meet the Thringstone Fault at Whitwick. The local basal Triassic conglomerates contain dolerite pebbles. A K-Ar age of 243 ± 11 Ma (Meneisy and Miller, 1963, pp.516–517) indicates that the intrusion took place in mid Permian times.

The Whitwick Dolerite was first described by Hull (1860, pp.64–66) and further details were given by Fox-Strangways (1907, pp.33–35). From these accounts, it is clear that the dolerite is intrusive and not a lava flow. It only causes limited thermal metamorphism of the surrounding Coal Measures. Hornfels extends for only 0.4 m below the dolerite in the Church Lane Borehole [4304 1585] and narrow, chilled margins to several sills are preserved in the Upper Grange Farm Borehole [4411 1181]. Abundant mudstone inclusions are incorporated into the dolerite in the cores of the latter borehole, but are not encountered elsewhere. Recent boreholes show that the dolerite is less extensive than was indicated by Fox-Strangways (p.34), except north of Whitwick Colliery. Both the earlier writers placed the northern limit of the dolerite between Whitwick No.1 and No.2 shafts. According to Hull, it is 18.1 m thick in

No.2 Shaft [4288 1475] but absent in No.1 Shaft [4291 1477]. Unfortunately none of the original records of these shafts has survived. Recent boreholes have proved that the dolerite sill extends up to 1 km to the north and north-west of Whitwick Colliery, so its absence from No.1 Shaft now seems less plausible.

Usually the dolerite occurs as a single sill. It reaches a maximum thickness of 24.7 m in Whitwick No.6 Shaft [4313 1450], although it is generally much thinner, even where the overlying beds are preserved. East of Ellistown, the sill is split and in the Upper Grange Borehole, there are 9 sills, from 0.2 m to 12.0 m thick and totalling 26.3 m.

The dolerite lies only 14.5 m above the Excelsior seam in the South Leicester Shaft [4312 1182], but as much as 51 m in the Hermitage Road Borehole [4303 1558]. Because the original top of the intrusion is generally eroded, it is not possible to distinguish between lateral thinning, or splitting of the intrusion, or transgression across the bedding as the cause of these variations. No dolerite is encountered in boreholes close to the Thringstone Fault east of Whitwick and it presumably comes to incrop beneath the Triassic before reaching the Fault, since the beds here dip steeply to the west.

Although several seams have been extensively worked beneath the Whitwick Dolerite, its feeder has never been located. Fox-Strangways suggested (1907, p.33) that the magma rose along the Thringstone Fault and this seems likely. Kaolinitic 'porcellanous breccias', with clasts of altered dolerite, occur close to the Thringstone Fault or within the steeply dipping, oxidised Coal Measures associated with the fracture, in the Etna Heading at Desford Colliery [463 058] and in the Merry Lees Drift [468 059]. This dolerite is petrographically similar to, and probably consanguineous with, the Whitwick Dolerite (Butterley and Mitchell, 1945, p.708; Horton and Hains, 1972, pp.58, 60, 76). It appears to have been intruded as a sill, locally into an horizon about the Lower Main or Yard. It is now highly altered and its igneous origin is barely recognisable.

Hydrothermal alteration, followed by oxidation and reddening beneath the sub-Triassic unconformity, have extensively altered the Whitwick Dolerite, and many of its original textures and structures have been destroyed. The rare examples of fresh rock show it to be ophitic olivine-dolerite (Fox-Strangways, 1907, p.34).

Thin sections reveal widespread alteration, particularly of the ferromagnesian minerals. Olivine is completely serpentinised or chloritised, and the ophitic augite is completely altered to green chlorite (E 48073). Plagioclase laths are strongly zoned with oligoclase rims, and are partially sericitised. Abundant needle and hair-like apatite prisms are enclosed by the plagioclase. Opaques occur as euhedral to subhedral grains, acicular prisms and spongy masses in a chlorite host. Rounded areas of calcite were probably vesicles.

The cross-cutting mineral veins prominent in hand specimens are of quartz, chert, hematite, goethite, siderite, calcite, baryte, and chlorite, commonly in successive crusts of a single mineral species (E 48074).

CHAPTER 8

Pleistocene: glacial deposits

Glacial deposits cover much of the eastern part of the district but are less common in the west. All the evidence suggests that they were produced by a single glaciation. The stratigraphy of the deposits is described below, based on the recognition of local varieties of till, and glacial sands and gravel. A sequence of deposits has been built up (Table 9) which applies across most of the district. In places, however, it is difficult to draw boundaries between the different varieties of till or to assign outlying patches of drift to the general succession. The drift stratigraphy has not been shown, therefore, on the 1:50 000 map and the detailed descriptions of the drift, given below, are based mainly on lithologies rather than on stratigraphical formations. However, the broad distribution of these formations is summarised in Figure 22.

STRATIGRAPHY

The glaciation which produced the glacial deposits of the district preceded the Ipswichian, or last Interglacial, which is locally represented by low-level river terrace deposits (the Beeston Terrace of the River Trent and its tributaries), near Derby. Consequently, the deposits are dissected, as a result of erosion throughout the Ipswichian and Devensian stages.

In the present standard sequence of British Quaternary stages (Mitchell and others, 1973), the deposits are assigned to the Wolstonian glacial stage, immediately preceding the

Ipswichian (Shotton, 1976). The relative ages of Midlands glacial deposits are, however, far from settled. For example, Perrin and others (1979), on the evidence of the heavy mineral content of a sample from Huncote, thought that the chalky Oadby Till of the present district was deposited by ice from the east, that invaded the Midlands via the Wash, and which also surged southwards into East Anglia to deposit till which forms the type of the Anglian stage, the glaciation preceding the Wolstonian. The different views are summarised by Catt (1981).

The earliest stratigraphical treatment of the local drift was by Deeley (1886). He distinguished between an earlier, or Pennine Boulder Clay which had a reddish brown matrix of Triassic derivation and was laid down by ice coming from the north, and a later, Great Chalky Boulder Clay, deposited by ice from the east. He assigned the associated deposits of sand and gravel to two levels, one within the Pennine Boulder Clay and the other below the Great Chalky Boulder Clay. He also recognised 'aqueous boulder clay', which was material deposited in water near an ice-front, in the vicinity of Market Bosworth. Harrison (1898) saw a wider significance in these last deposits, suggesting that they were laid down in a considerable ice-dammed lake, south-west of Charnwood, extending from Ashby-de-la-Zouch to Market Bosworth and Hinckley. Shotton (1953) traced these lake deposits far to the south of Hinckley and proposed the name Lake Harrison for the body of water in which they accumulated. The sequence of events during the formation and

Table 9 Glacial deposits of the Coalville district and their correlatives elsewhere in the Midlands

Deeley 1886	Shotton 1953	Rice 1968	Shotton 1976 Rice 1981	Douglas 1980	present account
Great Chalky	Dunsmore Gravel		Dunsmore Gravel	Flinty gravel	flinty gravel
Boulder Clay	Upper Wolston Clay	Upper Oadby Till	Upper Oadby Till	Chalky Till	Oadby Till
		Lower Oadby Till	Lower Oadby Till	Pennine Till	
Melton Sand	Wolston Sand	Wigston Sand and Gravel	Wolston Sand and Gravel	Cadeby Sand and Gravel	Cadeby Sand and Gravel
Middle Pennine Boulder Clay	Rotherby Clay	Bosworth Clays and Silts	Bosworth Clays and Silts	Bosworth Clays and Silts	Bosworth Clays and Silts
Quartzose Sand	Lower Wolston Clay				
Early Pennine Boulder Clay		Thrussington Till Glen Parva Clay	Thrussington Till	Basal Till	Thrussington Till
	Baginton Sand Baginton–Lillington Gravel	Thurmaston Sand and Gravel	Baginton Sand Baginton–Lillington Gravel		Baginton Sand and Gravel

filling of the lake, and then its over-riding by the Chalky Boulder Clay ice, has been developed in a succession of papers (Bishop, 1958; Posnansky, 1960; Shotton, 1953, 1976, 1977; Douglas, 1980; Rice, 1981). Stratigraphical nomenclature developed in these accounts is shown in Table 9.

The classification by Rice (1968) is followed in the main in this account. Two main tills, the Thrussington Till below and the Oadby Till above, are separated by lake clays and by sand and gravel. The Thrussington Till is mainly derived from Triassic rocks and the Oadby Till is mainly a chalky boulder clay, but neither is exclusively of one lithology; for example, the Thrussington Till, immediately south-east of the present district, has a dominantly 'Triassic' facies in the west and a chalky facies in the east (Rice, 1981). Similarly, the Lower Oady Till (Rice, 1981) includes material derived from Triassic lithologies. Where the intervening lake deposits and sand and gravel are absent, it is difficult to assign a boulder clay stratigraphically. For example, in the southern part of Charnwood Forest, Bridger (1975) recognises three tills of which the youngest, the Anstey Till, being chalky, correlates with the Oady Till.

The presumed extent of Lake Harrison in the present district is shown in Figure 22. The top of the lake deposits here, as in the area south of Hinckley, is generally around 122 m (400 ft) above OD. Before it was glaciated, the whole area from Bredon Hill in Gloucestershire, north-eastwards to beyond Leicester, is thought to have constituted the drainage basin of a northward-flowing Proto-Soar, to which the stream in the Hinckley Valley, flowing past Market Bosworth, was a tributary. Part of the buried Proto-Soar valley crosses the south-eastern corner of the present district (Figure 22). With ice blocking the Proto-Soar north of Leicester, the River Tame east of Birmingham and the Severn valley to the south-west, their waters were impounded against the Jurassic escarpment of the Cotswold Hills to the south-east and south. The heights of the few available gaps in the Cotswold escarpment, which acted as outlets for the lake, governed its water level at about 122 m above OD.

The uppermost part of the Hinckley valley within the present district contains patches of boulder clay and glacial sand and gravel along the Gilwiskaw Brook valley between Snareston and Packington, and possible lake clays around Arlick Farm [353 127]. A branch may have extended north-westward from Snareston along the line of the present Mease valley.

Besides the Hinckley and Proto-Soar valleys, there are two other subdrift valleys (Figure 22), the Lullington valley in the west and the Unthank valley in the east. Both may, in part at least, be subglacial furrows. Rice (1981, fig. 10) described as the Thurlaston Furrow a depression filled with irregularly disposed, glacigenic and waterlaid materials, extending due south from the junction of the Unthank and Proto-Soar valleys (Figure 23) between the two outcrops of Baginton Sand and Gravel at Huncote (Figure 23, EF).

The Unthank valley seems to have had tributaries flowing south-eastwards from Bagworth and south-westwards from Glenfield, along the valley which Rothley Brook now drains north-eastwards. The head of the Glenfield tributary may have been near Thurcaston, in the Leicester district, since Rice (1968, p.497) noted that Rothley Brook beyond Thur-

caston seems to have no pre-glacial ancestor.

Baginton Sand and Gravel

This is the basal drift deposit in the Soar valley. Rice (1968), using the name Thurmaston Sand and Gravel, suggested that it accumulated on the floor of the Proto-Soar valley at a time when an ice-sheet from the north-west had blocked its outlet and diverted its drainage eastwards past Melton Mowbray. The deposit at Huncote (Rice, 1981) comprises up to 5 m of fine gravel (Baginton-Lillington Gravel), composed predominantly of 'Bunter' pebbles with a few Carboniferous pebbles, overlain by up to 6 m of cross-bedded, well sorted, red sand (Baginton Sand), its bedding delineated by coal grains.

The surface height of the sand near Leicester is 67 to 73 m above OD. In the present district, it occurs near Huncote, with its top at 75 to 80 m above OD, and is probably represented by a valley-bottom deposit of sand around Kirby Muxloe cemetery [518 046], where the present surface height is also about 80 m above OD.

Thrussington Till

Continued advance of ice from the north-west, to about as far south as Warwick and Leamington (Shotton, 1976, p.248), covered the basal sands and gravels with a layer of Thrussington Till. Near Leicester, it is more than 15 m thick in places; in the present district, it also thickens in places rather than forming merely a veneer over the pre-glacial topography, as postulated by Douglas (1980).

In the north-west, the Thrussington Till occurs around Church Flats and Grange Wood, where it was deposited against the rising ground of the Bromsgrove Sandstone outcrop towards Netherseal. It is much dissected hereabouts.

The boulder clay spreads around Newton Burgoland and Twycross, and farther south around Shenton are a red sandy clay with scattered 'Bunter' pebbles, and fragments of green-grey siltstone, sandstone and coal. Around Nailstone, by contrast, the boulder clay is predominantly chalky; Charnian erratics are apparently absent, suggesting that the ice came here from the east.

Thrussington Till derived from Triassic rocks fills much of the Unthank valley northwards from near Newhall Park Farm [506 003]. An isolated patch of till [482 996] at 91 m above OD south of Peckleton is probable Thrussington Till in a pre-glacial tributary to the Unthank valley. Another part of the fill of this tributary valley may be represented by a till that caps the ridge east of Earl Shilton church (Figure 23, EF).

Bosworth Clays and Silts

According to Rice (1968), the Thrussington Till ice-sheet retreated from the present district, though it still blocked the Trent Valley at the mouth of the Proto-Soar, while a chalky boulder clay ice-sheet blocked the river's earlier outlet at Melton Mowbray. Under these circumstances the ponding of Lake Harrison apparently continued.

West of Shenton, on the Upton road [380 002], a thin wedge of gravel lies between the lower Thrussington Till and

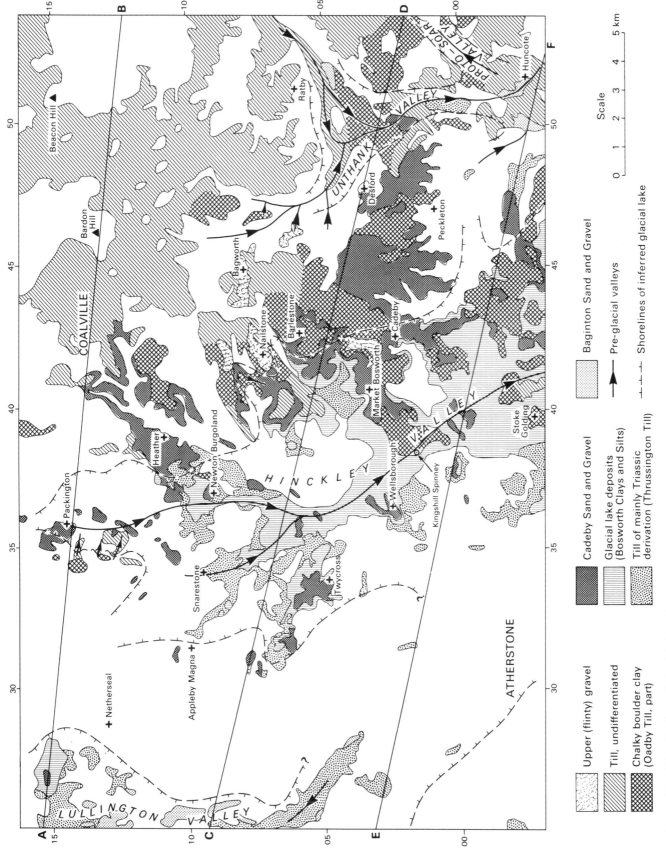

Figure 22 Distribution of glacial deposits

Legend:

Upper (flinty) gravel

Till, undifferentiated

Chalky boulder clay (Oadby Till, part)

Cadeby Sand and Gravel

Glacial lake deposits (Bosworth Clays and Silts)

Till of mainly Triassic derivation (Thrussington Till)

Baginton Sand and Gravel

Pre-glacial valleys

Shorelines of inferred glacial lake

Scale: 0 1 2 3 4 5 km

Place names: Beacon Hill, Bardon Hill, COALVILLE, Packington, Netherseal, Heather, Newton Burgoland, Snarestone, Appleby Magna, Twycross, ATHERSTONE, LULLINGTON VALLEY, HINCKLEY VALLEY, Wellsborough, Kingshill Spinney, Stoke Golding, Market Bosworth, Cadeby, Barlestone, Nailstone, Bagworth, Ratby, Desford, Peckleton, Huncote, UNTHANK VALLEY, PROTO-SOAR VALLEY

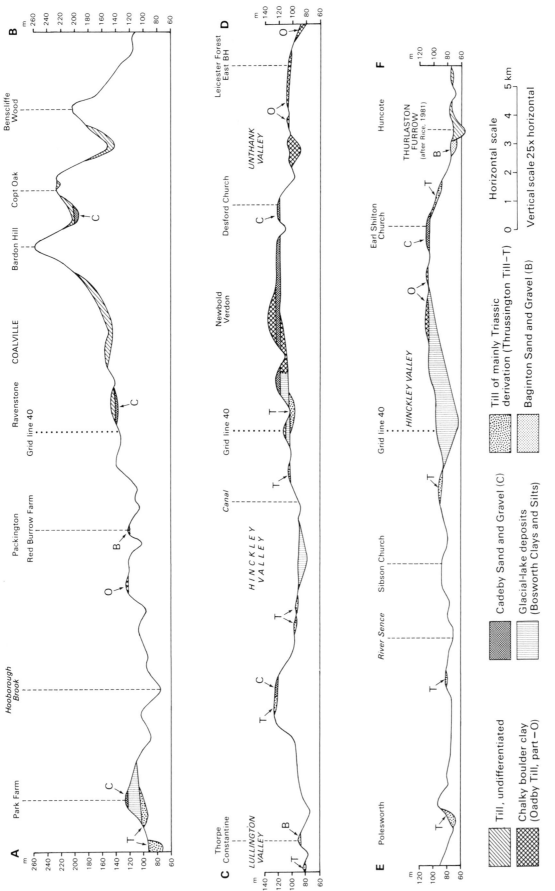

Figure 23   Sections showing the relations of the glacial deposits along the lines on Figure 22

an overlying lake clay. It consists dominantly of 'Bunter' pebbles in a red clayey sand matrix, and is almost certainly outwash from the Thrussington Till ice, possibly laid down during the initial stages of ponding of the lake.

At Appleby the lake deposits interdigitate with the underlying boulder clay, but eastwards, they pass beneath a thick covering of Oadby Till and gravels. A distinctive sandy facies, interpreted as marginal within the lake, occurs around Barton in the Beans [393 063], Newton Burgoland and in two outliers west of Stoke Golding. Lake clays in a temporary exposure west of Newton Unthank [490 043], at about 100 m above OD, suggest that an arm of the lake extended up the Unthank valley.

The highest surface outcrops of lake deposits are at 128 m above OD, south-west of Norton House [318 067]; the lowest is at 79 m above OD near Shenton. In the Hoo Hills No.2 B30 Borehole [3709 0371], lake clay was proved to about 70 m above OD, the base not being reached, and in the Kingshill Spinney Borehole [3814 0154], the base, with less certainty, was at 48 m above OD. This gives a height range of at least 58 m and possibly up to 80 m for the deposit.

Cadeby Sand and Gravel

In the Leicester district, the Wigston Sand and Gravel of Rice (1968, p.473) overlies the Glen Parva and Rotherby lake clays. Shotton (1976) correlated these lake deposits with the Bosworth Clays and Silts. Comparable sands and gravels are well developed around Market Bosworth, particularly in gravel pits near Cadeby (Douglas, 1974, 1980). Near Twycross and between Sibson Wolds [365 040] and Wellsborough, the sand and gravel directly overlie lake clay. At Twycross, there is a sandy gravel consisting dominantly of 'Bunter' pebbles with Triassic sandstone and siltstone fragments, but also including some flints.

At Wellsborough, the base of the deposit dips slightly to the north-east and the patches of sand and gravel 1 km farther to the north-east are probably remnants of the same deposit; they lie 5 to 10 m lower, but Douglas (1980, p.271, figs. 7, 8) invoked compaction of the underlying lake clays to explain this. More widely, from Newton Burgoland southwards, most of the gravels rest on surfaces sloping to the south-west, particularly where they lie on non-compacting boulder clay or Mercia Mudstone. The gravels which rest on lake clay have bases which slope generally into the 'Bosworth Arm' of Lake Harrison. This is probably due to an inward dip of the upper depositional surface of the lake clay towards the centre of Lake Harrison.

Whilst it is possible that some gravels may have been derived from the west, the presence at Twycross of flint and other erratics from the chalky boulder clay precludes such an origin for that particular deposit, and suggests derivation from the east. Indeed, near Market Bosworth, the sand and gravel is interbedded with chalky boulder clay.

Around Heather and Newbold Verdon, and between Desford and Peckleton, the wide spreads of sand and gravel contain predominantly quartzite pebbles, and were proved to be 7.6 m thick in The Fields No.3 [4517 0425] and Newbold Road [4546 0404] boreholes.

Oadby Till and flinty gravel

Rice (1968, p.474), defining the Oadby Till in the Leicester district, noted that it normally has a matrix derived from the Lias clay and a high content of chalk, flint and Jurassic rocks. He described a lower subdivision with some of the characteristics of a lodgement till, since it contains more abundant Liassic than Middle Jurassic and Cretaceous erratics, and locally at the base even a Mercia Mudstone matrix. This deposit merges upwards into a much thicker, chalky boulder clay resembling an englacial deposit and containing a high proportion of chalk and flint erratics. The ice which laid down the Oadby Till is held responsible for the folding and thrusting of earlier glacial deposits seen in M69 motorway cuttings near Hinckley (Shotton, 1976, p.249; Rice, 1981).

Between Dadlington and Stoke Lodge [4085 9732], the glacial sequence is complex, with deposits of bedded clay, sand and boulder clay of limited extent showing rapid lateral, and probably vertical, variation. It has not been possible to separate the individual outcrops on the map, and the whole has been grouped as Boulder Clay. The assemblage may have formed in a complex depositional environment which existed at the edge of an ice-sheet that temporarily paused hereabouts. Alternatively, the complexity of the sequence may result from folding and thrusting as seen along the M69 motorway, the Dadlington drift thus representing the north-western extremity of a belt of glacitectonics.

Some patches of flinty gravel, occurring as residual spreads on hill tops in the eastern part of the area around Bagworth and at Bull in the Oak [423 053], were probably formed as outwash during the final retreat of the ice and correspond to the Dunsmore Gravel of Shotton (1953).

BCW,RAO,KA,MGS

LITHOLOGIES

The glacial deposits are subdivided on the 1:50 000 geological map into boulder clay, glacial sand and gravel and glacial lake clay. Pebbles of liver-coloured quartzite and white vein quartz (hereafter 'Bunter' pebbles) are of common occurrence. Mapping of these deposits has been supplemented by a programme of boreholes up to 20 m deep, drilled by a lorry-mounted B30 power auger (Institute of Geological Sciences, 1976) and referred to as the B30 boreholes.

Boulder clay

The boulder clays comprise heterogeneous, stony clays and are likely to include both ground moraines (lodgement tills) and flow tills. There are two main varieties, each of which locally exceeds 15 m in thickness. One, of north-westerly or northerly derivation, has a stiff, red-brown, clay matrix derived chiefly from the Mercia Mudstone; its erratics mainly consist of 'Bunter' quartzite and quartz, Triassic sandstone and siltstone, Coal Measures sandstone, shale and coal and, more rarely, Carboniferous Limestone. Charnian rocks are common, even as large boulders, in Charnwood Forest. Where erratics are few, this boulder clay looks very similar to weathered Mercia Mudstone in hand-auger samples.

The other main variety is chalky boulder clay, derived from the north-east or east. This has a dark grey, Jurassic (probably Liassic) clay matrix with abundant chalk pebbles and flints, as well as erratics of Jurassic limestone and the more robust Liassic fossils such as *Gryphaea* and belemnites. Charnian erratics are common. In general, the emplacement of the boulder clay derived from Triassic rocks preceded that of the chalky boulder clay, but in places the two interdigitate and there are deposits of an intermediate character difficult to assign with confidence to one type or the other.

Details

Lullington – Shuttington – Polesworth

In the Coalpit Lane Borehole [2525 1494], Dr. J. M. Slater of the National Coal Board logged the cuttings as:

	Thickness m	Depth m
Sand, dull brown, medium-grained, with a few pebble fragments	10	10
Clay, dull red-brown, with a few pale green patches	3	13
Sand and gravel; predominantly fine-grained sand with pebble fragments of vein-quartz, quartzite, pale green sandstone of 'Keuper' aspect, and dark grey and black mudstone of Coal Measures aspect	3	16
Clay, dull brown, with many pebble fragments, predominantly of vein-quartz and 'Keuper' sandstone	6	22
Mercia Mudstone	—	—

The considerable thickness of these deposits suggests that they fill either a pre-glacial valley or a subglacial channel. The sand and gravel in the borehole may correlate with that which crops out southwards at Lullington. Around Church Flats Farm are small exposures of red or reddish brown till, in places sandy and very pebbly, with erratics, including igneous boulders and a few flints. To the north, this till probably passes beneath the stoneless clay which forms the lower slopes of the Park Farm hill. The top of the hill is capped by glacial gravel (Figure 23, AB). Till derived from Triassic rocks occupies the lower slopes of the valley south of Lonkhill Farm [253 077] which is cut into Mercia Mudstone; the till appears to fill a channel. Between Lonkhill Farm and Shuttington, the till augers as stiff grey to brown clay, forming heavy, brown clay soils. There may also be glacial lake clays in this channel. Beyond Shuttington, the channel trends south-eastwards as far as Bramcote Hall. Further in this direction, the base of the till rises to Warton and then ends. The channel may trend southwards from Bramcote Hall beneath the wide spread of boulder clay north of Polesworth, where brown, sandy clay with 'Bunter' pebbles and flints occurs. The boulder clay at Dordon Farm [278 003], on the lower slopes of the Anker valley, may be a remnant of a formerly more extensive tract of channel deposits. A temporary section by the farm buildings showed brown, stony clay with coal streaks.

Willesley – Measham House

Much of the patch of boulder clay [347 149], ENE of Willesley has been dug in a pit, now overgrown, 4.5 to 6 m deep in grey to brown stiff clay with quartz pebbles, flint flakes and some chalk pebbles. This is presumably an outlier of Oadby Till and was probably dug

as 'marl' for spreading on nearby sandy fields. On the larger spread of till to the south, the brown clay soil with numerous flints again suggests chalky boulder clay. The eastern part of that spread extends into a tributary valley of Gilwiskaw Brook.

Farther south [346 131] sandy till with quartzite pebbles and many flints caps a hill; a little to the east and at a lower level, a ditch section [3490 1326] shows dark brown, stiff, laminated clay with silt laminae and small race nodules. Grey-brown clay around Arlick Farm, on the lower slopes of the valley of the Gilwiskaw Brook, may be a weathered lake clay, in the lower part of a north–south pre-glacial valley, to which the two smaller channels mentioned above, and a third one forming a westward prolongation at the south end of the Arlick Farm spread, were tributaries (see Figure 22).

The spread of till south-west of Measham House contains flints and shelly, oolitic limestone pebbles, together with quartzite pebbles, in a red sandy clay matrix, a mixture of chalky boulder clay and Triassic erratics.   BCW

Appleby Magna – Newton Burgoland – Carlton

The narrow ridge trending south-east from Appleby Magna is capped by boulder clay with interdigitating lake clays. The till thickens north-east of Culloden Farm [336 082], where several trenches exposed red-brown, sandy clay, sticky and soft in places, with scattered 'Bunter' pebbles, siltstone fragments, and thin lenses of chocolate-brown lake clay. The till is correlated with the Thrussington Till. In the stream flowing north-north-east from the Twycross Zoo entrance [3194 0631], more than 2 m of red-brown clay with green-grey mottling, siltstone fragments and 'Bunter' pebbles, are exposed. Similar sections are exposed in nearby ditches and also south of Gopsall Hall Farm, where the till includes sand pockets [3610 0576].

East of Newton Burgoland, sandy, pebbly, boulder clay derived from Triassic rocks is seen to underlie chalky boulder clay in a ditch [3755 0956]; slight mixing of the two types can be seen. In a brick pit [384 095] SSW of Heather, boulder clay is well exposed. At the south-west end of the north face, 1.76 to 2.6 m of boulder clay rest on an irregular surface of Triassic mudstone. In the eastern face, it is cut out by alluvial deposits. The boulder clay is dark red-brown and structureless, with many fragments of Triassic siltstone and sandstone, and scattered 'Bunter' pebbles, concentrated, in places, in pockets. The boulder clay thickens north-westwards along the north face to about 3 m, with gravel wash above, and is noticeably more sandy in the north-west corner. Here, a 40 cm-deep channel filled with red sand, and with pebbles in the lowest 10 cm, cuts the till.

In contrast to these Thrussington Till occurrences, there are also several exposures of Oadby Till. For example, a boulder clay wedge 1 to 2 m thick, is preserved above lake clay and below sand and gravel at [3274 0617]; it consists of red-brown sandy clay with siltstone fragments and 'Bunter' pebbles and is therefore similar in lithology to the Thrussington Till, but is younger as it overlies the lake clay. A similar till was proved to be 1.8 m thick in the Lagos Farm B30 Borehole [3362 0572] but dies out beneath gravel to the south-east. Another lens crops out beneath the gravel SSW of Castle Farm [363 061], but is of small lateral extent as it is absent from a B30 borehole [3673 0593] nearby.

Grey-brown Oadby Till, with chalk fragments and a few gravel lenses, crops out north-east of Newton Burgoland and its junction with the underlying Thrussington Till has been described above. Westwards, the chalky boulder clay appears to grade into lake clay, and northwards towards Swepstone it becomes locally Triassic-rich. Many Charnian erratics up to 30 cm across, together with flints, 'Bunter' quartzite pebbles and coal fragments are found in dark brown, silty clay with sand pockets, north-east of Odstone, on the north side of the Ibstock road. An exposure [3950 0816] shows about 1.2 m of red to grey-brown sand and soft clay, with coal fragments. To the south-west, the boulder clay becomes more Triassic-rich, as seen in a ditch [3895 0755]. Other Charnian erratics occur in the till ESE of Odstone, and west of Barton.

Around Barton in the Beans and Carlton, the till is predominantly dark brown to blue-grey, silty to sandy clay, frequently weathering orange-brown, and the main erratics are 'Bunter' quartzites and coal with subordinate ironstone and chalk pebbles. Typical chalky boulder clay up to 1.5 m thick, with abundant chalk, flints, 'Bunter' quartzites and coal, rests on lake clay in a ditch NNE of Carlton Church [3967 0499]. Locally, however, the till is red-brown and Triassic-rich.

Wellsborough – Market Bosworth – Cadeby

The Thrussington Till hereabouts is generally a red clay, commonly sandy, and with 'Bunter' pebbles, angular fragments of Triassic siltstone and sandstone and pieces of coal. The matrix varies from a soft red clay to a clayey sand, and can closely resemble weathered Mercia Mudstone. Commonly, the till appears brecciated, with hardened, red marl pellets in a matrix of softer and, in places, sandier material. Many of the spreads of till form gentle topographical features. The maximum thickness of till locally is about 3 m. Exposures are rare. On the north side of the Coton road south of Wellsborough, orange, sandy clay with angular pebbles is seen in a ditch [3628 0193]. A B30 borehole [3597 0255] proved 1.2 m of till beneath lake clay and resting on Mercia Mudstone; 3 m of similar till occurs in a similar sequence in a borehole [3809 0344] south-west of Friezeland Farm.

In a new pit [4410 0136] east of Kirkby Lodge, about 2.5 m of Thrussington Till lies directly on Mercia Mudstone and below the worked sand and gravel. The till contains lenses of clayey, red sand and, near the top, coal-rich bands up to 0.05 m thick. The till crops out west of Kirkby Lodge but thins to the south and north, where it appears to be overstepped by the sand and gravel.

Another small outcrop of this till occurs [4105 0226] 200 m west of the Old Monument; other outcrops are shown on Figure 22. In the Cadeby gravel pits around Naneby Hall Farm [4365 0259], several faces showed good sections in Oadby Till overlying sand and gravel (Douglas, 1974, fig. 2), although parts of the till contain Triassic erratics. For example, the west face of one pit [4314 0277] showed:

	Thickness m
Boulder clay. chalky, with irregular lumps of Triassic-derived till	to 1.2
Triassic-derived till and chalky boulder clay, interbedded	0.71
Boulder clay, chalky	0.38
Triassic-derived till	0.41
Gravel, cemented (Cadeby Sand and Gravel)	to 0.50

Douglas (1974, 1980) found that the stone count of the Triassic-derived till resembled that of the subjacent Cadeby Sand and Gravel. He considered that the interbanded Triassic-derived and chalky types of till are flow tills. A similar succession was proved in Fields No.2 Borehole [4387 0468] where 5.49 m of 'boulder clay' rested on 5.18 m of sand and gravel.

In Sutton Cheyney, the logs of two water boreholes [4187 0066] proved:

	Thickness m
Sand and gravel	0.9
Boulder clay	1.6
Sand and gravel	0.4
Boulder Clay	6.7

Sand and Gravel	4.7
Lake clay	5.2
Mercia Mudstone	—

Field mapping confirms that this is the local glacial sequence. The lower boulder clay is of Triassic derivation at the base but becomes chalky at the top. The higher boulder clay is predominantly chalky, although patches of red, apparently Triassic-derived, material occur throughout.

Shenton – Stoke Golding – Stapleton

Outcrops of Thrussington Till occur immediately south-west of Shenton and, more extensively, west of Stoke Golding, capping the higher ground. The till is typically a red-brown, sandy clay with 'Bunter' quartzite, Triassic sandstone and siltstone and some coal pebbles.

Small outcrops of chalky boulder clay of Oadby type occur in the Stoke Golding area. Disused workings [3939 9728] show more than 2 m of chalky, brown-grey clay.

Around Dadlington and Stapleton, boulder clay rich in material derived from Triassic rocks tends to occur in the lower part of the local boulder clay, or as lenses in the top of the underlying glacial lake deposits. Pockets of this till appear to persist upwards within the chalky boulder clay which outcrops extensively on the higher ground.

Heather to Charnwood

Red Thrussington Till covers much of the ground north-east of a line between Ellistown [430 110] and Sinope [400 153], whilst chalky, grey Oadby Till lies south-west of this line. Exceptionally, small outcrops of chalky boulder clay occur within the tract of red till at Coalville and Charley Knoll, but only rarely are the tills seen in contact. In the Charley Knoll cutting [487 160], chalky boulder clay overlies red boulder clay (Poole, 1968, p.146), but a small patch of chalky boulder clay underlies 3 m of red boulder clay in a pit at Ellistown [4397 1077].

A grey, Liassic-derived, non-chalky boulder clay ('the Raunscliffe Till') was recorded in a cutting on the A50 at Raunscliffe [487 107] by Bridger (1975, p.193).

Many areas of the Mercia Mudstone outcrop in Charnwood Forest, previously mapped as drift-free, are now known to be covered by Triassic-derived till, much new information coming from cuttings along the M1 motorway (Poole, 1968). Dark red boulder clay up to 3 m thick, resting on Charnian and overlain by head, was encountered during the construction of a roundabout [4965 1022] at Field Head. The boulder clay contained numerous Charnian fragments and 'Bunter' pebbles together with coal fragments and ironstone nodules. Towards the base of the section were lenses of very muddy sand and gravel, up to 0.3 m thick.

Chalky boulder clay up to 8 m thick is exposed on the east face of the main pit at Ibstock Brickworks [4177 1075]. The boulder clay rests directly on Mercia Mudstone and fills a shallow depression. It is red-brown in the upper 2 to 3 m, but passes downwards into blue-grey. There are a few lenses of buff, gravelly and silty, false-bedded clay up to 0.3 m thick.

About 2 m of dark brown, grey-streaked, chalky till overlie sand and gravel at Heather Minerals gravel pit [4008 1080]. At Hugglescote, chalky boulder clay and sand and gravel appear to interdigitate, and surface mapping suggests that the drift boundaries are steeply inclined locally. A B30 borehole near Hugglescote Church [4265 1280] proved interbedded boulder clay and sand and gravel.

Nailstone – Bagworth – Thornton

Some of the till underlying the Cadeby Sand and Gravel hereabouts, includes chalky lithologies more characteristic of the Oadby than the Thrussington Till. In the valley around Nailstone Gorse, for example, the chalky boulder clay beneath the gravel apparently includes no Triassic-derived material although it rests directly on Mercia Mudstone. At the north-east corner of Nailstone Gorse [410 076], the till extends into the valley bottom, possibly filling a pre-drift valley oblique to the present-day valley. East and north-east of Carlton [405 055], the boulder clay is sandy and gravelly in places and includes both chalky and Triassic-derived types. South and south-west of Barlestone, chalky boulder clay underlying the Cadeby Sand and Gravel extends well below the level of the lake deposits to the west. Evidently this till, or its parent ice-sheet, was present during the deposition of the clays in the lake to the west.

Where the Cadeby Sand and Gravel is absent, the Oadby and Thrussington tills can be difficult to distinguish in the field. The Oadby Till usually contains abundant Charnian erratics, commonly as large boulders.

Around Belcher's Bar [408 085], the boulder clay is chalky and rather sandy but around Barton in the Beans [400 665], the till is mainly derived from Triassic lithologies but with pods of chalky boulder clay.

Around Nailstone Wiggs and towards Ellistown, the boulder clay is again Triassic-derived, as in a ditch [430 092] showing 2 m of red boulder clay with scattered 'Bunter' pebbles and flints. In other places, debris from small excavations for electricity poles shows red sandy clay, or red clay with large fragments of Triassic siltstone. Despite the abundance of flint, no chalk fragments are found.

Around Bagworth, the boulder clay is similar but with some chalk pebbles and gravelly patches. About 1 km ESE of Bagworth, a tongue of boulder clay extends down the valley slope, perhaps filling a valley in the pre-drift surface.

The broad tract of till north and east of Barlestone [440 660] is dominantly chalky, and similar spreads containing abundant Charnian erratics also occur south-east of Stanton under Bardon, and between Markfield and Botcheston, forming a plateau falling gently to the south and south-east.

Chalky boulder clay caps the southern part of the Thornton ridge. At its southern end [470 070], a small lobe of till extends into the valley south of Thornton Reservoir. A similar lobe of Triassic-derived boulder clay west of Merry Lees [467 059] was proved by the Merry Lees Drift (Butterley and Mitchell, 1945, pp.705–706) to contain pebbles of coal, quartzite and flint and boulders of Mercia Mudstone; the till also included beds of gravel. It extended down to 81.4 m above OD [4670 0586] in No.1 Drift, well below the floor of the valley to the east lying at about 100 m above OD. A third lobe of boulder clay occurs about 1 km to the south [469 051]. The presence of these till lobes suggests that the valley south of Thornton Reservoir was excavated in pre-drift times, the Merry Lees lobe occupying a tributary valley entering from the west. A gravel deposit, apparently filling a channel near the bottom of the valley at Bagworth Park [453 087], may be a remnant of an early glacial deposit, so this valley may also be a re-excavated pre-drift valley.

Desford – Earl Shilton

Patches of till west of Newton Unthank [487 043] mark the continuation of the pre-drift, Unthank valley southwards from Thornton. In temporary exposures south of Botcheston, dark grey-brown, mottled, smooth, 'glacial-lake' clay, sandy, red clay derived from Triassic lithologies and brown, chalky boulder clay are intermixed within this valley; its southern boundary is set by a temporary section [4869 6414] (Bridger and Rice, 1969) that showed Mercia

Mudstone at 4 m depth immediately beneath alluvial deposits. A tributary of the pre-drift valley, along the present Rothley Brook valley east of Newtown Unthank, might account for the 12.5 m of till and gravel proved beneath 2.6 m of river gravel in the Newtown Unthank Borehole [4947 0447] (Fox-Strangways, 1907, p.345) and for the boulder clay proved to 12.1 m in the Barron Park Farm Shaft [5098 0454].

The main buried valley is believed to extend south-eastwards beneath the spread of chalky boulder clay covering the plateau around Park House Farm [497 035] and then below Newhall Park Farm [506 003] where boulder clay forms the west side of the present-day valley. The boulder clay hereabouts is mainly of Triassic derivation, but a ditch section 200 m north of the farm exposes smooth, red-brown, silty clay, possibly a glacial-lake deposit, and chalky boulder clay caps the ridge west of the valley [508 014].

In an isolated patch of low-lying till at Long Spinneys, a temporary excavation [4812 9962] showed, at 1.5 m depth, red clay with mostly rounded pebbles 25 to 50 mm across, of chalk, shelly, oolitic limestone, coal, quartz, quartzite, red mudstone and fine, green-grey, Triassic sandstone, together with angular flints. In its top 1 to 1.2 m, the deposit was decalcified to a red-brown, sandy clay with quartz and sandstone pebbles.

Chalky boulder clay covers the plateau around Lockey Farm [464 028] and Stocks House, although around the fringes of the deposit, where it abuts against spreads of glacial sand and gravel, red-brown, sandy boulder clay of Thrussington type can be seen in a few places. The Stocks House Borehole proved till to 12.2 m, logged as 'red clay and stones'. The ridge from the Earl Shilton cemetery to Barrow Hill [488 971] is capped by sandy Thrussington Till. By contrast, the ridge to the south, through Huit Farm, is capped by chalky Oadby Till.

Newtown Linford – Huncote

The ridge between Newtown Linford and Groby is covered with reddish brown till, mainly derived from Triassic rocks but with many flints, which is locally more than 9 m thick. In places, there are many erratics of Charnian and igneous rocks, and other patches are rich in coal and chalk fragments. For example, red till with many coal fragments is exposed west of Newtown Linford in a 1.5 m deep ditch [5133 0974], whilst in the northernmost face of the Bradgate Hill Quarries [5148 0931], 4.5 m of stiff, brown boulder clay rest on 1.5 m of purple boulder clay on diorite: these tills contain flint, quartzite and igneous rock erratics.

In the motorway cutting [508 068] in the southern part of Martinshaw Wood, 6 m of red till derived from the Mercia Mudstone is mixed with beds of purple till rich in Coal Measure fragments and grey, chalky till with many flints. Small pockets of loose, brown sand are interbedded with these tills which rest directly on Mercia Mudstone in the south-eastern part of the cutting. Farther to the south-east, in the cutting [5123 0656] north-east of Ratby, 6.1 m of brown till with lenticular brown sand bands overlie 4.3 m of stiff, grey, chalky till which rest in turn upon 1.5 m of stiff red Triassic-derived till. The chalky till thins to the south-east so that, in the motorway cutting below the Ratby-Groby roadbridge [5160 0636], 6 m of red, Mercia Mudstone-rich boulder clay with many flints, quartzites, Charnian and igneous rock erratics rest directly on Mercia Mudstone, but it reappears towards the southern end of the cutting [5185 0614] where grey chalky boulder clay rests upon a 3 m thick lens of red, glacial sand. The remainder of this cutting is in purple and grey, chalky till. A section [5125 0605] in Ratby shows 3 m of red-brown, Thrussington till containing erratics of Triassic sandstone and siltstone, coal, black shale, weathered ironstone nodules and small quartz and quartzite pebbles, but no flints.

In the motorway cutting north-east of Kirby Fields, a red, Triassic-derived till, up to 2.4 m thick, lies in the south-western bank [5312 0446] below the Ratby Lane bridge. In the opposite bank [5314 0435], red sand overlies the till and expands southwards into the western bank of the cutting [5310 0426] where 2.1 m of purple and grey Oadby Till with chalk and flints rests on 1.8 m of red, medium-grained sand with coal streaks, on 5.2 m of stiff, red, Thrussington Till. Farther to the south-east, and to the west of the Airman's Rest Inn [5325 0408], the cutting is entirely in till consisting of 1.2 m of yellowish brown weathering, chalky boulder clay on 6.1 m of bluish grey, chalky boulder clay mixed with reddish purple, Mercia Mudstone-rich boulder clay and purplish grey, drab boulder clay with Coal Measure fragments, abundant chalk, quartzite pebbles, flints and a few Charnian mudstone and igneous rock fragments. Beneath the Hinckley Road bridge [5358 0337], pink and grey boulder clay was exposed to the floor level of the 6 m-deep cutting and southwards, the cutting continues in dark purple-grey, sandy boulder clay with erratics of chalk, Liassic fossils, quartzite, flint, Coal Measures sandstone, shale and coal, Carboniferous Limestone and local Charnian slate and igneous rocks. Coarse, brown sand and gravel, mainly of Coal Measures and Triassic derivation, lies on till in the west bank of the cutting [5367 0301]. Farther south, pockets of red till and bluish grey, chalky till are mixed with drab, purple clay, just north of the Service Area.

Excavations for a new electricity station [5301 0035] north-west of Enderby, revealed up to 3 m of mixed red and grey tills with many flints and 'Bunter' pebbles.

In the south-west corner of the disused Enderby Hill Quarry [5323 9953], 3 m of pale bluish grey Oadby Till is banked against the diorite, but in the railway cutting [530 993] west of Enderby, the boulder clay is entirely of Thrussington type.

In the sand pit west of Huncote Cemetery [5151 9782], 4 m of pinkish brown, coaly, Baginton Sand is overlain in turn by 0.3 m of pink and brown laminated, calcareous clay with scattered stones, 3.7 m of bluish grey, chalky boulder clay and 2.4 m of mixed, red, Mercia Mudstone-rich, and bluish grey, chalky, boulder clay that weathers to a sandy, brown till in the uppermost 1 m. Rice (in Shotton, 1977, pp.23–24) described the section as reddish, marly Thrussington Till overlain along an irregular boundary by blue-grey chalky Oadby Till, the two types interdigitating to produce distinctive 'banded tills'. The heavy mineral assemblage of the Oadby Till from this section (Perrin and others, 1979, p.543) resembles those of the other chalky boulder clays of eastern England, whereas the assemblage from the 'red till' differs, showing high percentages of mica group minerals, mostly chlorite, and of titanite.

Glacial sand and gravel

Deposits of sand and gravel occur at three main stratigraphical levels (Table 9). The lowest are gravels and sands worked near Huncote [515 975] in the south-east of the district. These are well sorted, water-laid deposits of Triassic derivation, more than 10 m thick, and have been correlated with the Baginton-Lillington Gravel and the Baginton Sand (Rice, 1981), a name here shortened to Baginton Sand and Gravel.

At about the middle of the glacial succession, the Cadeby Sand and Gravel of Douglas (1980) comprises 5 to 10 m of sands and gravels. This deposit contains flints as well as 'Bunter' pebbles, though it is less flinty than its likely correlative, the Wigston Sand and Gravel (Rice, 1968) of the Leicester area. The deposits form spreads at Heather, Nailstone, Newbold Verdon and Desford, interpreted by Douglas (1980) as parts of a sandur, or outwash plain, perhaps from the Oadby Till ice.

Lastly, at the highest level, some residual spreads on hill tops near Bagworth [450 080] probably formed as outwash

during the final retreat of the ice, and are 10 m thick at Bagworth. They contain pebbles representing all the erratics found in the tills including 'Bunter' pebbles, Triassic sandstones, Charnian rocks (absent from the lower gravels), flints and probably Chalk, though this last tends to be leached out near the surface. The gravel is usually quite clayey and commonly ochreous, in contrast to the more sandy, earlier gravels.

Details

Lullington – Warton

A pebbly soil with a few flints and fragments of Carboniferous Limestone is developed on the patch of gravel [252 138] north of Lullington. Around Lullington village, there are signs of small disused diggings up to 2.5 m deep in sandy gravel.

The gravel ridge between Clifton Campville and Thorpe Constantine seems to have been laid down in a subglacial channel, since it does not follow the present drainage and is cut by the valley north of Thorpe Hall. Small disused pits [2572 1055, 2573 0963], up to 4.5 m deep, show gravel of quartz and quartzite pebbles up to 50 mm across in a red-brown, loamy to clayey sand. No flints are recorded here, nor in the small patches of gravel farther south, at Lonkhill Farm and between Shuttington and Warton.

Measham – Packington

At Measham Station [333 118], about 2 m of gravelly, coarse sand is exposed. The banks of the canal that crosses the gravel spread north of Snarestone showed at one point [3475 1043] 2.5 m of gravel resting on Bromsgrove Sandstone. This, and similar flat spreads along the Gilwiskaw Brook valley might be regarded as river terraces but the amount of gravel seems too large for such a small valley. Around Snarestone church, gravel at a similar level (99 m above OD) lies at the base of the glacial drift sequence, so the Gilwiskaw Brook valley deposits are, therefore, considered to be relics of an early glacial fill, presumably comtemporaneous with the Baginton Sand and Gravel.

Twycross – Newton Burgoland – Barton in the Beans

At Twycross, Cadeby Sand and Gravel forms a plateau with its base at a maximum height of 120 m above OD In a disused pit [3413 0526] north-east of Twycross Church, 1.5 m of gravel is exposed with a further 0.5 m unexposed, resting on lake clay. The gravel consists mainly of rounded 'Bunter' quartzite pebbles and with angular flints, Triassic sandstones and a few coal-rich lenses, in an unbedded matrix of medium-to coarse-grained, orange to red-brown, quartz sand with clay-rich pockets. Pebbles constitute 20 to 40 per cent of the deposit and most are less than 2 cm across. A few large pebbles were found, up to 28 cm (flint), or 18 cm ('Bunter' pebbles and Triassic sandstone). The same deposit is at least 0.9 m thick, in a ditch [3303 0574] south-west of Lagos Farm, resting on lake clay, the interface being cryoturbated with lobes over 0.5 m in amplitude.

A lens of soft, brown sand with pebbles crops out about 300 m E of Lagos Farm [334 057]. Its thickness of 1.7 m was established in a B30 auger borehole [3362 0572], and it lies beneath thin boulder clay. The latter is overlapped eastwards by an upper gravel which comes to rest directly on the sand.

The maximum thickness of gravel at Twycross is about 6 m and it is plateau-like in form, dipping gently to the north-east.

The gravel patches near Twycross Zoo [318 062] and Elms Farm [311 078] are only thin cappings, and are probably outliers of the gravel at Twycross. They are marked at the surface by a high pebble-density, and augering shows a red-brown and yellow-grey, medium- or fine-grained sand which is clayey in parts.

About 6 m of sand, resting on lake clay, was found in a B30 borehole [3573 0593] south-east of Gopsall Hall Farm. The sand is medium grained, red-brown, soft and clayey with 'Bunter' pebbles, siltstone fragments and pockets of yellow and grey sand.

To the south of Heather, the gravel deposits between Newton Burgoland, Odstone, and Ibstock show much evidence of former workings. The bases of all the local spreads dip to the south-east, suggesting deposition from the north-west. About 1 m of gravelly, orange sand is exposed in a small bank, 90 m east of Newton Burgoland Post Office. It contains about 20 per cent gravel, mainly of rounded quartzite pebbles.

The sand deposit encircling Barton in the Beans is generally more clayey, and poorer in pebbles, than others in the area. A ditch [3986 0713] north of the crossroads, exposes about 1 m of red-brown, clayey sand with coal fragments.   KA

Heather – Ellistown – Charnwood

Cadeby Sand and Gravel forms extensive spreads from Hugglescote [427 128] and Ibstock, westwards to Measham Hall and Packington. East of Heather, the sand and gravel underlies Oadby Till of both Triassic-derived and chalky facies and rests directly on rockhead. West of Heather, boulder clay in many places lies in hollows topographically below the sand and gravel and although the till may be earlier, it is more likely to be later and filling hollows in the sand and gravel.

The extensive spreads of sand and gravel die out rapidly east of Hugglescote and only small lenses of sand and gravel occur below the till, as at Hugglescote Grange [438 124] and south and west of Stanton under Bardon.

The probable feather-edge of the Cadeby Sand and Gravel was encountered beneath boulder clay in the workings of Ellistown Pipeworks Ltd [4350 1050], where the section is:

	Thickness m
Boulder clay, red-brown, Triassic-derived with 'Bunter' pebbles, flint and Charnian erratics	3.0
Silt, brown, poorly bedded, with pebbly sands including coal fragments	0.7
Clay, brown, silty	0.1
Clay, sandy, very pebbly; possibly reworked boulder clay, poorly exposed	to 1.0

The top of the underlying Mercia Mudstone is intensely cryoturbated to a depth of 2 m.

Two distinct gravels are exposed in a disused gravel pit near Highfield Farm, 1.3 km ENE of Heather Church [402 114] as follows:

	Thickness m
Sand and gravel, very muddy, roughly bedded and containing abundant angular flints	2.5
Sand, coarse, pebbly, fairly clean, with false-bedding dipping westwards; some coal pebbles	0.6
Sand and gravel, coarse, well rounded, with many pebbles of soft sandstone and ironstone	6.0
Sand, brown, coarse, muddy, false-bedded with black oxide-rich laminae	0.3
Gravel, sandy and silty, with thin ribbons of black oxide-coated sand	0.9 (seen)

The upper angular gravel is absent in the pit worked by Heather Minerals Ltd only 400 m to the south [401 108] where the succession is:

	Thickness m
Boulder clay, chalky	2
Sand, gravelly, fairly clean, with coaly streaks; false-bedding dipping westwards	1.5
Mud, red, gravelly	0.1
Sand, buff	0.03
Sand and gravel, coarse, well rounded, passing into sand up to 5 m thick in the north-east corner of the pit	6.0

The lowest bed in this section correlates with the bed of the same thickness in the Highfield Farm pit.

Gravel forms a small outcrop on either side of the railway line at Alton Hill [3945 1535], where it fills a depression in the upper surface of the Mercia Mudstone, and at least 2.5 m of coarse gravel are exposed in an overgrown pit on the south side of the cutting. A very hard, indurated band of gravel up to 0.5 m thick lies towards the base of the gravel exposed in the pit and it contains an unusual suite of pebbles including abundant, rounded and angular chalk, flint, and Triassic sandstone and marl, together with Charnian and quartz pebbles and Jurassic limestone fragments. The indurated band also appears on the north side of the cutting. Its pebble content suggests that it is associated with the chalky boulder clay, the nearest outcrop of which is 200 m to the west.

Small areas of sand and gravel resting on boulder clay occur in Charnwood Forest. A NW–SE ridge, 600 m long, occurs at Ulverscroft Lodge [499 131]. Like the small patch of sand and gravel at Bushy Field Wood [499 112], this deposit is rather muddy and not easily separated from the surrounding boulder clay.

At the north-west end of Bardon Hill Quarry, the Trias is overlain at about 200 m above OD by a pocket of glacial sediments, largely obscured by slipping of the overlying head. A measured, but not necessarily typical, section [454 130] reads:

	Thickness m
Gravel, very silty and with a clayey, brown matrix, with well rounded pebbles of flint, quartz, quartzite, Carboniferous-type sandstone, clay ironstone concretions and agglomerate	1.0
Sand, brown, fine, very silty, thin-bedded, with some clay bands.	0.5
Clay, brown, silty	0.1
Mercia Mudstone	—
	RAO

Nailstone – Bagworth – Thornton – Ratby

Gravels assigned to the Cadeby Sand and Gravel occur in the west of this area and are generally overlain by boulder clay. No Charnian rocks have been found in them. They are generally clean and sandy, and their bases are flat with erosive contacts. The main occurrence is around Nailstone and Barlestone. They crop out along the south-eastern slopes of the valley near Osbaston Hollow, forming a prominent feature, especially in the south-west where they overlie lake clay. In a road cutting at Osbaston Hollow [4164 0602], most of the pebbles are Triassic sandstone ('skerries') and 'Bunter' quartzites; debris from an old gravel pit [4192 0624] includes, in addition, flints and rare, Jurassic ironstone pebbles. In the top of the Osbaston Lount Brickpit [410 056], Fox-Strangways (1894, field map) recorded 4 m of gravel, the pebbles consisting mainly of sandstone and chalk. The gravels around Barleston [430 056] locally incorporate lenses of sandy boulder clay.

Gravel crops out on both sides of the valley between Nailstone Gorse and Belcher's Bar. On the south side [408 073], the sandy Cadeby Sand and Gravel is in part overlain by clayey, flinty gravel. On the north side of the valley [407 080], the gravel forms a strong feature. It evidently pinches out northwards since it does not crop out in the valley 0.5 km to the north. East of Belcher's Bar [414 083], the lower gravel is again overlain directly by flinty gravel.

Upper, flinty gravels form broad, hill-top spreads around Nailstone Grange [418 085], Nailstone [418 073], Bagworth [445 079] and west of Merry Lees [461 061]. The gravels at Thornton [462 082], and at two small areas NNE of Barlestone [432 068 and 436 064], were probably deposited coevally with these.

The two spreads of gravel at Ratby Burroughs [490 061 and 495 061] overlie chalky boulder clay on their northern margins, but cut down southwards to rest on Mercia Mudstone. They may represent a remnant of a gravel fill in an outwash channel. MGS

Wellsborough – Market Bosworth – Cadeby – Newbold Verdon

Two small areas of sand and gravel lie beneath lake deposits. A bed of sandy 'Bunter' pebble gravel 1 km north of Temple Mill [357 045] underlies lake clay and is up to 1.5 m thick. Sandy gravel with many rounded flints [378 003] SSE of Stubble Hills Farm, overlies till but is covered by lake clay. Its thickness is about 7.5 m. A small patch of sandy gravel [320 021] south-east of New House Grange, is probably of the same age.

Beds of Cadeby Sand and Gravel directly overlie lake clay between Wellsborough [362 023] and Sibson Wolds [365 040]. At Wellsborough, 4.6 m of sand were recorded in a water borehole [3607 0235], and along the south side of the Market Bosworth road there are a few overgrown workings.

South-east of Cadeby, in a gravel pit [4421 0160] east of Kirkby Lodge, the sequence reads:

	Thickness m
Gravel, mainly of 'Bunter' pebbles with red, sand matrix	1.2
Sand, yellow, pebbly, becoming gravelly near the top	0.3 to 1.9
Gravel, medium-grained, poorly sorted, mainly 'Bunter' pebbles but with many subangular flints in a deep red, sand matrix	0.6 to 1.8
Sand, red, medium- to coarse-grained, with yellow, brown, black and grey colour bands which show cross-bedding dipping S and SE	0.2 to 0.5
Gravel, fairly coarse, poorly sorted, predominantly 'Bunter' with some Mercia Mudstone siltstone and sandstone fragments and weathered coal; sandy near the top	0.8 to 2.3
Sand, red, coarse, quartzose, rounded grains with pebble bands, cross-bedding dips S	2.2 to 3.3
Gravel, medium to coarse, mainly 'Bunter' pebbles in a coarse, red, sand matrix with some coal, sandstone and subangular flint fragments	0.7 to 1.1
Thrussington Till	about 2.3
Mercia Mudstone, not exposed but proved in excavations in pit bottom	—

Similar sand and gravel has been exploited around Naneby Hall Farm [4635 0259] (Douglas, 1974, fig. 2), where 3.5 m of yellow to red sand overlain by 0.7 m of gravel is seen above water level in a pit [4315 0257] south of the road. Poorly developed cross-bedding in the sand dips east; however, a rose diagram by Douglas (1974, fig. 3), based on 25 observations, shows a mainly westward dip of the

cross-bedding. In the pits north of the road, 0.5 m of sand is overlain by 2.0 m of gravel and interbedded sand, capped by 3.0 m of red till (p.93). On the western side of the pit [4314 0277], 0.5 m of cemented gravel is overlain by the boulder clay; the cement is calcareous, and hardest near the gravel boulder clay junction. Pockets of sand and gravel in the till at the top of the face presumably result from periglacial activity and post-date the melting of the ice. Douglas (1974, p.60) remarked that these pockets seem to have developed more readily on the surface of Triassic-derived till than on that of chalky till.

In disused pits [4345 0270] north of the farm house, a face exposes 1.9 m of gravel with horizontal sand layers, overlain by 0.7 to 3.5 m of coarse, poorly sorted gravel with a sandy matrix; the pebbles are mainly 'Bunter' quartzites and weathered flints with a few large blocks of Mercia Mudstone siltstone. Patches of Mercia Mudstone visible in the bottom of the pit suggest that the gravel directly overlies bedrock.

From borehole evidence, sand and gravel appears to be continuous under chalky boulder clay between Osbaston and Newbold Verdon. The Fields No.2 Borehole [4387 0468] records 5.18 m of sand and gravel beneath chalky boulder clay, and Newbold Verdon Borehole [4444 0414] records 9.1 m of sand on Mercia Mudstone, suggesting that the lower sand at the Naneby Hall Farm pits has thickened at the expense of the overlying gravel.

The gravel outcrop around Bull in the Oak [4223 0344] appears to be mainly flinty gravel up to 9 m thick. There are three disused pits along the A447; none affords exposures, but only the higher part of the deposit appears to have been worked. The lower part is sandy and near Bull in the Oak, the contact between the sand and underlying chalky boulder clay gives rise to a marked seepage line. From field debris, the gravel appears to be coarser and richer in Charnian fragments and flints than the gravels below, but the main constituent is still 'Bunter' pebbles. JB

Shenton – Stoke Golding – Stapleton

Two outcrops north-east of New Barn [391 995] are almost certainly parts of the same deposit. A ditch [3928 9975], in a slight hollow between them, entered lake clay with overlying pockets of gravel, separated by a cryoturbated interface. The eastern outcrop, the base of which dips at about 2° to the south-west, was exposed in a ditch [3978 9975] running from the top of the railway embankment, north of the canal overbridge, down to the canal. Up to 50 cm of red-brown, clayey sand with 'Bunter' pebbles and flints was exposed.

At Stoke Golding, a ditch exposure [3982 9687] south-east of the road showed 1.9 m of red-brown sand with scattered pebbles, with a 0.35 m, pebble-rich band dipping at 5° to the south-east. The pebbles include 'Bunter' quartzites, flints, coal and Triassic siltstones. A disused working, just east of the canal in Stoke Golding, may have been worked partially for gravel, though the remaining face consists only of chalky boulder clay. KA

The sand and gravel deposits of the Dadlington to Stapleton area can be separated into two groups on the basis of their grain size. The older deposits are generally fine grained, reddish brown sands, clayey sands and silts. They occur within the Bosworth Clays and Silts but generally lie at the top of the sequence. They appear to have accumulated in a quiescent, fluviatile or lacustrine deltaic environment. The Cadeby Sand and Gravel deposits, on the other hand, are of much coarser grade with pebbles of chalk, flint, 'Bunter' quartzites and other Triassic types, in a varying proportion of quartz-chalk sand matrix. They are commonly associated with deposits of chalky boulder clay, and occur above the lake deposits, as at Dadlington where they fill a channel at least 10 m deep. They may extend down to the bedrock at a level comparable with the upper surface of the glacial lake deposits, as for example north of Kirkby Wood [445 000]. Coarse gravel and sand at Greenhill Farm [4065 9920] has a base at about 91 m above OD, which is below the general level for this lithology. It forms a lens within the lake deposits, and may represent either a distributary channel within the lake or a later episode of fluviatile deposition. AH

Desford – Kirby Muxloe – Earl Shilton – Huncote

Orange-brown sand and clayey sand [519 047] north of Kirby Muxloe village, forms the southern slope of the valley of Rothley Brook and abuts against the First Terrace at the foot of the slope, possibly passing beneath the terrace deposits.

A disused gravel pit [457 976] south-west of Westfield Farm, Earl Shilton, is 4.5 to 6 m deep; small, partially obscured sections show yellow to reddish brown medium-grained sand, with cross-bedding dipping between south-east and WSW, and some pebbly streaks along the bedding including small pebbles of quartz, Triassic sandstone and flint. The sand is overlain on the west face by re-sorted reddish brown clay about 1.8 m thick, possibly a flow till derived from the Mercia Mudstone; at one point inclined pebbly streaks show cross-bedding inclined to the south.

Abandoned sand pits west of Stony Bridge [503 972], and a temporary section on the west side of the M69 motorway [4992 9746], show up to 6 m of Baginton Sand, consisting of reddish brown sand with cross-bedding dipping north to north-west at one point, overlain along a horizontal junction by 1 to 2 m of till. In the most southerly of the sections, the till is a reddish brown, Trias-derived clay, and in another, 300 m to the north, it is a chalky boulder clay with streaks of reddish brown clay at the base. A patch of glacial gravel [494 985] crops out on a spur at 91 m above OD, on the east side of the Thurlaston Brook valley. Rice (1981, pp.390, 398) listed the erratic pebbles from a shallow trench hereabouts; though the deposit resembles Baginton Gravel in composition, he considered it to be a localised outwash laid down in front of the advancing, Thrussington Till ice.

In a pit west of Huncote Cemetery [5145 9785], 4 m of cross-bedded, medium-grained, pinkish brown sand with laminae rich in small coal fragments and with a few, small, angular, flint fragments occur at the eastern end of the face. Westwards, the sand is very gravelly with many large cobble-size fragments of Triassic sandstone, flint, Lias limestone and ironstone; no Charnian or igneous rocks are present. Rice (1981), describing a pit farther to the northwest [513 982], found an absence of flint, which he considered differentiated the Baginton Gravel from other local gravel deposits. EGP

Glacial lake clays

Patches of chocolate to greyish brown, laminated clays, the Bosworth Clays and Silts, are common throughout the central part of the district, especially within the pre-glacial Hinckley Valley. They also occur elsewhere in the district, for example near Park Farm, Overseal [268 153]. They are uniform clays, with only sporadic, quartz-silt laminae, and they reach a maximum thickness of 30 m near Market Bosworth.

Exposures of the clays commonly show strong, vertical joints, probably caused by shrinkage during periods of dry weather, extending to 6 m or more below the ground surface. On the joint faces, the clay is commonly pale greenish grey where it has been reduced by weak humic acids produced by recent rootlets. Other clay samples from near the ground surface are mottled green and brown, where only partial reduction has occurred. Small, whitish, calcareous nodules or 'race' are common in the weathered clay.

In Charnwood Forest, sections along the M1 Motorway revealed evidence of an isolated glacial lake in which gravel and sand accumulated near Birch Hill [478 136], and sandy

silt with rootlets at White Hill [497 122] (Poole 1968; Bridger 1975).

Details

Netherseal

On the slopes of the hill capped by glacial sand and gravel at Park Farm [269 153], dark-brown and grey mottled, stiff clay with a few erratics, probably glacial-lake clay, crops out. Fox-Strangways noted on his field map a section at Netherseal Colliery [2748 1536], showing 'clay with chalk flints over laminated clay'. The Netherseal [2678 1521] and Wynne's [2715 1526] boreholes, starting on the thin glacial gravel that caps the hill, proved 21.8 and 18.3 m of drift respectively. The local thickness of lake clay, therefore, is probably about 20 m. It original extent is not known but is unlikely to have been great (Figure 23, AB). It overlies the red till that occurs around Church Flats Farm.

Swepstone

Around Swepstone, silts and clays locally lie at the base of the glacial sequence. Laminated, brown, silty clay was dug at the roadside [3718 1081] north-east of Swepstone Church, and soft, brown clays were augered in fields around Upper Fields Farm [373 111]. Similar silty clays occur south of Swepstone [369 101].

Twycross, Newton Burgoland, Barton in the Beans, and Carlton

Around Norton House, laminated clay is exposed in a ditch on the south side of the A444 road [3162 0648 to 3138 0675], where more than 1 m of reddish grey-brown clay with pockets of red-brown sand is exposed. A gas pipe-line trench dug in 1972, north-east-trending across the outcrop, showed that the clay overlies till immediately south of the road, and thickens to more than 1.3 m to the northwest. The exposed clay showed many slickensided surfaces, laminations and scattered 'Bunter' quartzite pebbles and flints.

Laminated clay occurs beneath gravel (Cadeby Sand and Gravel) near Twycross. About 500 m south-west of Lagos Farm, a ditch [3303 0547] shows 2 m of clay with a cryoturbated top, beneath sand and resting on Mercia Mudstone. The clay contains a few Mercia Mudstone fragments. The bed thickens eastwards, 4 m being proved to a depth of 7.4 m in the B30 Lagos Farm [3362 0572] Borehole, and 11.6 m to a depth of 17.7 m in the B30 borehole in Gopsall Park [3573 0593]. Coal debris, Triassic siltstone and 'Bunter' quartzite pebbles are present in the clay hereabouts. The basal 0.5 m of the clay was cored in the Lagos Farm Borehole; it was markedly calcareous and showed some stratification, with slightly gritty lenses and red marly patches.

West of Bilston, a line of B30 auger holes [3552 0522, 3556 0523, and 3559 0523] crossed the edge of a spread of clay where it rests on till. The junction dips east at 1 in 7. Flow tills may be present within the laminated clays locally.

Around Shornhills [344 075] and east of Culloden Farm [336 082], the clay seems to interdigitate with till. In the stream near Culloden Farm [3406 0808], the clay fills a hollow cut in till beneath the alluvium.

North of Shackerstone, the clays gradually become coarser grained and this may indicate a margin to the glacial lake in this direction. About 500 m north of the disused railway line [366 079], fragments of coal, Triassic siltstone and mudstone are present in the clay and northwards, the deposits become increasingly sandy and pebbly.

By the stream west of Newton Burgoland, a lens of red to orange-brown, sandy, variably clayey and pebbly, laminated clay persists for about 1 km within the lake deposits [362 092 to 370 100]. North-east of the village, the laminated clay appears to grade laterally eastwards into chalky boulder clay (Oadby Till); the boundary being taken at the incoming of chalk debris.

The coarser, marginal facies, which is very variable, has also been recognised at Barton in the Beans, and near Carlton where chocolate-brown clay passes up into chalky till (Oadby Till) in a ditch [3892 0541] NNE of the church.

Barlestone

Lake deposits in the valley west of Barlestone consist of smooth, stoneless, blue-grey or mottled grey and brown clays, commonly with white patches and small race nodules in the weathered zone. Lenses of silt and sand occur locally. The clays were formerly dug for brick-making at Osbaston Lount [410 056], where the clays were said to have been proved for 12 m below the floor of the valley (Fox-Strangways, 1900, p.38; 1907, p.71).

On the west side of the valley [404 053], the lake clays are underlain by a smooth, red-brown till with sparse pebbles.

Wellsborough – Market Bosworth – Sutton Cheney

Hereabouts, the lake clays are brown, plastic and homogeneous, and commonly weather to blue-grey. There was little evidence of lamination in auger samples and none in the few ditch sections.

Drainage ditches near Botany Spinney [3775 0300] and east of Shenton Gorse [3769 0154] expose 1.5 to 2 m of lake clay with a variable cover of sandy wash. The clay contains lenses up to 0.05 m long, of subangular, medium-grained, red sand and a few small fragments of Triassic siltstone.

A B30 auger hole [3709 0371] north-west of White House Farm proved 17.1 m of clay (to 70 m above OD) without reaching the base. As the top of the lake clay at Wellsborough, 1.1 km to the south-west, lies at about 115 m above OD, the lake in this area may have been 45 m deep. Another hole [3721 0356] in this area proved:

	Thickness m
Clay, brown, becoming silty towards base, with a band of Triassic mudstone debris at the base	3.4
Sand, fine, silty and waterlogged	3.0
Clay, brown	1.2 base not seen

The sand must be strongly lenticular as it does not appear to crop out in the adjoining fields. The Kingshill Spinney Borehole [3814 0154] records 38.4 m of 'boulder clay and sands' which may be entirely lake deposits (see Douglas, 1980, p.263).

North-east of Market Bosworth, the lake clay contains abundant 'race' concretions, so that it superficially resembles chalky boulder clay. A lenticular body of Triassic till, up to 400 m wide, interdigitates with lake clay 300 m south of Carlton. Farther south, the lake clays were formerly worked for brick-making at Brick Kiln Farm [4225 0145], and at Stapleton Brick Yard [4295 0094] NE of Sutton Cheney, but no sections remain.

Shenton to Earl Shilton

The main spread of glacial lake clays does not extend much west of Stoke Golding and Shenton. Numerous exposures, largely in stream banks, show up to 0.5 m of smooth, plastic, dark chocolate-brown, nearly stoneless clay beneath alluvium. There are four patches of lake deposits west of the main spread, however; two of them, 500 m west and 1.2 km south-west of Shenton respectively, consist of chocolate-brown clay. To the south, a third patch [375 971] WSW of Foxcovert Farm consists of partly laminated, red-brown and grey sandy clays and clayey sands. About 800 m farther south, a larger patch consists of brown, silty and grey, plastic clays, which resemble alluvial clays in colour and texture. These last two patches are possibly a marginal facies.

Around Sutton Cheney, Dadlington and Barwell, the lake deposits are chocolate to greyish brown clays, laminated with pale fawn or reddish brown, silt or sand partings. Seams and lenses of silt and sand are present throughout whilst beds of reddish brown, pebbly clay also occur. The lowest beds are more uniform lithologically but upwards the variety of interbedded lithologies increases; for example, near Poplars Farm [4133 9867], elongate lenses of chalky boulder clay and sand and gravel occur within plastic, greyish brown clays. The lake deposits hereabouts also include red sands, sandy clays and pebbly, sandy clays which, however, are too thin and impersistent to map separately.

Excavations south of Earl Shilton [4607 9867] show 2 to 3 m of dark red to dark brown, silty clay with many listric surfaces; sparsely scattered, quartz pebbles (up to 10 mm) and some smaller erratics (6 mm) including coal fragments, are present.

Charnwood

Sand, silt and gravel, probably deposited in a local glacial lake separate from that in the Hinckley Valley, occur between Birch Hill [478 136] and Ulverscroft Grange [486 115] (Poole, 1968, pp.143, 146, figs.3, 4). The deposit grades from gravel and sand around Birch Hill to sandy silt farther south. It is largely concealed beneath the surface deposits of till and head, and lies at about 200 m above OD between Bardon Hill and Copt Oak (Fig. 23, AB). In the western bank [4791 1217] of a motorway cutting near White Hill, 5.5 m of Mercia Mudstone-rich till rests upon 0.6 to 0.75 m of brown sandy silt with abundant rootlets, which in turn rests upon a lenticular (0 to 0.02 m) bed of stiff, red, Mercia Mudstone-derived clay with many small stones and roots, and upon 1 m of brown, sandy silt with a few roots. Bridger (1975, p.193) described this section as showing over 3 m of interstratified clays and silts with rootlets, which he named the White Hill Clays and Silts.

The road bridge to White Hill is built upon a former island of bedrock in the glacial lake (Poole, 1968, fig. 4), since it is surrounded by deposits of sandy silt. These gradually thin southwards against the Mercia Mudstone.

CHAPTER 9

Late Pleistocene and Recent: post-glacial deposits

The term 'post-glacial' is used here for the period that has elapsed since the area was last glaciated. It spans in time the Ipswichian interglacial and Devensian glacial stages, as well as the Flandrian, or post-Devensian glaciation, stage. All the deposits described here relate to the present-day drainage, and so an account is given of how this is believed to have developed after the final withdrawal of ice sheets from the district.

RIVER DEVELOPMENT

The district lies within the catchment of the River Trent, but the eastern streams reach the Trent via the River Soar, while the central and western streams flow via the River Mease or River Tame (Figure 24). Beneath the glacial deposits lies a different drainage system, and if, as seems most likely (Douglas, 1980), this was fluvial rather than glacial, then at least two-thirds of the district must have drained into the Soar, by way of the Hinckley valley, before the onset of the ice sheets. The Hinckley valley (Figures 22 and 23) was more closely related to the geological structure than is the present drainage system. It followed the principal belt of soft rocks in the district, in the middle part of the Mercia Mudstone outcrop, between the higher ground of Charnwood Forest and that formed by the skerries in the lower Mercia Mudstone in the west. The buried Lullington valley, in the extreme west, also followed the strike of an easily eroded belt of Mercia Mudstone.

At present, the line of the Hinckley valley is marked by low-lying ground through Shackerstone and west of Market Bosworth and Stoke Golding. It is crossed by the westerly-flowing River Sence and Sence Brook. Small, north- or south-flowing 'subsequent' tributaries of these streams flow across the soft clays that fill the depression, and part of the earlier drainage has been captured by the River Mease near Shackerstone.

Some of the sediments filling the Lullington valley are glacial gravels which have proved resistant to erosion. The River Mease at Clifton Campville and a minor stream to the north of Thorpe Constantine cut across the ridge capped by this gravel.

Such contrasts between the former and present drainage patterns show how much they were changed by glaciation. Glacial deposits choked many of the existing valleys and caused new valleys to develop. For example, many streams rise on the high ground of Charnwood Forest, and a radial drainage from the area above 200 m above OD, similar to that of today (Figure 24), probably existed in pre-glacial times. The present River Sence, however, whose source lies near the summit of Charnwood, seems to be one of the new valleys; it flows diagonally across the line of the Hinckley valley where it is filled by glacial lake deposits to reach the ground to the west.

Subsequently, the drainage basin of the River Tame was, perhaps, more rapidly eroded than that of the River Soar, so that the 'consequent' line of drainage of the Sence was established as a main element of the drainage. Shotton (1953, p.258) suggested that the Tame extended its drainage basin in post-glacial times at the expense of the Soar in the Coalville district because its main stream was still at about its pre-glacial size, whilst the headwaters of the Soar had been captured by the Avon. Alternatively, the present Tame and Soar drainage basins could have resulted more from the immediately post-glacial topography than from a contest in erosion between the Soar, Avon and Tame. The valleys are generally wider along the Tame than along the Soar. The Tame and its tributaries have cut down mainly into glacial-lake deposits with little or no protective capping of Oadby Till or associated gravels, and then into relatively soft Mercia Mudstone. The Soar basin, however, is largely filled by thick deposits of tenaceous chalky boulder clay so valley-width appears to reflect the nature of the material being eroded. Each river has incised its main valley to a similar depth, as shown by the position of the 50 m contour in Figure 24. The Tame Valley, however, particularly below Tamworth, is practically free from glacial drift, while the slopes of the Soar valley are still largely formed of chalky boulder clay.

Present-day streams are able to cut down with surprising speed; for example, mining subsidence along Saltersford Brook in the South Derbyshire Coalfield, produced small lakes in 1967. Two of them, one sited where the brook is crossed by the Oakthorpe-Donisthorpe road [322 137] and the other 600 m upstream [328 139], were caused by subsidence above workings flanking a belt of unworked ground left along a north-trending fault. In the Main, this belt was 40 m wide but it was 150 m wide in the underlying Woodfield, Stockings and Eureka. At least 5.5 m of coal has been removed. At the surface, therefore, along a fairly narrow tract crossing the stream, the ground remained at its original level, while on either side it subsided by several metres. The effect on the stream was that of a ridge rising in its path providing the conditions for 'antecedent drainage'. The Main workings date from 1869 to 1873; the latest in the Eureka, date from 1955 to 1964. The brook had by 1967 incised a gorge whose winding course followed the meanders originally present on its flood-plain. At one point [3270 1378] the gorge was 3 m deep and its steep sides exposed the former alluvium overlying Coal Measures mudstone and sandstone. The stream had therefore, cut down 3 m in less than a century.

If the main base level of the district, represented by the River Trent, were to be lowered, then down-cutting would be correspondingly rapid along all tributaries. This seems to have happened in the past; for example, the gradient of the Second Terrace of the Mease, which is parallel to the present stream profile seems to record a general drop in base level. This view is supported by the occurrence of a knickpoint on

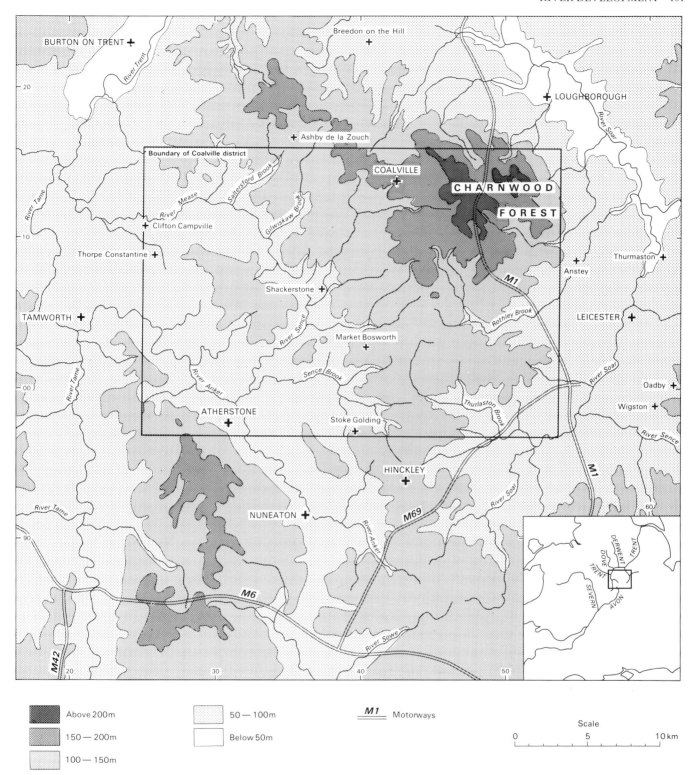

Figure 24 Relief and drainage of the Coalville district and surrounding areas

the river at Netherseal, where the river crosses the outcrop of an unusually thick sandstone near the base of the Bromsgrove Sandstone.

About 400 m south-east of Shackerstone Fields Farm [358 077], there is an example of river capture, the headwaters of the Mease having cut back to capture a tributary of the Sence.

Many streams on the lower slopes of Charnwood Forest, notably in the northern and eastern parts of the Forest, run discordantly across outcrops of pre-Triassic rock. Several instances of discordance are located at heights around 122 m above OD but others are found at lower altitudes. For example, the Ulverscroft stream at Newtown Linford swings north-eastwards to cut a gorge 300 m long through hard diorite in Bradgate Park [530 100]. Watts (1903, 1905, 1947) attributed some of these instances to exhumation of Triassic valleys, others to the superimposition of drainage developed on a formerly complete cover of Mesozoic rocks. Bridger (1968, 1981), however, has stressed the importance of glaciation suggesting that some of the diversions or narrow valleys might have been formed by meltwaters flowing beneath ice or diverted by spreads of drift deposits.

HEAD

Head is a post-glacial deposit formed by solifluction under periglacial conditions. It is generally poorly structured but can retain relic shears in some circumstances. These may adversely affect its geotechnical properties and be a hazard to construction. In Charnwood Forest, head is principally debris which formed in post-glacial times around Charnian crags and has moved downslope by solifluction over boulder clay or Mercia Mudstone. It forms smooth aprons on the hill slopes and is typically marked by large numbers of Charnian blocks in the soil and by the absence of rounded pebbles from the boulder clay. The boulder content of the head, however, clearly depends on the up-slope outcrop from which it was derived, so that it varies considerably from one hill slope to the next. Generally, the head is better drained than either the boulder clay or Mercia Mudstone. The lower margin of the head, as mapped, is rather arbitrary. No attempt has been made to map solifluction deposits that have moved downslope from the boulder clay or Mercia Mudstone.

The head is thickest where banked against its source outcrop. Thicknesses of 5 to 8 m are common in working quarries, like those on Bardon Hill and Cliffe Hill. Downslope, the head gradually thins, although it can continue for 500 to 600 m from the nearest outcrop. Its matrix is an orange-brown or buff, silty clay, the proportion of which increases downslope relative to the boulder content. In some quarry sections, the base of the head contains an admixture of Mercia Mudstone or glacial drift.

The mapped head is entirely post-glacial in origin; it overlies boulder clay at the entrance to Croft Quarry and fluvio-glacial drifts on the south side of Bardon Hill Quarry. No instance of head lying beneath glacial drifts was observed during the survey, but subdrift head was recorded by Bridger (1975, pp.191–192) in excavations for the Markfield A50 by-pass at Raunscliffe [487 107] and for the M1 Motorway at White Hill [470 121].

Head, formed by downwash from sand and gravel, adjoins the lower slopes of the Sence valley between Kelham Bridge and Heather (Institute of Geological Sciences, 1976, p.6). It is exposed along the western face of a disused brick-pit at Heather [3915 1045] as 2 to 3 m of unbedded sandy, pebbly, silt.

In the Snarestone-Shackerstone area, the head is limited in extent and rarely exposed. The mapped deposits are more than 1 m, but rarely more than 2 m, thick, except where they have accumulated in valleys.

Deposits east [395 090] of Odstone Hill Farm and about 1 km south of Heather [389 100], are both derived from gravels. Those in a brick-pit [384 095] are too thin to map; up to 0.7 m of a gravel of small, 'Bunter' pebbles in a pale to dark orange-grey, sand matrix are seen. The base is lobate, with 'pipes' extending 1.1 m into the underlying boulder clay. The form of the deposit east of Odstone Hill Farm suggests that the gravels on the ridge were originally continuous with those at Ibstock. Red, plastic clay, 600 m south-east of Lea Grange Farm [322 054], is probably derived from the outcrops of Triassic mudstone that form the surrounding higher ground. A deposit of yellow-brown, sandy clay [354 063] south-east of Gopsall Hall Farm, lies below a capping of gravel on lake clay, and is probably derived from these.

Head deposits lie in shallow hollows 500 m north-east of Gopsall Hall Farm [356 068] and 700 m SSW of Shornhills [344 075]; both are derived from boulder clay.

Accumulations of head lie on valley sides 600 m SSE of Snarestone Church, 700 m S of Shackerstone station, and 600 m SSW of Beanfield House. They are derived from till and Mercia Mudstone outcrops and may, in part, be waterlaid.

Outcrops to the east of Lodge Farm [3353 0046] are of sandy clay, probably derived from the hillslope to the south. Two short north–south valleys east of Sibson [355 008] have sandy, pebbly wash in their floors, whilst head appears to underlie lacustrine alluvium in the valley nearer to the village.

RIVER GRAVELS

Patches and small spreads of terrace gravels occur along most of the streams. A First and Second Terrace can generally be distinguished. The Second Terrace is now much dissected and consists of scattered patches of gravel with their upper surface at 5 to 10 m above the adjacent floodplains. Such a height suggests correlation with the Beeston Terrace of the River Trent (Posnansky, 1960), which Shotton (in Mitchell and others, 1973, p.22) dated as Ipswichian on the basis of its *Hippopotamus* fauna, though deposition of these terrace gravels may well have continued into the early Devensian.

The First Terrace generally lies 1 to 1.5 m above the floodplain of each stream, and is probably late Devensian to early Flandrian in age.

The Second Terrace, in particular, has been locally worked on a small scale for gravel but all the terrace deposits are thin, averaging less than 2 m in thickness. Their clasts are largely derived from the local glacial gravels.

River Anker

River gravel occupies much of the Anker valley downstream from Polesworth. The pebbles are mainly of 'Bunter' quartzite, with many yellow and white flints, dark igneous rocks, and Coal Measures sandstones and ironstones, in a reddish brown sand matrix. Many of the gravel spreads have been affected by mining subsidence and some lie in hollows below river level. Lower and higher terraces cannot be differentiated, except for two patches of higher gravel which are probably remnants of the Second Terrace. One at Waven Farm [2423 0542], just west of the present district, is about 6 m above the alluvium and consists of numerous flints in addition to 'Bunter' quartzite and igneous pebbles. The other, at a similar height, occurs near Pooley Hall Colliery [2609 0369].

Farther upstream, the First and Second terraces can be recognised at Grendon. The Second Terrace forms a plateau of 'Bunter' quartzite-flint gravel above the village and 10.5 m above the alluvium [2930 0095], with a smaller patch at a similar height at The Mount [2880 0207]. The First Terrace, 1 to 1.5 m above the alluvium, is narrow around Grendon but widens southwards where the river skirts Atherstone. It consists mainly of 'Bunter' quartzite and flint pebbles, together with fragments of Coal Measures and Triassic sandstones and many dark, igneous pebbles. Three boreholes sunk for bridge foundations [3027 9818], where the Atherstone by-pass crosses the canal, proved 3.3, 5.6 and 6.1 m respectively of First Terrace deposits consisting of sand, sandy clay, and gravel. Some large nodules of flint occur. A patch of gravel rises to 2.4 m above the alluvium at Birches Barn [2805 0618] and to 3.7 m south of the barn in High Royal Plantation, and may represent an intermediate terrace.

River Sence and Sence Brook

In the upper reaches of the River Sence and its tributaries, small terraces, usually no more than 1 or 2 m above the level of the floodplain, occur sporadically. They usually consist of silt and clay with gravelly seams. A 0.2 m bed of peat, as yet undated, lies within First Terrace deposits [4133 1203] 750 m upstream from Kelham Bridge.

Downstream, between Heather and Congerstone, the Sence has two terraces, about 1 m and 4 to 5.5 m respectively, above the alluvium. Second Terrace deposits exposed south-east of Newton Burgoland, in a ditch by the road [3742 0823], consist of red-brown, clayey gravel up to 1 m thick, with mainly 'Bunter' pebbles and a few flints. About 2 m of similar gravel are exposed [377 075] immediately south of Tivey's Farm. On the Sence Brook, the terrace gravels south of Shenton are exposed [3870 9961] 600 m WNW of New Barn, on the north side of the river. Up to 1 m of gravel here consists of mainly 'Bunter' pebbles, with some angular flints, set in a brown, sand matrix with blue-grey, clay pockets.

Downstream from Congerstone, both the First and Second Terraces are present and no difference in content between the gravels is found. It is also difficult, commonly, to differentiate between Second Terrace gravel and glacial gravel, as Fox-Strangways (1900) noted. The glacial gravels are sandier and generally contain more Triassic sandstone and siltstone than river gravels, but a pipe-line excavation [3261 0135] crossing the river at Sheepy Magna encountered over 3.5 m of 'running sand' in river gravel deposits.

The Second Terrace is well developed between Sheepy Magna [325 013] and Pinwall [308 001], where it is about 10.5 m above the flood plain. The Terrace has been much eroded and only an area 500 m north-east of the crossroads at Pinwall shows a typical, flat terrace surface. There are no good exposures, although disused workings are common.

The First Terrace rises to a maximum of 3.5 m above present river level and is extensive around Sheepy Magna and Pinwall. A ditch section [3164 0007] showed 2 m of coarse, 'Bunter' quartzite gravel in a sandy matrix. On the north bank of the Sence [3407 0266] 360 m south-east of Overfield Farm, an overgrown section in a small gravel pit suggests that the First Terrace gravel is here at least 3 m thick. Angular Charnian fragments are a common constituent of the field brash on First Terrace gravel [3358 0244] south-west of Lovetts Bridge.

River Mease

Along the River Mease below Snarestone, the gravels are unlikely to exceed 3 m. South of Netherseal, a former course of the river, to the south of the present course and separated from it by a sandstone hill [290 125], is marked by a narrow depression floored by a clayey deposit mapped as First Terrace loam. Deposition of First Terrace gravel across both the entrance and the exit to this depression seems to have caused its abandonment.

Charnwood Forest streams

A gravel terrace occurs south of Black Brook [480 153] below Cat Hill Wood. It stands about 0.5 m above the floodplain and extends about 700 m southwards towards Charley Hall. Only part of the terrace is in contact with the flood plain of Black Brook, and it may be partly lacustrine in origin.

Stony clay forms the First Terrace of the Ulverscroft stream near Newtown Linford, and occurs on both banks at about 1.2 m above the alluvium. West of Groby, the Slate Brook contains a well developed First Terrace, about 1.2 to 1.5 m above the alluvium; it is composed of gravelly loam mostly derived from the boulder clay, with a few local Charnian and igneous rock fragments. Similar small patches of terrace deposits occur in the valley north-east and east of Groby.

Rothley Brook

First Terrace deposits at Merry Lees consist of brown, clayey sand and silt, containing 'Bunter' pebbles and subrounded pebbles of flint and Charnian rocks, the latter up to 0.2 m long. The poor sorting of the deposits and the composition of the pebbles, which are the more resistant lithologies found in the local glacial drifts, indicate that they are largely of reworked boulder clay or glacial gravel.

For 2 km or so downstream from Newtown Unthank, First Terrace gravel north of the brook consists of a coarse gravel predominantly of 'Bunter' quartzite pebbles up to 150 to

200 mm across, in a clayey sand matrix. A large spread of such coarse material is unexpected along such a small stream, and the terrace, which abuts against spreads of low-level boulder clay, may represent a re-working by the stream of glacial deposits preserved within a buried channel beneath the Unthank valley (Figure 22).

The First Terrace is well developed on both sides of the valley west of Glenfield at heights of 1.2 to 1.8 m above the alluvium, and extends up tributary streams near Ratby. The deposit is generally a thin, gravelly loam with many flints and 'Bunter' quartzite pebbles.

Thurlaston Brook

Both the First and Second terraces are present along Thurlaston Brook. The Second Terrace consists of four, scattered, small patches of pebbly gravel, at 3 to 7 m above the stream. The First Terrace is more extensive and lies not more than 1 m above the floodplain. Much of the Terrace appears to have originated as fans where small tributaries enter the brook, for example around [455 993], south of Kirkby Mallory. South of Dairy Farm, Normanton [4965 9838], a deposit of First Terrace loam occupies a hollow, evidently a former loop of the brook abandoned when its inlet was choked by the First Terrace gravels that now separate it from the floodplain.

In the valley between Huncote and Thurlaston, narrow tracts of First Terrace gravels occur on both sides of the small brook at 1 to 1.5 m above the alluvium, and similar tracts extend westwards along either side of Thurlaston Brook at 1.5 m above the alluvium.

ALLUVIUM

The floodplains of streams thoughout the district are formed of clayey alluvium. Meadows on alluvium commonly show a dark, silty, loamy soil and much of the alluvium is still subject to flooding after heavy rainfall, especially along the tributaries of the Sence.

While the floodplains of the major streams tend to be linear, with widths corresponding to those of the meander-belts, many of the smaller streams are misfits with floodplains too wide to be formed by the migration of meanders. The deposits in some of these wide, alluvium-floored hollows have been distinguished as Lacustrine Alluvium, and are described separately.

One of the most extensive alluvial tracts in the district is at Austrey Meadows [283 059], where the soil is a dark grey to brown clay with scattered, quartzite pebbles. The few sections suggest that the alluvial deposits, probably clay with a little sand or gravel at the base, do not exceed 2 m in thickness.

To the south lies a smaller alluvial flat south-east of Warton. Ditches reveal up to 1.5 m of blue-grey silty clay with carbonaceous patches, possibly root remains, and scattered quartzite pebbles between dark, alluvial clay above and Mercia Mudstone below. The alluvial flat drains into the Anker by a narrow outlet [288 015] just south of The Mount, where ditch-deepening shows that the blue, silty clay does not extend into the outlet but that thin, alluvial silt is underlain by

a gravel of 'Bunter' quartzite and Charnian pebbles resting on Mercia Mudstone. This gravel can be traced into the First Terrace, while the alluvium is graded to the present river floodplain. It seems that the flow of this tributary of the Anker has been ponded back by the main stream, forming a bar of First Terrace gravel across its outlet. A similar mechanism may explain some of the anomalously wide alluvial tracts along the tributaries of the Sence and Sence Brook. Along the main valleys of the last two streams, the alluvium consists mainly of a dark blue or yellow-grey, silty clay which is variably sandy and pebbly. It is generally about 1.0 to 1.5 m thick but is thinner towards the margins of the floodplains. In the north, the streams commonly have pockets of gravel beneath the clay. Drainage channels, and ploughing across the Sence floodplain north-east of Shackerstone, show the gravels to be lenticular.

On the Culloden Farm approach road [3405 0807], 0.3 m of gravel underlie 0.7 m of clay. Upstream, the gravel dies out and more than 1 m of alluvial clay is present. In a stream [3635 0640] north of Castle Farm, 0.47 m of alluvial clay overlie 0.45 m of gravel. The gravel consists of a medium- to coarse-grained, orange-brown sand, with 70 to 80 per cent pebbles in the upper 0.2 m, decreasing to 50 to 60 per cent below. The pebbles are dominantly 'Bunter' quartzites, with some angular flints and rare Charnian rocks. They are poorly sorted, unstratified, and mostly less than 2 cm across.

In Charnwood Forest, the alluvium commonly is a mixture of gravel derived from head or boulder clay, and red-brown clay derived from the Mercia Mudstone. The gravel content decreases with increasing distance from the Charnian outcrops.

In areas underlain by colliery workings, mining subsidence has often disrupted the floodplains, so that the alluvial flats are not necessarily the only areas now liable to flooding. For example, large areas of ground covered with thin, dark grey-brown, loamy clay indistinguishable at the surface from alluvium, occur 1 km south-west of Bagworth and in the heads of the valleys just north-west and south-west of Bagworth Colliery [4403 0833 and 4410 0840]. They are probably the sites of temporary floods caused by mining subsidence in the past. In 1975, several small areas of flooding due to subsidence were present north and NNE of Bagworth.

Small spreads of alluvium flank most of the streams around Bagworth and Desford. They consist of dark grey-brown to red, clayey loam, underlain by gravel composed of pebbles derived from the glacial drifts. Towards the heads of valleys, the alluvium is commonly very pebbly.

LACUSTRINE ALLUVIUM

A roughly circular area of lacustrine alluvium about 700 m across occurs at Beaumanor Park [535 155]. The lake, in which the alluvium was laid down probably formed in a depression in the boulder clay immediately following the disappearance of the last ice sheet. The overflow from the temporary lake was probably to the north-east along the valley occupied by the present-day drainage. The lake seems to have been drained in two stages, since about two-thirds of the alluvium is formed of a flat terrace standing 2 m above the remaining area of alluvium to the east. The older alluvium is predominantly a muddy, red sand, finely interbedded with red clay. The more recent alluvium is predominantly red or orange-brown clay with scattered pebbles derived from the boulder clay.

A small deposit [387 057] about 1 km south-west of Barton in the Beans, mapped as lacustrine alluvium, is about 1 m thick; it is a soft brown-yellow silty clay which rests on glacial-lake clay, boulder clay and Mercia Mudstone. The origin of this flat tract of ground is uncertain, since it does not lie in a valley or hollow and is not obviously linked to the natural drainage system. It seems to mark the site of a small lake, whose existence depended on a dam to the south, though whether the dam was stagnant ice or glacial debris since removed, is unknown.

An area of lacustrine alluvium [316 013] about 1 km west of Sheepy Magna covers about 15 hectares of flat meadowland and was probably the site of a roughly circular temporary lake with a narrow outlet to the south.

Similar flat, low-lying areas [324 040, 362 010 and 359 999], although all now linked naturally or artifically to the present drainage system, were not deposited by the small streams now in existence; it seems probable that they too were formerly shallow lakes.

PEAT

Small spreads of peat occur along seepages emerging from the base of the glacial gravels [406 080] north of Nailstone Gorse and at Osbaston Hollow [418 062]. The peat contains disseminated calcareous tufa, presumably derived from the solution of chalk and limestone in the gravels.

Other small patches of peat, formed at springs issuing from glacial gravels, occur [471 017] south-west of Broomhills Farm and at [491 017] south of Alder Hall. The deposits are only about 0.5 m thick.

CALCAREOUS TUFA

Deposits of calcareous tufa along the eastern margin of Austrey Meadows and of the alluvial tract [295 028] south-east of Warton, have been formed where springs issue from fine-grained sandstone beds in the basal part of the Mercia Mudstone. The tufa is interbedded with peat. A pipe-line trench [2945 0528] showed 0.5 m of brown, alluvial clay, resting on 0.2 m of yellowish white, powdery tufa, on a peat lens up to 2 m wide and 0.1 m thick, which in turn rests on clayey tufa with streaks of peat, seen for 0.3 m. Nearby, some blocks of a hard, white, tufa rock with gastropod shells were excavated from the trench.

LANDSLIP

A small landslip has developed in the Mercia Mudstone [503 156] about 1 km north-west of Beacon Hill. At the toe of the landslip is a bank 2 m high, mainly of mudstone with a thin veneer of head. The east side of the slip is marked by a gully 1 m deep and 6 m across. Across the top of the slip and down its west side is a slight depression, which shows that the top has dropped by about 0.5 m. There is a record of this landslip in the Harleian MS at the British Museum, reproduced in Juke's geological Appendix to Potter (1842, pp. 29 – 31). The slip occurred in 1679 and the top showed a drop of 'half a yard'. According to the tenant farmer, the slip is still slowly moving and the fairly fresh appearance of the bank at the toe lends support to this view.

North-north-east of Lount Farm [4118 0576] and just east of the road [4176 0620] at Osbaston Hollow small landslips occur where glacial-lake clays have been weakened and lubricated by water draining from the overlying gravel. Another landslip farther up the valley [4262 0667] may similarly be due to the presence of a small pocket of lake clay beneath the gravel.

CHAPTER 10

Structure

PRECAMBRIAN ROCKS

The Charnian rocks are folded into the south-easterly plunging Charnwood Anticline (Figure 2), a structure which was outlined by Sedgwick (1834) and described in more detail by Watts (1947), Evans (1963) and Moseley (1979). The fold is asymmetrical with a steeper north-east-facing limb, and has a complex hinge, although this part of the structure is poorly exposed. Evans (1963) concluded, from stereographical analyses of the bedding planes, that the anticline is conical rather than cylindrical in form. The axial trace of the anticline is sinuous. In the north, the trend is about 120°, changing to about 160° in the central area and swinging back to as little as 95° at the eastern end of Bradgate Park.

Minor east- and south-west-trending folds have been described by Evans (1963) and Poole (1968), but poor exposure does not allow them to be delineated in detail. The east-trending folds are considered by Evans (1979) to be contemporary with folds of the same trend in Cambrian rocks at Merry Lees and Swithland Reservoir [555 135]. The northerly dipping lavas and underlying tuffs of Bardon Hill are isolated from the rest of the Charnwood outcrops and the relationship between the Bardon dips and the main Anticline is not clear. The easterly trend of the beds at Bardon Hill may be due to minor folding with this orientation.

The cleavage affecting the Charnian varies from a penetrative, slaty cleavage in the argillaceous Swithland Greywackes, to a poorly developed, fracture cleavage in the coarser parts of the more siliceous Maplewell Group. Previous workers have noted the average 280° strike of the cleavage (Jukes, 1842, p.23; Watts, 1947, p.86). The most conspicuous feature of the Charnian cleavage is that it is oblique to the main axis of the Anticline. In detail, the cleavage takes a slightly sinuous course, becoming more easterly in strike at the north and south of the Charnian outcrop (Evans, 1963).

Watts (1947, p.23) considered that the cleavage developed after the formation of the Charnwood Anticline by renewed Precambrian movement on the same axis. Evans (1963, p.76) supported this conclusion by an analysis of the cleavage directions, which he proved did not converge towards the anticlinal core, as would be expected if the two sets of structures had formed under the same stresses. Subsequently, K:Ar age determinations on the Charnian have yielded ages of 413 ± 19 Ma for the tuffs at Beacon Hill, 374 ± 13 Ma for the cleaved Lubcloud Dyke [4737 1716] (Meneisy and Miller, 1963) and 417 ± 16 Ma and 398 ± 16 Ma for the Swithland Greywacke (Evans, 1979). It has also been observed that the cleavage post-dates the intrusion of the North Charnwood Diorites (p.74). Evans, revising his earlier opinion, considered that these dates supported the view that the cleavage is late Silurian in age, as had been proposed by Jones (1927). The view is based on the assumption that the K:Ar systems were reset by complete recrystallisation during the Caledonian earth movements. An alternative view favoured by Pharoah and others (in press (a)), based on white mica crystallinity studies, is that a metamorphic hiatus exists between the more highly metamorphosed Charnian and the Cambrian. On this view the Charnian has undergone Precambrian metamorphism.

The faults cutting the Charnian have two main trends, one subparallel to the fold axis and the other roughly perpendicular to it. Many of the faults previously mapped (Watts, 1947, fig. 40) have not been confirmed by the present survey. All the fractures cutting the Charnian appear to be high-angle, normal faults. Some of their throws are very large; for example, the Abbot's Oak Fault throws about 1000 m in its south-western part and the Newtown Lindford Fault throws about 750 m.

CAMBRIAN ROCKS

Neither the structural nor the stratigraphical relationship between the Cambrian and Charnian rocks is proved. In the Nuneaton inlier, however, just south of the present district, there is an angular unconformity of about 10° between the late Precambrian Caldecote Volcanic Series and the overlying Lower Cambrian Hartshill Quartzite (Allen, 1968, p.15), and the basal beds of the Hartshill Quartzite contain fragments of Caldecote Volcanics (Eastwood and others, 1923). The Precambrian rocks had thus been tilted and eroded by Lower Cambrian times. The Precambrian and Cambrian rocks of Nuneaton were more strongly folded by Hercynian movements but these were insufficient to produce a cleavage (Allen, 1957, p.28). The most severe folds are localised around the edge of the Warwickshire Coalfield. Where Cambrian strata are encountered beneath Coal Measures, in boreholes in the central part of the coalfield, they generally have a low dip.

Cambrian strata east and north of the Nuneaton inlier are known from boreholes and mine workings where, in contrast to the exposed sequence, they show steep pre-Westphalian dips and a variable cleavage which, in places is strong. From this, Evans (1979) concluded that they had been subjected to more intense deformation, in a fault-bounded trench or aulacogen, than the rocks of the relatively stable cratonic area to the south-west. He also showed that the cleavage in the probable Cambrian rocks at Swithland Reservoir has the same trend as that in the Charnian, thus supporting the view that both are end-Silurian in age. On the other hand, cleaved Upper Cambrian mudstone from the Rotherwood Borehole yielded a K:Ar age of 477 ± 19 Ma which would suggest that the cleavage in the rocks is much earlier, if the K:Ar systems had been re-set by cleavage formation.

There is some evidence for a major, Precambrian fold

episode affecting the Charnian (see above) but analogy with Nuneaton suggests that gentle flexuring pre-dated the deposition of the Cambrian rocks. The Charnwood Anticline may have been formed at this time, or shortly after the Cambrian beds were laid down, or during the Caledonian orogeny. The minor folds and cleavage appear to be later in origin but their age is not certain.

The faults cutting the Charnian post-date the formation of the cleavage and some displace the North Charnwood Diorites, and South Charnwood Diorites. It seems reasonable, therefore, to assume a Caledonian age for the faults and this view is supported by the fact that the north-east-trending faults apparently pre-date the Thringstone Fault of Hercynian age.

The various events in the pre-Westphalian history of the district are summarised below.

EVENT	AGE
Deposition of Charnian and Caldecote Volcanics	Vendian
Intrusions of North and South Charnwood	late Precambrian
Minor folding of Caldecote Volcanics and ?Charnian	late Precambrian
Deposition of Cambrian	Lower to Upper Cambrian
Major folding of Charnian and Cambrian, followed closely by cleavage and minor folding of Charnian and Cambrian	Siluro-Devonian
Intrusions of South Leicestershire	
Faulting	?Lower Devonian

CARBONIFEROUS ROCKS

The detailed geological structure of the three coalfields within the district is shown in Figures 25 and 26. The Carboniferous rocks are preserved on the downthrow sides of the Polesworth, Netherseal, Boothorpe and Thringstone faults, but they were eroded from the central tract of the present district, which was uplifted in post-Carboniferous times. The upfold plunges northwards to bring in the Westphalian beds of the South Derbyshire Coalfield, but the central tract is underlain by an extensive outcrop of Cambrian beds. By basin inversion in Triassic times, the uplifted area became the Hinckley Basin, in which a thick pile of Triassic sediments accumulated.

Between the Boothorpe Fault and its projection southwards, and the Thringstone Fault, the thin Coal Measures and Triassic sequences of the Leicestershire Coalfield are thought to rest on Cambrian mudstones (Whitcombe and Maguire, 1981a). The seismic refraction traverse, upon which this interpretation rests, also suggests that there is a massive igneous intrusion, a possible continuation of the South Charnwood or South Leicestershire diorites, rising to 1 km or less from the surface beneath this tract. The presence of igneous rocks at such shallow depths would explain the relatively simple structure of the overlying rocks of the Leicestershire Coalfield. The South Derbyshire Coalfield is more complicated in structure (Figure 25); if, as seems likely, its basement consists of Cambrian rocks overlying Charnian rocks, then perhaps these beds offered less resistance to Hercynian deformation.

The Thringstone Fault is a fracture with a considerable reversed throw in Hercynian times, during which the Charnian rocks of Charnwood Forest were pushed over the eastern margin of the Cambrian and the postulated intrusion; unlike the Boothorpe and Polesworth faults, the Thringstone Fault did not move significantly in Triassic times.

That part of the Warwickshire Coalfield falling within the present district is the north-eastern corner of the main structural basin of the coalfield (Mitchell, 1942, pp.14–17; National Coal Board 1957). The detailed structure of this ground (Figure 26) is not well known because the earlier mining records are incomplete and there are large, unworked areas. The main synclinal fold gives southerly or south-westerly dips within the present district, with local minor flexures. A complementary anticline occurs along the eastern flank of the coalfield as the north-westerly prolongation of the Nuneaton Anticline, as proved in opencast workings between Grendon and Polesworth, and by mapping and deep mining farther north.

At its north-eastern end, the Warwickshire Coalfield is bounded by the Warton Fault. The Shuttington Fields Borehole [2642 0610], 0.9 km north-east of the Fault, proved Cambrian rocks of Tremadoc age beneath the Triassic. The absence of Coal Measures on the eastern side of the Fault proves that it must have thrown down to the west in post-Carboniferous times. Its post-Triassic throw is, of course, in the opposite direction. Mining records show that the fault is very steep to vertical.

The South Derbyshire Coalfield is a north-west-trending, flat-bottomed syncline with much minor folding and normal faulting. The three main fractures in that part of the coalfield within the present district (Figure 25) are the Netherseal Fault trending just east of north, the north-west-trending Overseal Fault, and the Boothorpe Fault, similarly aligned on the east side of the coalfield.

West of the Netherseal Fault, the concealed 'Western Extension' to the coalfield has not been worked and its structure is not known in detail. Between the Netherseal and Overseal faults, the Gunby Lea Fault trends ENE near Grangewood Hall. South of the Fault is a half-basin whose detailed structure is still uncertain, despite the many boreholes hereabouts. Two fractures have been proved within the half-basin in underground workings (Mitchell and Stubblefield, 1948, p.26); one is aligned parallel to the Overseal Fault and throws 88 m down to the north-east [300 144] near Seal Pastures; the other crosses from this fracture to the Overseal Fault, with a sinuous course proved in workings in the Eureka seam. Between this last fault and the Overseal Fault is a down-faulted tract of Upper Coal Measures cropping out in Hooborough Brook, west of Donisthorpe. North of the Gunby Lea Fault, dips are gentler. An anticline west of the Netherseal Shaft, and a syncline east of it, both trend northwards and are cut by small cross-faults.

The Overseal Fault can be traced at outcrop from near Overseal, south-eastwards as far as the western end of the

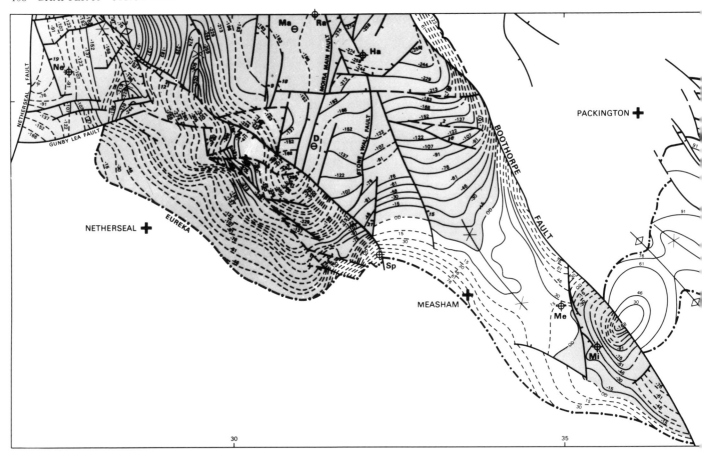

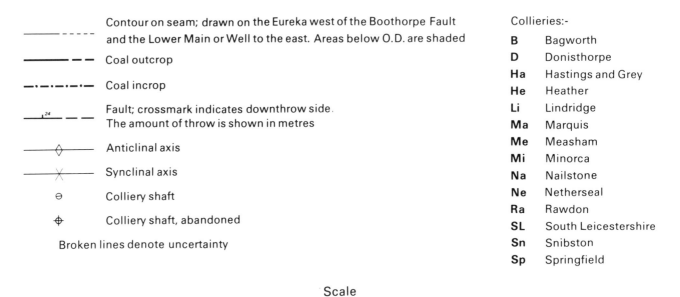

Contour on seam; drawn on the Eureka west of the Boothorpe Fault
and the Lower Main or Well to the east. Areas below O.D. are shaded

Coal outcrop

Coal incrop

Fault; crossmark indicates downthrow side.
The amount of throw is shown in metres

Anticlinal axis

Synclinal axis

⊖ Colliery shaft

⊕ Colliery shaft, abandoned

Broken lines denote uncertainty

Collieries:-

B	Bagworth
D	Donisthorpe
Ha	Hastings and Grey
He	Heather
Li	Lindridge
Ma	Marquis
Me	Measham
Mi	Minorca
Na	Nailstone
Ne	Netherseal
Ra	Rawdon
SL	South Leicestershire
Sn	Snibston
Sp	Springfield

Scale

0   1   2   3km

Figure 25   Structure of the South Derbyshire and Leicestershire coalfields

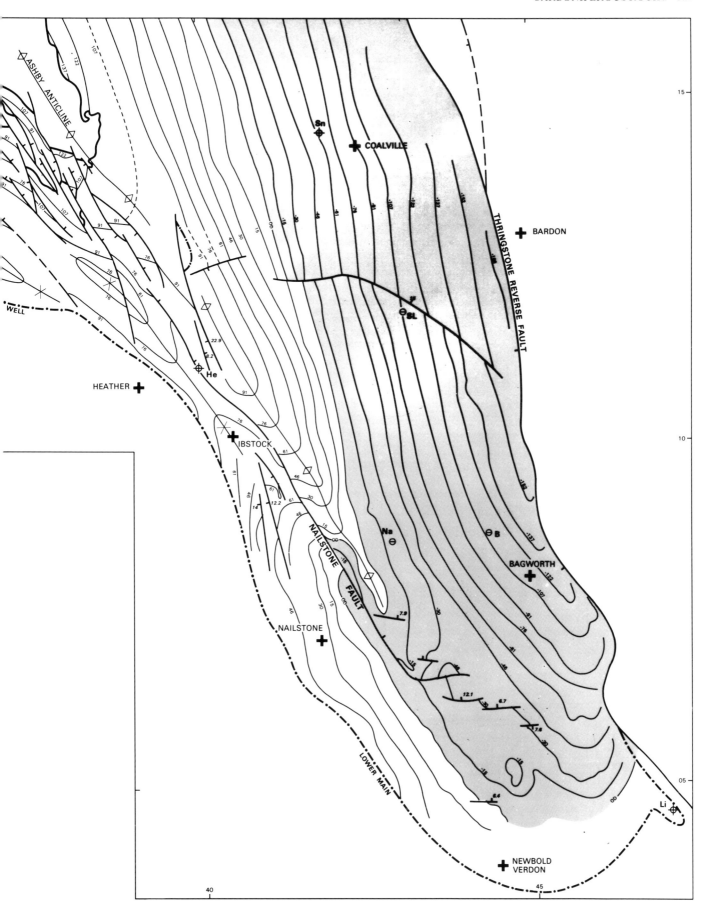

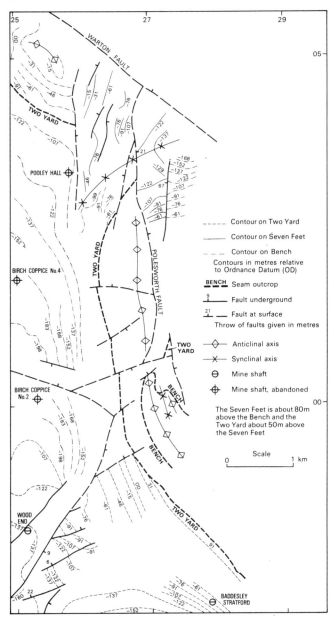

Figure 26   Structure of the north-eastern part of the Warwickshire Coalfield

throwing down 107 m east and 28 m west respectively. Another prominent fracture trends eastwards through The Furnace and throws down to the north about 32 m. In the west beneath Overseal, a well defined syncline and anticline trend north or NNW, with dips between 14° and 27° on their limbs (Figure 25). Farther east, an area of shallow dips culminates in a gentle NNW-trending anticline extending northwards through the Reservoir Shaft [3008 1645]. East of the Moira Main Fault the beds are folded into a gently north-plunging syncline, its eastern limb rising steeply against the Boothorpe Fault. A small basin is developed [358 117] beneath Triassic cover on the west side of the Boothorpe Fault, south-east of Measham Colliery, and the subcrop of the Main is repeated across the basin. Spink (1965, pp.71–72) thought the repetition was due to the Boothorpe Fault having a reversed throw, but this view is not tenable when the structure of the basin is recognised.

East of the Boothorpe Fault, the Eureka and higher seams crop out north of Willesley [340 148], but older rocks of Namurian age are brought to the surface in a small, up-faulted block near Valley Farm [344 154]; this structure passes eastwards beneath the Trias and its relationship with the Ashby Anticline in the district to the north is not known.

The Leicestershire Coalfield is a syncline whose eastern limb is truncated by the Thringstone Reverse Fault. Dips are generally to the east at about 5°, although along its western edge the coalfield extends onto the crest of the complementary Ashby Anticline (Figure 25). The folds trend generally NNW and plunge gently to the south-east. The main syncline is almost entirely concealed but its detailed structure is known from extensive colliery workings. Near the Thringstone Fault, the beds are sharply upturned and dip steeply to the west. For example, in the Lindridge Shaft [4720 0454], the Coal Measures dip at 45° to 60° (Fox-Strangways, 1907, p.107), and in the Ibstock No.1 Borehole [4732 0464], 150 m to the north-east and probably close against the Thringstone Fault, they proved near-vertical. The direction of dip in the shaft is not recorded but, assuming it is to the west, there is a minor syncline here, parallel to the Thringstone Fault, with the incrops of the Lower Main and Kilburn coals curved tightly around its axis.

The Ashby Anticline, in the north-west of the coalfield, has many small faults parallel to its axis and details of its structure are only well known where the coals have been worked opencast (Mitchell, 1948; Spink, 1965). Most of the faults are on the western limb of the anticline and throw down to the east, successively repeating the sequence. Details of faulting on the south-eastern limb of the Anticline at the Blower's Brook [395 125] opencast site were published by Spink (1965, pp.73–74). The Anticline has not been traced south of Heather, but another upfold occurs *en échelon* to the south-east between Ibstock and Nailstone (Figure 25). Both these folds are also much faulted, as proved in the old workings at Heather Colliery and in less well known ground, west of the Nailstone Fault and north of Nailstone.

The Thringstone Fault was shown as a normal fault on the primary one-inch geological map but is now known to have a reverse throw. The hade of the fault in the Merry Lees Drift, for example, was 40° to the north-east (Butterley and Mitchell, 1945) and this seems to be maintained farther north. Near Desford Colliery, the incrop of the Fault is offset by

Oakthorpe Meer opencast site [3225 1275]; beyond, its course must lie west of Chapel St. No.3 Borehole [3239 1247] but it does not seem to cut the Triassic rocks hereabouts. About 900 m west of Donisthorpe Shaft, a heading passed through the Fault from the Main Coal into the Upper Kilburn, indicating a throw of about 180 m, but the structure contours (Figure 25) suggest smaller throws to the north, of about 15 m near Short Heath, and to the south-west, of about 73 m near Oakthorpe. Near this last locality, a local basin on the downthrow side of the Fault [309 139] is not matched on the upthrow side.

Between the Overseal and Boothorpe faults, the main fractures are the north-trending Moira Main and Stonewall faults (Fox-Strangways, 1900, p.45; 1907, pp.88–89), about 600 m. This is not a cross-fault, since the Coal Measures are unfaulted hereabouts; the hade of this section of the fault is about 40° to the south-east. At Bardon, the throw of the Thringstone Fault is at least 550 m, assuming that 190 m of Carboniferous strata underlie the Lower Main, but this is a minimum estimate of the throw since an unknown thickness of Charnian has also been faulted out. Mine plans from Ellistown southwards, show many small, normal faults extending into the coalfield perpendicular to the Thringstone Fault. North of Bardon, there may be an eastern branch of the Thringstone Fault passing east of the Holly Hayes boreholes to separate the Coal Measures or Millstone Grit rocks in these boreholes from the Charnian of Spring Hill.

PERMO-TRIASSIC ROCKS

Contours on the base of the Permo-Triassic rocks are shown in Figure 27. The Hinckley Basin (Warrington and others, 1980, fig. 4) is the most conspicuous structure, though the Needwood Basin, lying north-west of the district, is just as deep.

Thickening of the Triassic rocks into these basins (Figures 14 to 16) shows that they were subsiding during Triassic times, generally along growth-faults. The Hinckley Basin is a clear example of basin inversion, since it developed above an area of post-Carboniferous uplift which had exposed an extensive outcrop of Cambrian rocks by Permian times, as proved in the Shuttington Fields, Twycross, Kingshill Spinney, and Dadlington boreholes.

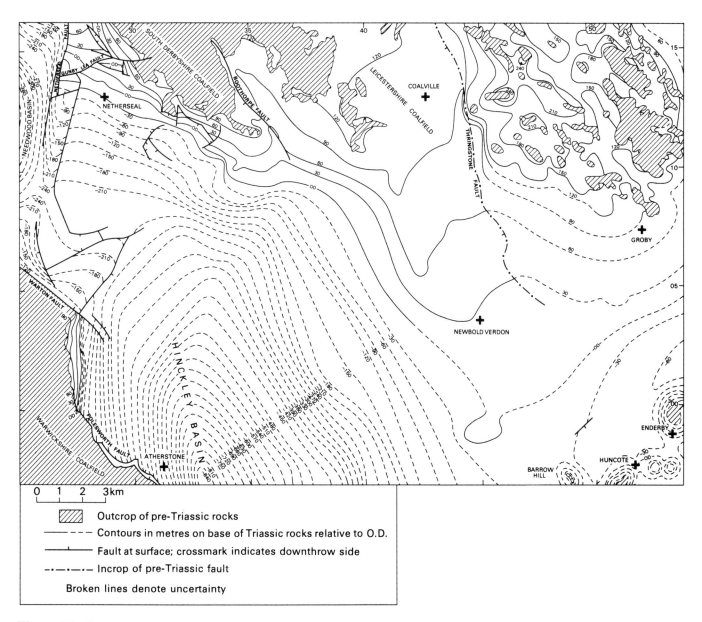

Figure 27 Structure contours on the base of the Triassic (including Hopwas Breccia and Moira Breccia)

The south-western margin of the Hinckley Basin has a steep slope which is partly formed, as seismic evidence shows (p.122), by the Polesworth Fault. The Basin appears from gravity measurements and limited borehole evidence to be separated from the Needwood Basin by a col extending between the Warwickshire and South Derbyshire coalfields.

The Triassic rocks along the southern fringe of the South Derbyshire Coalfield are much affected by faulting. In some cases, the throws in the Triassic rocks are in the opposite direction to those in the Coal Measures. For example, a narrow fault-trough of Bromsgrove Sandstone [284 157], 1 km west of Overseal, lies vertically above a fault in the Coal Measures that appears to be a steeply-hading continuation of the Overseal Fault. The throw is about 40 m down to the west in the Coal Measures (Figure 25), but 45 m down to the east (Figure 27) in the Triassic beds. This is to be expected from the quite different stress patterns operating in late-Carboniferous and Triassic times.

The most notable faults affecting the Triassic rocks are the Polesworth Fault, which throws the Triassic outcrops about 300 m vertically (Figure 27), but has a total throw of about 1 km (Whitcombe and Maguire, 1981a, fig. 4) when the growth element is included, and the Neatherseal Fault, which has a throw of about 150 m in Triassic, and about 450 m in the Carboniferous, beds. The Netherseal Fault probably grew during the deposition of the Polesworth Formation, since the sequence is 180 m thick on the downthrow side and only 50 m thick on the upthrow side (Figure 15).

CHAPTER 11

Geophysical investigations

The wide variety of rocks within the district, with their greatly varying physical properties, produces several large Bouguer gravity and magnetic anomalies. Interpretation of these anomalies is particularly valuable in understanding the geological structures hidden beneath the extensive cover of Triassic sediments. Several of the anomalies reflect variations in the thickness of the Triassic cover rocks, but others are due to structures within the Palaeozoic and Precambrian rocks. The latter group of anomalies extend for considerable distances beyond the limits of the present district and their interpretation, therefore, has widespread application. On the other hand, the anomalies marking outcrops of the older rocks provide some control on the interpretation of similar anomalies elsewhere in the Midlands, where the basement rocks are entirely hidden.

The first Bouguer anomaly data for the present district were included on the 1:253 440 scale gravity overlay maps (sheets 15 and 11), published by the Geological Survey in 1954 and 1956. The data were obtained from several sources, such as Cook and others (1951), who included the Warwickshire Coalfield in their survey. They pointed out that the Cambrian and Precambrian ridge at Nuneaton gives rise to a Bouguer anomaly high and suggested that the coalfield rests on a basement of Lower Palaeozoic rocks with a density of 2.71 Mg/m³.

Bullerwell (in Stevenson and Mitchell, 1955) summarised the geophysical data for the Burton-upon-Trent (140) district, north-west of the present district. The near-surface geology around Burton is dominated by the Needwood Basin where the Triassic rocks are at least 700 m thick, but an east–west, elongated, Bouguer anomaly high is revealed after the correction for the Triassic cover has been made. The relationship between the known thickness and density of the Triassic rocks, and the Bouguer anomaly values indicates a density contrast of 0.26 Mg/m³ with the underlying rocks.

Spink and Strauss (1965) reviewed the geophysical evidence, mainly Bouguer anomaly data, in the area just north-east of the present district and concluded that there might be a concealed coalfield east of Loughborough.

In an attempt to trace the boundaries of the Mountsorrel Granodiorite, Davies and Matthews (1966) carried out detailed magnetic, gravity and seismic surveys which showed that a hidden extension of the intrusion to the west beneath Charnwood Forest was unlikely.

Howell (in Taylor and Rushton, 1971, pp.63–65) described the results of radiometric and resistivity logging of three Geological Survey boreholes at Merevale, near Atherstone, which penetrated Carboniferous, Devonian and Cambrian sediments.

Evans and Maroof (1976) used geophysical data from Charnwood to support their suggestion that mineralisation in several areas of Britain was spatially related to concealed granites. They suggested that intrusions with physical properties similar to the Mountsorrel Granodiorite (Table 10b) might not be revealed on Bouguer anomaly maps, because they closely resembled the basement host-rocks in density, but that they could be the source of strong aeromagnetic anomalies, as yet unexplained.

A palaeomagnetic study of the Leicestershire diorites (Duff, 1980) revealed at least two components of magnetisation, one of which was consistent with magnetisation during late Precambrian or Cambrian times, in line with their radiometric ages.

The deeper structure of the Charnwood area has been examined in recent years using long seismic refraction profiles, mainly recording quarry blasts (Whitcombe and Maguire, 1980, 1981a, 1981b, Maguire and others, 1981).

PHYSICAL PROPERTIES OF ROCKS

To supplement data available from published sources, rock samples were collected from local quarries and outcrops and their physical properties have been determined (Table 10). A comparison of the saturated densities shows that there is a density contrast of approximately 0.3 Mg/m³ between Triassic sedimentary rocks, with densities between 2.15 and 2.57 Mg/m³, and rocks of Dinantian age or older, with densities varying from 2.62 to 2.80 Mg/m³. There is a smaller density contrast of about 0.2 Mg/m³, between Westphalian and Namurian sediments with a mean saturated density of 2.48 Mg/m³, and the older rocks. The Triassic and Upper Carboniferous samples have consistently high porosity values, while the igneous rocks have the lowest porosities and highest densities of all the samples obtained. For the Charnian, Maroof (1976) reported an average density of 2.80 Mg/m³. This value seems too high for the metasediments and volcanic rocks in Charnwood Forest. Whitcombe and Maguire (1980) reported densities of 2.70 Mg/m³ and 2.63 Mg/m³ for rocks of the Maplewell Group and the Blackbrook Formation respectively. These rocks form a major part of the outcrop in Charnwood Forest and have a substantial thickness, so an average density for the Charnian of 2.70 Mg/m³ is probably more realistic.

The following densities were adopted for the elevation correction when calculating Bouguer anomaly values:

Trias	2.40 Mg/m³
Westphalian and Namurian	2.50 Mg/m³
Dinantian, Cambrian and Precambrian	2.70 Mg/m³

Data on the magnetic properties of the rocks have also been collected from several sources (Table 10) and include measurements of the susceptibility, remanent magnetisation and Q-values (ratio of remanent magnetisation to the magnetisation induced in a field of 36 A/m). As expected, the sediments are generally non-magnetic but susceptibility measurements indicate a weak magnetisation in some of the

Table 10a Physical properties of local sediments

Age	Rock type	Location	Grid ref.	No. of samples	Mean density Mg/m³ Saturated	Grain	Effective porosity %	Magnetic susceptibility SI units (×10⁻⁵)	Sonic Velocity km/s Dry	Saturated	Source
Recent	Clays	M69 motorway sites	—	948	2.10						5
Trias	General	Sproxton Borehole	[845 239]	10	2.40±0.09						3
	Bromsgrove Sandstone	Near Grendon	[280 000]	3	2.23±0.01	2.64±0.02	24.90±4.00	16±1	2.63	2.69	1
	Bromsgrove Sandstone	Leicester Forest East Borehole	[524 029]	3	2.42±0.08	2.74±0.04	18.1±4.1				1
	Mudstone in Bromsgrove Sandstone	Near Grendon	[278 002]	1	2.57	2.70	7.55				1
	Polesworth Fm.	Near Polesworth	[275 015]	4	2.54±0.03	2.72±0.01	11.00				1
	Sandstone	Near Measham	[342 120]	2	2.15±0.02	2.65	31.77				1
Carbonif-erous	Westphalian	Sproxton Borehole	[845 239]	5	2.47±0.14						3
		Coventry Colliery	[315 847]		2.48±0.07						4
	Namurian	Derbyshire	[267 532]	3	2.38±0.04	2.66	8.40±8.00	3±0			1
		Sproxton Borehole	[845 239]	4	2.55±0.14						3
	Dinantian	Derbyshire	[284 543]	3	2.65±0.03	2.69	0.78			6.22±0.01	1
		Cloudhill Borehole	[411 217]	46	2.70±0.03						3
Cambrian	Stockingford Shales	Merry Lees Drift	[460 068]	7	2.72±0.02						3
	Purley Shales	Near Oldbury	[300 964]	2	2.70±0.03	2.92±0.04	11.03±3.30	83	4.50±0.30	4.92±0.20	1
	Oldbury Shales	Near Oldbury	[315 950]	1	2.72	2.77	3.03	88	5.42	5.55	1
	Cambrian Mudstones	Rotherwood Borehole	[345 155]	7	2.76±0.04	2.81±0.04	3.30±1.90	65±21	4.67	4.67	1
	(?)	Leicester Forest East Borehole	[524 029]	3	2.69±0.04	2.82±0.02	6.71±1.58				1
Pre-cambrian	Swithland Slate	Near Groby	[500 085]	2	2.66±0.01	2.73	4.63±0.40	30	5.42	5.57	1
	Phyllitic Shale	Sproxton Borehole	[845 239]	8	2.84±0.09						3
	Bradgate Tuff Fm.	Near Newtown Linford	[530 110]	3	2.69±0.06	2.73±0.06	2.54±0.40	69	5.67	5.98	1
	'Slate Ag-glomerate'	Markfield			2.78±0.02						3
	Beacon Hill Tuff Fm.	Near Newtown Linford	[520 110]	1	2.72	2.76	2.52	53	5.24	5.65	1
	Beacon Hill Tuff Fm.	Beacon Hill	[515 120]	2	2.71±0.06	2.73±0.06	0.55	82±54	5.93	6.13	1
	Beacon Hill Tuff Fm.	Near Newtown Linford	[520 111]	1	2.83	2.83	0.32	57	6.30	6.48	1
	Blackbrook Fm.	Near Blackbrook	[463 171]		2.63±0.03						3

Sources: 1–BGS, 2–Leeds University (private communication from B. A. Duff), 3–Maroof 1973, 4–Cook & others, 1951, 5–Cementation Ltd, M69 motorway samples

Cambrian samples. The igneous rocks generally show higher susceptibilities, which can vary, even for samples from the same intrusion. The granodiorites at Bradgate, Cliffe Hill, Markfield, Groby East and Mountsorrel are particularly magnetic, but the remainder have lower susceptibilities, similar to those of the more magnetic sediments. The degree to which the magnetic properties reflect differences in the original compositions of the intrusions, is partly obscured by their secondary alteration. The age of the alteration is not always known, weathering was widespread beneath the sub-Triassic unconformity, and was also extensive in Tertiary times. The result of the alteration is widely differing magnetisation and density values. At Markfield, for example, one group of two samples gave mean values of 2.84 Mg/m³ and 4 × 10⁻² SI units for the density and susceptibility, respectively, and a second group of four samples gave 2.72 Mg/m³ and 5 × 10⁻⁴ SI units. The contrast probably arises from the secondary alteration, in the second group, of the opaque and ferromagnesian minerals.

Estimates of the velocities of the main rock types are

Table 10b Physical properties of local igneous rocks

Age	Rock type	Location	Grid ref.	No. of samples	Mean density Mg/m³		Effective porosity %	Magnetic susceptibility SI units ($\times 10^{-5}$)	Q	Velocity km/s		Source
					Saturated	Grain				Dry	Saturated	
Cambrian	Lamprophyre	Nr Atherstone	[300 970]	2	2.71±0.04	2.75±0.02	2.42	72		4.97	5.61	1
Caledonian	Diorite	Barrow Hill N	[488 972]	6	2.74	2.74	0.72					1
		Barrow Hill S	[488 970]	5	2.72	2.73	0.66					1
	Diorite	Enderby	[539 002]	1	2.74	2.75	0.71			5.92	5.60	1
		Enderby	[538 001]	9				68±235	18			2
		Bradgate Park	[520 105]	1	2.77	2.82	2.83	679		4.73	5.28	1
		Bradgate Quarry	[510 086]	8				136±206	0.46			2
		Cliffe Hill	[472 106]	1	2.72	2.74	1.04	58				1
		Cliffe Hill	[474 107]	6				1883±1810	0.37			2
		Markfield	[486 102]	4	2.72±0.03	2.75±0.03	1.81±1.80	49			4.94±0.28	1
		Markfield	[486 103]	2	2.84	2.84	0.07	4100		6.38	6.25	1
		Markfield	[486 103]		2.80±0.0?							3
		Markfield	[486 103]	5				376±292	1.05			2
	Quartz-diorite	Sapcote	[492 941]	10				70±30	0.25			2
		Croft	[512 968]	8				28±6	0.21			2
		Croft	[519 962]	3	2.70	2.70	0.23±0.04	16±2		5.47±0.02	5.78±0.03	1
	Diorite	Groby	[525 082]	3	2.71±0.01	2.74	2.12±1.26			4.94±0.27	5.00±0.55	1
		Groby West	[524 083]	6				53±5	0.14			2
		Groby East	[528 082]	7				790±685	0.63			2
		Groby			2.72±0.01							3
	Granodiorite	Mountsorrel	[576 149]	2	2.67±0.01	2.68	0.22±0.02	2638		6.0	5.5	1
		Mountsorrel	[576 149]	60	2.65±0.01							3
		Mountsorrel	[576 149]	12				771±678	0.22			2
		Newhurst	[486 179]	1	2.91	2.91	0.12	113		6.36	6.42	1
	Porphyroid	Nr Shepshed	[483 179]	3	2.80±0.09	2.84	1.95±0.50				6.27	1
		Spring Hill	[445 159]		2.70±0.08							3

Sources: see Table 10a

available from seismic refraction surveys and geophysical logs of boreholes. The mean values recorded are:

Cambrian (Stockingford Shales)	1,2	3.80–4.40 km/s
Precambrian (Charnian)	2	5.40–5.65 km/s
(Blackbrook Formation only)	2	5.60 km/s
Intrusive rocks: Diorite	2	4.70–4.80 km/s
Granodiorite	2	4.90 km/s

Sources: 1 NCB, 2 reported by Whitcombe and Maguire (1980).

JDC,JMA

GRAVITY SURVEYS

Gravity surveys of the northern part of the district were carried out by BGS between 1950 and 1973. These formed a basis for further work during 1976, which completed the regional survey and added detailed traverses across local gradients and anomalies. This later survey used a base at Cadeby [425 023] which was linked to the National Gravity Reference Network (Masson Smith and others, 1974) at Melbourne [364 242] in the Loughborough district. All gravity data were reduced to sea level, adopting the densities given above, and inner zone terrain corrections were applied. The outer zone terrain corrections were omitted as negligible.

The Bouguer anomaly contours, relative to the 1967 International Gravity Formula, are shown in Figure 28 and the map forms part of the 1:250 000 scale Bouguer anomaly map of the East Midlands (IGS, 1982). A principal feature of the Bouguer anomaly map is a ridge of high values, trending south-eastwards from a maximum at Ashby towards Desford, where a borehole at Stocks House [468 021] found igneous rocks 150 m below the surface (Fox-Strangways, 1907). The ridge near Ashby extends northwards, to the west of Derby, to coalesce with the high over the Derbyshire Dome (Maroof, 1976). In the Derby district, the ridge seems to be due to Lower Carboniferous and older rocks close to the surface (Cornwell in Frost and Smart, 1979), and this is also likely to be true in the present district.

In the south-western corner of Figure 28, a positive north-west-trending anomaly extends along strike over the Warwickshire Coalfield, and also marks the near-surface occurrance of older, denser rocks.

In the north-east of the district, the area of low Bouguer anomaly values over Charnwood Forest appears to be a step or positive interruption in the otherwise steep regional gradient towards the Bouguer anomaly low of the Lincolnshire area. Just to the north of the area of Figure 28, this low marks the depositional basin of Namurian and older sediments of the Widmerpool Gulf (Kent, 1966). The absence of a pronounced Bouguer anomaly high over the Precambrian rocks of Charnwood is unexpected, but may be due to the preponderance of sediments in the succession,

which have a density slightly lower than the numerous igneous intrusions. The western boundary of the Charnwood rocks, the Thringstone Fault, is represented by a north–south bend [440 150 to 440 090] in the Bouguer anomaly contours. West of the fault, the increasing Bouguer anomaly values over the Leicestershire Coalfield correspond with the decreasing thickness of lower density Westphalian sediments below the Trias. The major Bouguer anomaly low is centred over the Hinckley Basin of thick Triassic sediments. The low trends to the north-west and is separated from the Warwickshire Coalfield and the Ashby ridge by steep gradients. The gradient on the north-eastern side, possibly a continuation of the Boothorpe Fault, appears to separate into two smaller step-like features near [36 09], which merge again before entering the South Derbyshire Coalfield. The gradient on the south-western side of the low coincides with the Cambrian outcrops flanking the eastern side of the Warwickshire Coalfield.

As a first step in the interpretation of the gravity data, a regional field was computed for the area shown in Figure 28, with a 10-km extension on all sides using a second degree polynomial method. The main feature of the regional field is a broad closure centred about [30 10], with a decrease of 0.25 mGal/km towards all the boundaries of the district. Subtraction of this regional effect from the Bouguer anomaly values gives a residual map that differs only slightly from Figure 28.

To illustrate the geophysical responses of some of the rock sequences in the district, several anomalies are interpreted below, using two dimensional models, to produce geological cross-sections.

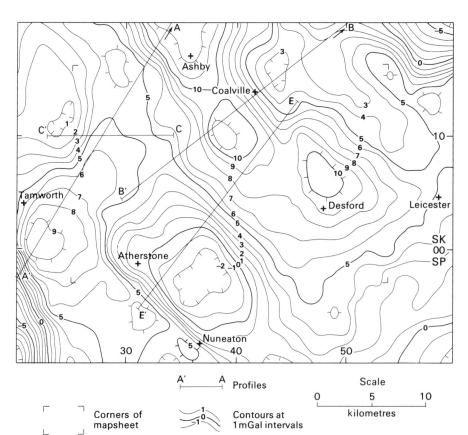

Figure 28 Bouguer anomaly map and location of profiles A′A, B′B, C′C and E′E

Warwickshire Coalfield—Ashby Ridge Profiles (A′A and B′B)

The Bouguer anomaly profile A′A (Figure 29i) [200 975 to 425 335] incorporates both the Warwickshire Coalfield and Ashby anomalies. The Western Boundary Fault of the Coalfield and the Boothorpe Fault are marked by steep gradients and part of the low over the Widmerpool Gulf, to the north of the present district, is apparent. The Thringstone Fault lies just east of the Ashby ridge and another sharp decrease in values occurs 2.5 km farther to the northeast. Further analysis of the anomalies requires the removal of the gravitational effects of the low density sediments of Triassic, Westphalian and Namurian age.

An upfold, bringing Namurian and Lower Carboniferous rocks close to the surface, is postulated between the Thringstone Fault and the Widerpool Gulf, to account for the local Bouguer anomaly high on the main gradient, and the outcrops of Dinantian and Namurian rocks, as at Breedon on the Hill [407 233], show that there is indeed upfolding hereabouts. The Thringstone Fault itself has no gravity response on the profile since there is little density contrast between the Namurian and Westphalian sediments on either side of the fracture. The main gradient farther to the northeast, appears to be due to faulting along the margin of the Widmerpool Gulf. The profile obtained after correcting for the Bouguer anomaly effect of the Upper Carboniferous and Triassic rocks (Profile 4, Figure 29i) shows two distinct gravity highs, which must therefore be due to features in the pre-Namurian 'basement'. The high near Ashby suggests uplift of the pre-Carboniferous basement.

It seems likely that the main gravity 'highs' on Profile 4 represent two separate blocks, one beneath the Warwickshire Coalfield and the other beneath the South Derbyshire Coalfield. Assuming an arbitrary density contrast of $0.1 \, Mg/m^3$ within the basement, and that the base of each block lies at 3.0 km below OD, then the Warwickshire block would rise to about 1 km below OD and the Derbyshire one to about 0.7 km below OD. These blocks would have to be composed of rocks comparable with the $2.80 \, Mg/m^3$ Precambrian basement described by Maroof (1976) and comprise more basic intrusions and indurated sediments than occur at

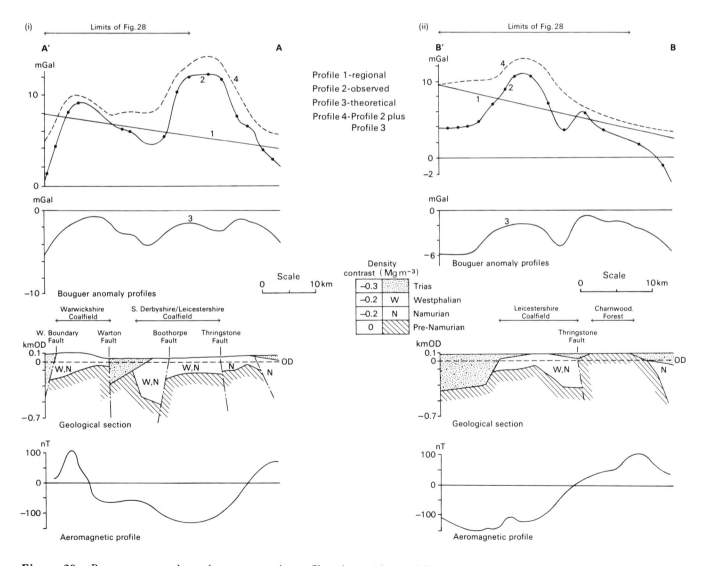

Figure 29 Bouguer anomaly and aeromagnetic profiles along A′A and B′B on Figure 28

the surface.

Whitcombe and Maguire (1981b) have described the results of a long seismic refraction profile across the Ashby ridge and its extension northwards. A refractor, with a velocity similar to that of the Charnian rocks, increases in depth just north of Charnwood Forest to lie at about 0.6 km below OD beneath the Trent valley. A borehole located on the Ashby ridge at Rotherwood [346 156] proved Cambrian mudstones with an average density of 2.76 Mg/m³ at a depth of 64.4 m below OD, and these sediments are dense enough to contribute to the anomaly.

A second profile (B′B, Figure 29ii) [297 045 to 550 238] also crosses the Ashby ridge. The Thringstone Fault and the wedge of Westphalian rocks present in the Leicestershire Coalfield are apparent on the geological cross-section. It also shows a steep gradient on the west side of the Ashby ridge, which probably marks Triassic growth-faulting along the presumed south-westerly continuation of the Boothorpe Fault. The Triassic sediments extend down into the Hinckley Basin below 300 m below OD. When these local effects have been removed, a positive anomaly remains which indicates a high density ridge in the basement between Ashby and Desford.

Hinckley Basin Profile (E′E, Figure 34)

The Bouguer anomaly low near Hinckley lies between 3 and 6 m east of Atherstone, and marks an area of thick Triassic sediments. The most likely explanation of the low, which has an amplitude of about − 9 mGal, is that it is due entirely to thick, low density Triassic sediments; the alternative, that the low is due to a combination of Triassic and Westphalian sediments, may be discounted following the unsuccessful search by the National Coal Board for concealed Coal Measures around Hinckley (Jones, 1981). Seismic reflection surveys were followed by the drilling of the Dadlington [3993 9909] and Aston Flamville [458 931] boreholes, which proved Cambrian sediments directly beneath the Trias on the flanks of the Basin. An explanation of the Bouguer anomaly low, in terms of variations in the thickness of the Trias alone is only partly supported by the borehole evidence. The calculations assume a density contrast of − 0.3 Mg/m³ between the Triassic cover and a basement of Cambrian sediments. Thus, the Dadlington Borehole proved only 314 m of Triassic sediments where a greater thickness would be needed to account completely for the Bouguer anomaly. Furthermore, the Kingshill Spinney Borehole [3814 0154] proved even thinner Triassic sediments, despite being situated, like Dadlington, in the zone of steep Bouguer anomaly gradients. Finally, the Twycross Borehole [3387 0564] proved much thicker Trias, down to 381 m below OD, in an area where the Bouguer anomaly is less pronounced (cf. profile B′B Figure 29).

It is probable that the thickness of the Trias does not relate exactly to the Bouguer anomaly values because the local background Bouguer anomaly values are not accurately known. These difficulties arise because the Bouguer anomaly highs, on either side of the basin, are increased by an unknown amount by the presence, at depth, of high density bodies (cf. Figures 29 and 34). It seems likely, on balance, that the maximum thickness of Trias occurs near the western margin

of the Hinckley Basin, and that the basin remains deep northwards, beneath the site of the Twycross Borehole.

Netherseal Fault Profile (C′C, Figure 30)

In the north-west of the district, a north-trending step-like feature on the Bouguer anomaly map coincides with the Netherseal Fault. The westward decrease of about 4 mGal can be explained by the thickening of both Triassic and Westphalian sediments on the downthrow side of the Fault. The fracture appears to be underlain by a magnetic body, discussed below.

Barrow Hill Intrusion

A detailed gravity traverse along the M69 motorway [393 824 to 536 004] revealed a sharp Bouguer anomaly high [490 968] near the Barrow Hill diorite quarries (Figure 31i). The residual anomaly has an amplitude of 1.5 mGal with a maximum gradient of 3 mGal/km on the western side, and is too localised to be apparent on Figure 28. The anomaly probably marks a mass of diorite forming a pre-Triassic hill thinly covered by Triassic sediments. Further work around the Barrow Hill quarries revealed a local Bouguer anomaly high, with a maximum amplitude of 1.7 mGal, associated with the exposed diorite.

The association of Bouguer gravity anomaly highs with buried hills of diorite led to a detailed gravity investigation of the southern part of the district, to prove other near-surface diorite intrusions (Allsop and Arthur, 1983).   JMA

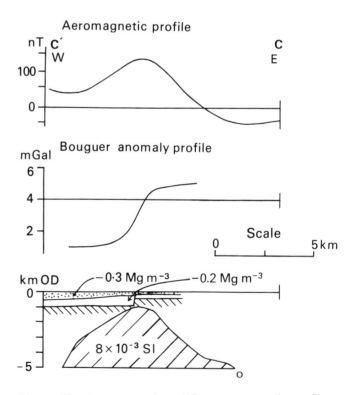

Figure 30   Aeromagnetic and Bouguer anomaly profiles along C′C

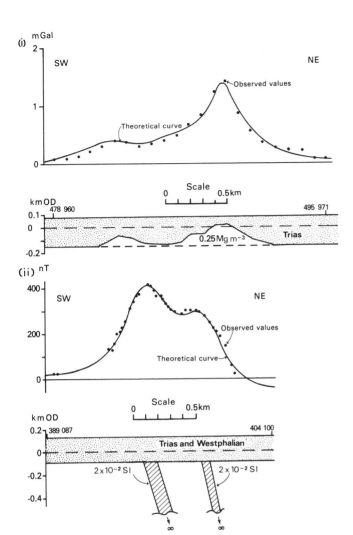

Figure 31 Detailed geophysical profiles

(i) Residual Bouguer anomaly profile (50 m station intervals) showing a topographical high in the sub-Triassic land surface due to a concealed extension of the Barrow Hill diorite.

(ii) Ground magnetic profile across concealed pre-Carboniferous magnetic bodies near Ibstock.

MAGNETIC SURVEYS

The district was included in an aeromagnetic survey flown for the Geological Survey in 1955. The east–west flight lines were one mile (1.61 km) apart and the north–south tie lines were spaced every 6 miles (9.65 km). North of about grid line 97, the magnetic data were recorded at a mean terrain clearance of 1000 feet (305 m) but south of this line, a constant barometric flight height of 1800 feet (549 m) was maintained. The results were published as a 1:250 000 scale map (Institute of Geological Sciences, 1980b).

The aeromagnetic map of the present district has a central zone of low gradients and negative values (Figure 32), flanked by two highly magnetic zones trending generally NW–SE. The limited areal extent and steep gradients of many of the anomalies indicate source rocks at, or near, the surface. The general trends of the magnetic anomalies broadly resemble those of the Bouguer anomalies but do not generally coincide with them.

The aeromagnetic flight records for the area include numerous anomalies of small amplitude, superimposed on the larger scale anomalies shown in Figure 32. These smaller anomalies usually indicate near-surface magnetic bodies but, in some cases, weak anomalies with amplitudes of 5 to 10 nanoteslas (nT) suggest deeper origins. The locations of the peaks of the smaller anomalies on individual flight lines are shown in Figure 32, and are joined if continuity is suspected.

Igneous rocks are responsible for some of the magnetic anomalies. For example, the elongated anomaly north-west of Leicester coincides with the outcrops of basic diorite at Groby and the diorite in Markfield Quarry. Samples from both sites (Table 10b) show fairly high susceptibilities. The aeromagnetic map suggests that the anomaly is due to a sheet-like body, with a NNW trend and a near-vertical dip. Detailed ground measurements, however, show that there are several separate anomalies hereabouts, some indicating unexposed magnetic bodies.

In the south-east (Figure 32) a circular anomaly marks the quartz-diorite quarried at Croft [512 962] and the large anomaly centred at Whetstone Pastures [564 940] is more tentatively correlated with the quartz-diorite proved in the Countesthorpe Borehole [567 955]. On aeromagnetic evidence, these two intrusions contrast with other exposures of diorite nearby, such as at Narborough [525 975] and Enderby [542 992], which appear to be non-magnetic or only weakly magnetic. The Enderby intrusion, which has a low magnetic susceptibility (Table 10b) seems to be associated with an anomaly seen on the aeromagnetic flight records but too small to produce a contour closure on Figure 32. The petrological classification of Le Bas (1972), and that described in this memoir, both group the magnetic Countesthorpe and non-magnetic Narborough intrusions as co-magmatic, so the difference in magnetic properties is puzzling. The contrast might be in the primary compositons of the intrusions; alternatively, the Narborough rock may have suffered a higher degree of alteration, resulting in a loss of magnetite similar to that found in the altered samples from Markfield Quarry (Table 10b).

The western end of the Croft intrusion is sharply truncated on the aeromagnetic map (Figure 32) but eastwards it could pass at depth into the larger igneous body beneath Countesthorpe. Although the area is built-up, which restricts the amount of magnetic surveying possible, ground magnetic readings taken around Croft show that the highest values occur near a small, disused quarry just north of the main working, and seem to suggest that the magnetite-bearing part of the intrusion has steep sides. A magnetometer traverse at Sapcote revealed an anomaly near disused quarries in quartz-diorite [492 941] (Table 10b), suggesting that it extends further beneath a thin Triassic cover.

A large magnetic high [575 150] marks the granodiorite outcrops at Mountsorrel. The centre of the anomaly is sharply peaked, and the broad zone of contours on the flanks indicates that the intrusion broadens downwards and extends down to a depth of several kilometres. The magnetisation of the Mountsorrel samples (Table 10b) is one of the highest for local rocks and is capable of producing anomalies of several

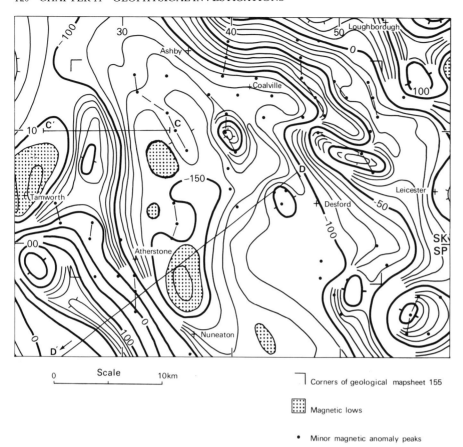

Figure 32 Aeromagnetic map with contours at 10nT intervals, and showing the locations of profiles C'C an D'D

0 ___ Scale ___ 10km

⌐ Corners of geological mapsheet 155

▨ Magnetic lows

• Minor magnetic anomaly peaks

⊢⊣ Profiles
D'   D

hundred nanotesla in amplitude. Davies and Matthews (1966) measured anomalies of up to 600 nT on the ground over the intrusion, and concluded that the magnetisation of the rocks was inhomogeneous, due probably to relatively basic phases within the intrusion. Alternatively, the variation may be due to patchy secondary alteration.

East of Ashby at the northern edge of Figure 32, is a large WNW-trending anomaly which could be another granodiorite intrusion, although the character of the anomaly is different from those described above. It occurs on the southern flank of the Widmerpool Gulf, and it is elongated parallel to the gulf axis. Indeed, the intrusion may have controlled the position of the gulf's southern margin.

The largest magnetic values in the district occur over the Warwickshire Coalfield and are truncated along the line of its Western Boundary Fault. Near the 200 nT closure in Figure 32 [211 997] there are only small outcrops of lamprophyre at Dosthill, and in-situ measurements show that these rocks have a low susceptibility, perhaps due to alteration effects. Therefore, it seems likely that the main magnetic anomaly is unrelated to the outcropping lamprophyres, and is due instead to a much larger diorite intrusion buried beneath the Warwickshire Coalfield. The main part of the anomaly indicates that the intrusion is asymmetrically shaped (Figure 33) with a steeper north-eastern slope and a shallower south-western slope extending at depth. The upper part of the body may be as shallow as 1 km below OD, but maximum depths of about 2.5 km

below OD are necessary if the magnetic material forms a sheet with a width less than its depth. The theoretical curve for such a sheet does not reproduce all the features of the observed curve and the sheet would need to dip steeply to the south-west. Uncertainties about the depth of the magnetic source and the thicknesses of Westphalian and older Carboniferous sediments beneath the coalfield persist, but it seems most likely that the magnetic body is a pre-Carboniferous, diorite intrusion.

The Warwickshire Coalfield anomaly includes on its flanks a few smaller anomalies which mark the outcrops of lamprophyre sills (Table 10b) in the Cambrian sediments. Compared with the main anomaly, they are small in amplitude, for example about 40 nT near [31 94], and are difficult to recognise unless the flight lines lie at right angles to the geological strike.

North of Polesworth, the Warwickshire Coalfield anomaly seems to extend in attenuated form as a north-trending high, with an amplitude of about 100 nT, beyond Newton Regis [27 08]. It coincides with the step-like Bouguer anomaly (Figure 28) along the Netherseal Fault. The origin of the magnetic anomaly is unknown but its continuity with the Warwickshire Coalfield anomaly suggests that it may be a part of it. The form of the anomaly is due to a slightly asymmetrical body, increasing in width down to an arbitary depth of 5 km below OD and with a susceptibility of about 8×10^{-3} SI units (Figure 30).

The central zone of low magnetic values in Figure 32 cor-

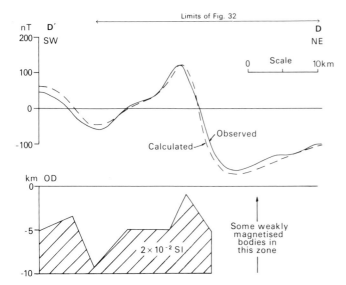

Figure 33 Aeromagnetic profile along D'D across the Warwickshire Coalfield

pre-Triassic basement rocks lie at depth here beneath the Needwood Basin, they also appear to be less magnetic than those elsewhere in the district.   JDC

SEISMIC SURVEYS

Davies and Matthews (1966) carried out shallow refraction surveys to locate igneous rocks and Lower Palaeozoic sediments beneath Triassic or drift cover, and the National Coal Board have employed high resolution, seismic reflection techniques to investigate the structure and extent of Westphalian rocks in the concealed parts of the coalfields (Jones, 1981).

A 24 km-long, seismic refraction profile (E'E Figure 28) was recorded jointly by BGS and the University of Leicester (cf. Whitcombe and Maguire, 1981a), to investigate the deeper structure of the district. Twenty-three seismic stations were set out along the profile, mainly to record blasts from two large quarries, at Mancetter [311 948] and Bardon [455 132]. The profile crossed the large Bouguer anomaly low over the Hinckley Basin. The study was intended to determine the depth and velocity of the basement beneath the Basin to supplement the interpretation of the gravity data.

The first arrival times from blasts in the two quarries were analysed using the 'plus-minus' method of Hagedoorn (1959), although 'two way' times were obtained only at stations 1 to 16 (Figure 34i); the equipment failed to record the smaller Mancetter blast at stations 17 to 23. The first three stations near the Mancetter quarry indicated a low apparent velocity of about 3.80 km/s but the corresponding value to station 20 from Bardon was 4.10 km/s. The 3.80 km/s velocity was recorded over the Cambrian inlier at Nuneaton, but the apparent velocity near Bardon may result from delays in arrival time due to the thicker Triassic and Westphalian sequences west of the Thringstone Fault. The 'minus' times (Hagedoorn, 1959) indicate a basement velocity of 5.76 km/s between stations 5 and 13 and a lower velocity (4.80 km/s) between stations 14 and 16. The change in velocity coincides with a decrease in the 'plus' times (Figure 34i), which indicates a major change in the basement surface between stations 12 and 14. The unreversed section of the profile between stations 17 and 20 indicates an apparent velocity of 5.1 km/s.

Whitcombe and Maguire (1981a, fig. 4b) present an interpretation based on the data recorded along the Mancetter–Bardon seismic line and also on quarry blasts at Croft (part of this line is shown in Figure 34ii). Using time-term analysis of the data, a more sophisticated approach than that described above, they postulate a thick sequence of Cambrian mudstones overlying Charnian rocks between stations 4 and 12 and interpret the 4.8 km/s layer as diorite intrusions. A second profile, ENE-trending from Bardon to Holwell [695 220] was interpreted to show, beneath cover, Charnian basement intruded by an igneous pluton between Woodhouse Eaves [530 165] and Barrow upon Soar [580 185]. It is thought to be an extension of the Mounsorrel Granodiorite.   JDC.

responds with the Hinckley Basin. Anomalies within it are generally small in amplitude, but they are sufficient to suggest that the basement rocks beneath the Basin are magnetic in places; they may be intrusions like those farther east. The anomalies also fix a lower limit to the base of the Triassic sediments, which are invariably non-magnetic. The weak, step-like feature, shown by northerly-trending contours around [40 00], seems to indicate the western boundary of a magnetic layer above 1 km below OD, although this anomaly merges with the magnetic low flanking the large magnetic high to the south-west. Northwards from about Nailstone [40 07], this boundary can be followed as a series of magnetic peaks (Figure 32), usually small in amplitude but including the large anomaly at Ibstock [398 098].

The source of the isolated Ibstock anomaly is concealed by Triassic, and possibly by Namurian, sediments. Detailed ground measurements reveal a double-peaked anomaly, 1.5 km long, of 400 nT amplitude, and elongated NW–SE. It is thought to be due to two magnetic horizons, dipping north-east at about 70° (Figure 31). Near Newbold Verdon [45 04], the aeromagnetic map (Figure 32) shows a small circular anomaly, indicated mainly by the closure of the −100 nT contour. On the aeromagnetic profile, two peaks, 1.3 km apart in an east–west direction, suggest comparison with the Ibstock anomaly.

Between Overseal and Shackerstone [30 16 to 36 08] is an elongated anomaly with a maximum amplitude of 20 nanotesla. Its source appears to lie less than 0.5 km below the surface, at or close to the base of the Westphalian.

Between Wellsborough [36 02] and Congerstone [36 05], the bend in the −150 nT contour indicates a local north-trending magnetic anomaly. Its source seems to lie at about 1 km below OD, beneath the deeper part of the Hinckley Basin (cf. Figure 34), and could be another igneous intrusion.

In the north-western part of Figure 32, an area of low magnetic values is devoid of minor anomalies. Although the

Figure 34 Seismic refraction
profile E′E

(i) Plus times for part of the
Bardon – Mancetter seismic
refraction line E′E (Figure 28)
(ii) Geological interpretation by Whitcombe
and Maguire (1981a).

Velocities in km/s

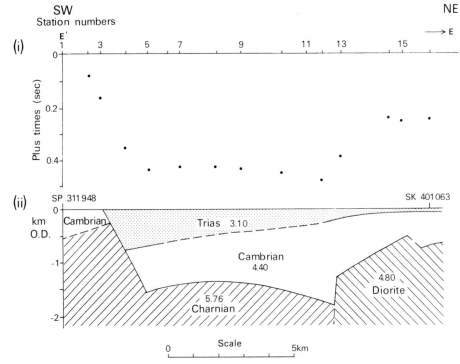

CHAPTER 12

Economic geology

COAL

The historical accounts of the three coalfields that follow rely on published annual lists and reports (Anon, 1882, 1883, 1891, 1948; Hunt, 1854–1881). Plans showing the extent of workings in each seam can be inspected at local British Coal offices. Maps showing how some seams vary within each coalfield have been published by the National Coal Board (1957, 1962).

Warwickshire Coalfield

In that part of the Warwickshire Coalfield which lies within the present district, eighteen coal seams have been worked, although many of these were exploited only in opencast sites working other seams.

Early working concentrated on the outcrops and gradually developed in depth. The parts of the exposed coalfield affected by the anticline near Polesworth (Figure 26), were not exploited on a significant scale by the early underground mines because the seams were too broken by faulting (Howell, 1859), although some of these areas have been recently worked opencast. The concealed part of the coalfield east of the Polesworth Fault was not entered until well into the twentieth century. Some of the seams are thick in places but vary rapidly along strike; splits and washouts are common. The combined Bench and Bench Thin seams, for example, reach a maximum thickness of 5.3 m at Orchard Colliery.

Mine abandonment plans and published lists of working collieries show that, up to the third quarter of the nineteenth century, exploitation was from small, shallow workings close to crop which, in total, extracted relatively little coal. By 1875, mining was concentrated at a few, larger and deeper collieries, which persisted into the second half of the twentieth century. Workings from these mines extended across the whole area and extracted much of the workable coal. After 1940, opencast operations were introduced to exploit seams close to crop, including some thinner seams which could not be profitably deep-mined.

Much of the early mining activity is not recorded. Baddesley Colliery appears to have operated the Speedwell shaft from at least 1854, and the owners of this mine had worked small drift collieries close by as far back as 1810. Farther north, there were shallow workings at this time, near Polesworth, served by several shafts.

Of the larger collieries, Alvecote worked between 1874 and 1951, and Pooley Hall from 1870 to 1966. In its last few years of operation, the latter combined with Alvecote and Amington collieries to form the North Warwick Colliery. Baddesley Colliery became established in about 1860 with the sinking to 328 m of the Stratford Shaft; this was subsequently deepened to 445 m and is still working. Birch Coppice, otherwise Hall End, Colliery was sunk in 1875 and

closed in 1986. It used subsidiary shafts at Wood End and Birchmoor (Cockspur Pit).

Shallow underground collieries include Machall No.1 and No.2 adit mines, east of Dordon, which worked the Seven Feet and Bench seams from 1920 to 1930. Nearby, Orchard Colliery worked the Double, Bench and Deep Rider seams from adits but these were superseded by opencast operations in 1941.

Thirteen seams have been deep-mined on a significant scale. They are all of high volatile, very weakly or non-caking type (British Coal Coal Rank Code 802 and 902).

South Derbyshire Coalfield

Twelve seams have been worked underground on a significant scale. The measures are more faulted than in the other coalfields (Figure 25) and worked areas are, consequently, less continuous.

Working in the coalfield possibly began in pre-historic times (Fox-Strangways, 1907) and continued intermittently on a small scale until the Industrial Revolution. Most of the early workings were near Measham, where the Main is at shallow depths beneath a large area.

In the late eighteenth century, mining became more intensive and by 1776, there were seven small pits in the Measham area. Transport of coal along the turnpike roads was difficult. Construction of the Ashby Canal and feeder tramroads in 1804 stimulated expansion earlier than in the Leicestershire Coalfield and led to widespread exploitation of the Main seam (Griffin, 1969). The sinking of Oakthorpe Colliery in 1787 and Donisthorpe Old Colliery (Brook Pit) in 1799 began this period of expansion. The Moira Colliery Company, for over a century the major operator around Moira, sank the Double Pits in 1802 to 41.5 m, and the Furnace Pit in 1812 to 184 m. The company sank Rawdon Colliery to 331 m in 1821, and Newfield Colliery (Hastings and Grey Shaft) to 379 m in 1832 (Fox-Strangways, 1907). The first published list of collieries (Hunt, 1854) suggests that those at Donisthorpe and Measham were idle at this time and that production seems to have been confined to the Moira pits. Oakthorpe Colliery came back into operation in 1856 and a mine at Measham appears in the lists from 1856 to 1861.

The 1870's were a further period of expansion; collieries were established at Netherseal and Donisthorpe, and the Moira Company opened Reservoir Colliery. In the next decade, this company established Marquis as a separate colliery. In 1878, a speculative shaft was put down at Snarestone into measures just below the Main seam. The venture was unprofitable as a colliery but the shaft was subsequently used to supply water to Snarestone Waterworks.

Working at Measham restarted near the turn of the century and a new shaft was sunk in the 1920's; the mine continued until 1986. During the present century, the mines were combined into larger, more viable collieries. By 1913, the working collieries were Rawdon, which now included the Marquis shaft, Netherseal, Donisthorpe, Measham and Reservoir. Of these only Rawdon is still in production.

The coals worked in this coalfield are mainly of high volatile, very weakly-caking type (Coal Rank Code 802), though there are some coals of non-caking type (Code 902) and a small amount of weakly-caking type (Code 702).

Leicestershire Coalfield

Most of the coalfield is overlain by Mercia Mudstone and Bromsgrove Sandstone, which give increasingly thick cover to the south. The deepest workings reach about 300 m below ground.

Early workings, from medieval times until the end of the eighteenth century, were largely confined to the northern part of the coalfield near Swannington and Coleorton, where coals crop out at the surface. By 1800, a mine was operating at Heather, but exploitation was generally restricted in the early nineteenth century by transport difficulties, and competition from other coalfields (Fox-Strangways, 1907; Griffin, 1969).

Workings beneath the Trias commenced with the opening of the Ibstock, Bagworth and Whitwick collieries between 1820 and 1826 and in 1832, Snibston Colliery was opened in conjunction with the new Leicester and Swannington Railway. The improved facilities offered by the railway, for the bulk transport of coal to the Leicester market, resulted in a significant increase in output from the existing collieries, and stimulated the deepening of Whitwick Colliery and the subsequent sinking of new shafts at Swannington and Coleorton in about 1845 (Griffin, 1969).

The first published list of coal mines shows workings at Bagworth, Coleorton, Ibstock, Snibston, Swannington and Whitwick (Hunt, 1854). Nailstone Colliery was opened in 1864. No workings were recorded at Heather until a new mine was opened in 1874, which operated until 1896.

The 1870's were a period of expansion, with the opening of the Ellistown (1874) and South Leicester (1877) collieries. About this time, a shaft was sunk at Lindridge in the south of the coalfield, but it was situated too close to the Thringstone Fault, where the measures are steeply dipping. The seams were proved in horizontal headings, but there were difficulties in pumping large quantities of mine water and the mine was a financial failure (Butterley and Mitchell, 1945). The southern area was not successfully exploited until Desford Colliery opened in 1902.

Prior to this, Swannington Colliery was extended southwards by the opening of Clink Shaft in about 1866 and Sinope Shaft in about 1891. Success was limited, since Sinope did not produce after 1902 and Clink closed even earlier. Between 1880 and 1933, workings from Coleorton Colliery extended southwards, between the Swannington Colliery take and the incrops of the lower seams, working the Lower Main, Nether Lount and Middle Lount seams. Most of the mines operating in the late nineteenth century continued in production until recently, except for Nailstone which closed in 1968, Ibstock which closed in 1929 and the Swannington collieries. The only new mines developed in

the present century are Desford Colliery (1902) and the Merry Lees Drift, which was driven in 1942 between Desford Colliery and the abortive Lindridge venture. This mine continued as a separate unit until 1968 when it was combined with Desford Colliery. In 1987 the only collieries still open were Bagworth and Ellistown.

In the faulted area around Packington, there was little working before 1940. Fishley Colliery worked the Middle and Nether Lount seams from a 20 m-deep shaft and was abandoned in 1945, while the Packington Coal and Clay Company had minor workings in the Yard Seam, which were abandoned in 1943. Otherwise, working has been confined to opencast pits exploiting the Lower Main, Yard, Nether Lount, Middle Lount and Upper Lount.

In evidence to the Public Inquiry into the proposed Vale of Belvoir Coalfield (day 10, p.72), the National Coal Board predicted that the entire Leicestershire Coalfield would be closed by 1990. The coals are all of high volatile, weakly- or non-caking type (Coal Rank Code 802 and 902), often with a high inherent moisture content.

IGNEOUS ROCK

The small outcrops of igneous rock in Warwickshire and Leicestershire provide some of the few sources of hard rock in the Midlands, suitable for aggregate and roadstone. They are of national importance since large areas to the south and east are devoid of comparable material. Consequently, they have been intensively quarried. In Warwickshire, basic minor intrusions occur in the Cambrian rocks near Atherstone. They are steeply dipping sills of lamprophyre, which yield high quality aggregate from large quarries at Griff and Mancetter. Further resources of comparable rock within the present district are small, though there are other occurrences around Nuneaton to the south.

Quarrying in Leicestershire for building and paving stone probably began before Roman times. Major developments did not occur, however, until the middle of the nineteenth century when the construction of railways and improved roads began to stimulate a demand for aggregates and to facilitate their transport.

Large-scale extraction began with the opening of quarries at Mountsorrel (1842), Markfield (1852), Bardon Hill (1857) and Croft (1868). Later, Enderby Quarry was opened in the 1870's to be followed by Cliffe Hill (1891), Charnwood (1892) and Whitwick (1893) (Anon, 1961, 1974). In 1895, when official production statistics were first recorded, output from the Leicestershire quarries had reached 860 000 tonnes. Records at this time show that Groby Quarry was also operational, together with seven smaller quarries at Sapcote and Stoney Stanton, two at Narborough, two at Huncote, and an additional quarry at Whitwick called Forest Rock. Originally, many of the quarries produced kerbs and setts but, more recently, this trade has been replaced by an increasing output of aggregates, which by the early 1970's exceeded 5 million tonnes a year.

The location of current quarries is given in Figure 35. The dacites and andesites worked at the Whitwick and Bardon Hill quarries, are commercially termed 'porphyry'. The other quarries work plutonic rock, sometimes referred to as

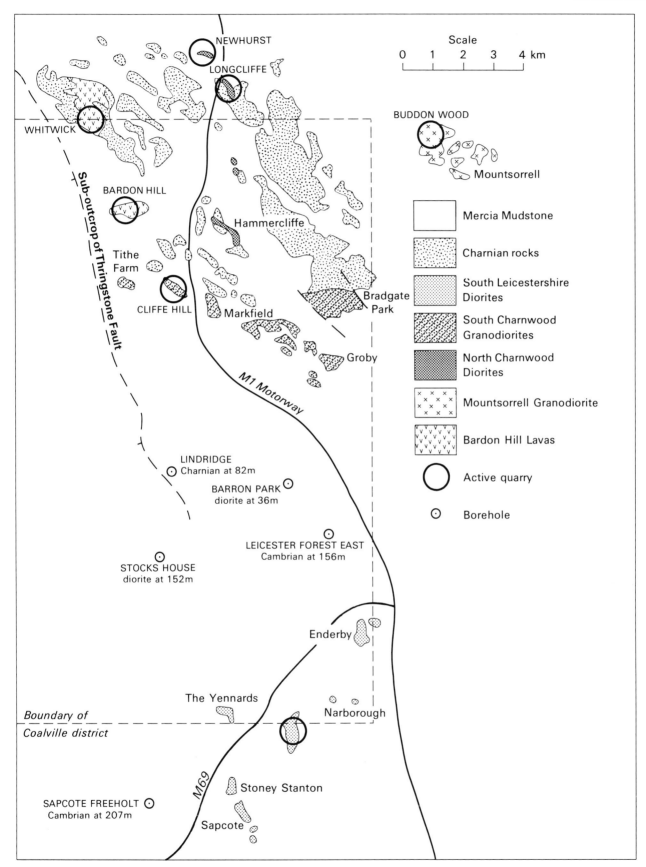

Figure 35 Distribution of igneous rocks and active quarries

Table 11   Physical properties of aggregates

	Buddon Wood	Charnwood	Cliffe Hill	Croft	Groby	Whitwick
Water absorption	0.75	0.75	0.4	0.95	0.4	0.19
Density	2.67	2.82	2.84	2.62	2.71	2.80
Strength (MN/m²)	—	217	—	194	302	243
Aggregate Crushing Value	18	17	11	20	15	16
10% fines value (KC)	200	260	370	200	330	280
Aggregate abrasion value	3.1	6.3	4.3	4.9	4.0	3.8
Aggregate impact value	18	15	13	19	14	14
Polished stone value	58	57	52–55	60	56	60

'Leicestershire granite or syenite', but in fact ranging in composition from granodiorite to diorite and quartz-diorite.

For completeness, the following list gives the rock type and operating company at working quarries. It includes the Buddon Wood and Charnwood quarries although they lie just outside the present district:

Quarry	Rock type	Operating Company
Bardon Hill	Andesite	Bardon Hill Quarries Ltd
Buddon Wood	Granodiorite	Redland Aggregates Ltd
Charnwood (Newhurst and Longcliffe)	Diorite and Charnian tuff	ARC
Cliffe Hill	Diorite	Tarmac Roadstone Holdings Ltd
Croft	Quartz-diorite	ECC Quarries Ltd
Groby (recently closed)	Diorite	ARC
Whitwick	Andesite	ARC

The physical properties of the aggregates, supplied by the operators of several quarries, are given in Table 11 and typical chemical analyses are provided in Table 7.

Although the rock in a quarry is commonly homogenous, its physical properties sometimes vary, so Table 11 is only a general guide to material properties rather than a precise specification. Nevertheless, the data indicate that the material is generally suitable for coarse aggregate in concrete, and for the main types of roadstone, although possibly not for the wearing course of very heavily used roads where, in some instances, the requirement for a minimum PSV (polished stone value) may not be attainable. In practice, the material is widely used for roadstone, and for railway ballast or fill. Smaller amounts are used locally for concrete aggregate, and minor quantities go for specific uses such as the surfacing of tennis courts or running tracks.

End-uses of igneous rock produced in Leicestershire and Derbyshire in 1985 were as follows:

	Thousand tonnes
Building stone	1
Coated roadstone	2923
Uncoated roadstone	2623
Railway ballast	1297
Concrete	428
Other construction purposes (e.g. fill)	1209

Since the Derbyshire output is restricted to one quarry, these totals may be reasonably taken to reflect the end-uses of Leicestershire rock.

In Leicestershire there is commonly a rapid increase in Mercia Mudstone cover at the edges of existing quarries. Typically, the old hills of resistant igneous rock have steep sides, against which the Triassic rocks accumulated until they buried the original topography. When a quarry, located in the top of one of these hills, expands laterally, it eventually reaches a zone where the overburden increases rapidly. This factor often controls the areal extent of the quarry, and has led to development downwards in many cases.

The recently closed quarry at Bradgate was effectively working three ridges of rock separated by areas of thick Mercia Mudstone overburden. Thick overburden affects one side of Cliffe Hill Quarry and another side is limited by hard Charnian tuffs. However, the Cliffe Hill diorite aggregates can admit up to 30 per cent of Charnian tuff without adverse effect.

Other quarries, which have developed in depth, include Groby 100 m deep, Croft 50 m deep, and Cliffe Hill, about 30 m deep. Enderby Quarry had also reached a considerable depth at the time of closure. In contrast, Bardon Hill and Whitwick quarries have been worked into hillsides on larger outcrops and do not yet need to work at lower levels. Although some companies have carried out programmes of shallow drilling on land adjacent to their quarries, accurate knowledge of the distribution of igneous rock beneath the cover of Mercia Mudstone is limited.

The rapid increase in demand for stone has led several operators to consider increasing the scope of their workings to obtain maximum extraction from a given site. Although the overburden may be thick, it can be profitably removed if the quarry is deep enough to obtain a sufficient volume of rock. Some operators regard the maximum workable ratio of overburden to rock in Leicestershire as about 1:2.

A limit to the ultimate depth of open-pit workings is imposed by considerations of rock mechanics, which require that the overall slope angle of a stable quarry face should be less than approximately 60° in this type of igneous rock. Much shallower angles are necessary in Mercia Mudstone.

The cost of pumping water from the local quarries is generally low, since the joints in the igneous rocks tend to be closed and little water enters the workings.

CLAYS

The major clay resources of the district occur within the Coal Measures of South Derbyshire, and the red mudstones of the Bromsgrove Sandstone and Mercia Mudstone sequences.

Within the South Derbyshire Coalfield, economically important fireclays occur within the Pottery Clays succession (p.40–43), where they are seat clays to the numerous thin coals in the 100 m or so of this sequence. The alumina content of the clays varies considerably, depending on the amount of silt impurity present. Some are extremely rich in kaolinite and therefore have high alumina contents. Mica is also present and contributes favourably to the vitrification properties of the clay. Analysis of the samples from the Hanginghill Farm Borehole (Worssam, 1977) showed that most contained over 25 per cent Al_2O_3 on a dry basis, with a maximum of 35.6 per cent Al_2O_3. The Pottery Clays represent one of the largest accumulations of high alumina clays in the United Kingdom.

The clays have been worked for centuries. In the early 19th century, they were used principally for pottery, and subsequently for sanitary pipes and refractory goods (Holmes, 1960). Except for small-scale workings at Moira [314 161] to produce clay for decorative and cooking stoneware, for studio pottery and for educational purposes, the pottery industry is no longer important. The major industries currently using the clays produce vitrified clay pipes and facing bricks, with a declining use in refractories. The clays and associated coals were worked together opencast, and large stockpiles of fireclay were produced which are still being exploited. The remaining reserves in the present district are limited and future prospects lie to the north within the Loughborough district.

The high-quality seat clay beneath the P31 coal, known as the Derby (or Main) Fireclay, which may contain up to 39 per cent alumina after calcination, historically formed the basis of the local refractory industry. Formerly, it was principally mined underground (Anon, 1920), and the last operating mine (Newfields No.5) [307 176] closed in the mid-1970s. The Deep Fireclay (P33) is another important refractory clay.

The Derby Fireclay and the underlying, and more siliceous, Derby Bottle Clay were formerly worked at crop for salt-glazed pipe manufacture. Prior to the Second World War, this area was the major centre of salt-glazed pipe production in the country. The use of seat clays for pipe production arose from the need for a fairly refractory clay, with a long vitrification range to withstand the considerable variations in temperature within the batch kilns then in use, and the high firing temperatures (1100° to 1200°C) which were required to volatilise the salt and allow it to react with the clay to produce the glaze. In order that the pipe did not shrink or distort at these high firing temperatures, a fairly refractory clay was required. The procedure was anomalous in that a refractory clay was used to produce a non-refractory product. The clays were also required to have sufficient green and dry strengths to retain their shape after extrusion, and to allow handling prior to firing. These properties are related to the high kaolinite content and fine particle size of the seat clays.

In the early 1960's, pipe manufacture changed and tunnel kilns replaced the old batch kilns and difficulties were then encountered in salt-glazing. The salt glaze was replaced by a ceramic glaze and finally the body of the pipe itself was made impermeable with the introduction of vitrified clay pipes. Better temperature control in tunnel kilns made possible the use of clays with shorter vitrification ranges. The need for a refractory clay was thus eliminated, and a much wider range of less refractory seat clays and mudstones is now used in the blended feed for pipe manufacture. Large, opencast mining techniques are employed to extract the coal and clay and, since clay extraction exceeds consumption, clay stockpiles are common. Clays are stored according to their stratigraphical horizon and quality, the higher alumina clays being used to make refractories, both locally and in other parts of the country. A layered stockpiling technique has been developed to aid blending of the clays and to ensure a uniformly consistent feed to the plant for pipe or refractory manufacture. Carbon and sulphur contents are monitored to keep them at a low level.

Seatearths elsewhere in the local Coal Measures have been worked in the past on a small scale. For example, the clay below the Yard Seam was worked at South Leicester Colliery (Mitchell and Stubblefield, 1948). However, they have not been exploited recently, and in view of the decline in demand for fireclays, both as refractories and for pipe-making, they seem unlikely to attract much future interest.

Mudstones in the Coal Measures have been worked for brickmaking in the past, principally at Polesworth and Dordon, but at present all local brickmaking uses Mercia Mudstone. This is exploited as a brickmaking clay at five sites within the district; they had a combined output of about 400 000 tonnes a year in 1986. Mercia Mudstone is worked principally for the manufacture of facing bricks at Ibstock [413 110 and 418 108], Desford [462 066] and Heather [386 094 and 395 100], but other products, such as land-drainage pipes, were formerly made, as at Ellistown [436 106]. Air bricks, chimney pots and roof ridges are made from clays within the Bromsgrove Sandstone at Measham [334 106].

Brick clays can have a wide range of composition. They must be sufficiently plastic, when wet, to be shaped and have sufficient green and dry strengths to retain their shape until the brick has been dried and fired. The clay should also vitrify sufficiently, at temperatures of 950° to 1100°C, to produce a brick with the desired crushing strength and porosity without excessive shrinkage or deformation. Colour and texture are also important and depend in part on the mineral composition of the clay. The clay minerals also confer the necessary plasticity, although other minerals of very fine particle size may also give this property. A high clay-mineral content can be a disadvantage, however, as it causes excessive shrinkage and distortion of the brick on drying and firing. An adequate quantity of non-plastic material in the clay is essential; this is usually quartz, but can also be dolomite in the Mercia Mudstone. The major fluxing agents which promote vitrification are soda and potash, contributed by micas and feldspar. The high proportion of reactive fluxes in the Mercia Mudstone allows the clay to vitrify over a short range of relatively low temperatures. Carbonate minerals present in finely divided form in the Mercia Mudstone are acceptable, as they react with the clay during firing. Coarser particles of carbonates, which remain as calcium or magnesium oxides after firing, can react with atmospheric moisture producing surface expansion and 'blowing' on the brick surface. A high carbonate content, mainly dolomite in the Mercia Mudstone, increases the refractory properties of the clay. Soluble salts such as gypsum and halite can cause efflorescence in bricks, but these minerals are not common enough in the local workings to cause problems.

Sandstone within the Mercia Mudstone is normally discarded during quarrying, but a proportion of sandstone may be included in the mix to reduce the plasticity of the clay and prevent excessive shrinkage on drying and firing. Chemical analyses of locally worked brick clays are shown in Table 12.

The presence of dolomite and calcite has a bleaching effect on the fired colour of a clay, and as the $MgO + CaO/Fe_2O_3$ ratio increases, the fired colour becomes progressively paler. The dolomite content of the Mercia Mudstone varies considerably and, by selectively quarrying specific horizons, a range of coloured bricks is produced. At Ibstock, three horizons are quarried to produce bricks with fired colours between pale buff and red. During the re-survey, the Mercia Mudstone from the three horizons worked at Ibstock was analysed by BGS, and dolomite contents ranging from 34 per cent in the uppermost horizon to 4 per cent in the lowest were identified; minor amounts of calcite are also present. In addition, the brick clays contain mica, with quartz, feldspar and chlorite. Compositionally, the brick clays vary, primarily in the relative proportions of carbonate minerals and quartz present. The fraction less than 2 microns consists of mica with subordinate quartz and chlorite. The limited plastic properties of the clays are attributed mainly to mica, and to a lesser extent, chlorite.

Table 12 Chemical analyses of Mercia Mudstone and Bromsgrove Sandstone clays

Quarry and references	Ibstock[1] [413 110, 418 108]	Ibstock[2] [413 110, 418 108]	Desford[3] [462 066]	Measham[4] [334 106]
SiO_2	36	54.6	47.1	66.5
Al_2O_3	11.00	12.1	11.3	11.5
Fe_2O_3	5.0	4.76	4.5	4.99
CaO	14.0	6.34	8.0	2.76
MgO	8.0	4.92	9.4	3.04
Na_2O	0.1	0.20	0.3	0.21
K_2O	5	5.04	4.4	4.22
TiO_2	0.5	0.64	0.6	0.58
Loss on ignition	20	10.9	12.6	5.89
Total %	99.60	99.50	98.2	99.69

Sources:

1 Upper light buff fired horizon. Ibstock Building Products Ltd
2 Middle buff fired horizon. Ibstock Building Products Ltd
3 Butterley Brick Ltd
4 Redbank Manufacturing Co. Ltd

Firing temperatures ranging between 980° and 1070°C are used for the local Mercia Mudstone clays, the higher firing temperatures being required for clays with high carbonate contents. The resultant bricks have crushing strengths ranging between 30 and 40 MN/m^2 and water absorptions of between 13.5 and 28 per cent, the high carbonate clays showing the highest porosity. Total linear shrinkage on drying and firing ranges between 1.5 and 5.5 per cent, the higher values coinciding with the more plastic clays.

Resources of brick clay within the local Mercia Mudstone and Bromsgrove Sandstone sequences are extremely large. On particular sites. the major geological constraints are the thickness of overburden and the amount of sandstone present. There is considerable variation in the dolomite content of the clay within a section, and within the Mercia Mudstone sequence as a whole.

SAND AND GRAVEL

The resources of the district consist of glacial sand and gravel deposits and Triassic conglomerates. The latter are at present unworked locally.

At the base of the Triassic sequence, the Polesworth Formation consists mainly of soft, red sandstone with only occasional pebbles; the conglomeratic beds are a minor part of the Formation and are of variable thickness and extent. The pebbles are mainly of white, vein quartz and brown quartzite, and the conglomerate can be processed to give a good-quality, coarse aggregate for concrete. However, in this area, borehole records suggest that actual pebble beds or conglomerates seldom constitute more than one third of the Polesworth Formation, and frequently fall well below that proportion (Table 13).

The nearest active working in these rocks is in the Weeford area, 8 km west of Tamworth, where the pebble beds (Cannock Chase Formation) yield an approximate 1:1 ratio of pebbles to sand over a considerable thickness. In contrast at a working at Acresford [29 13], near Netherseal, pebbles are sparse and the operator reported yields of 10 per cent gravel and 75 per cent asphalt sand, the remainder being rejected as slimes after washing. The poor gravel yield and the problems of slimes disposal have eventually rendered the operation uneconomic, and the pit has recently closed.

In a small, disused quarry [277 038] at Round Berry near Warton, a 3 m-thick pebble bed dipping at 20° was formerly worked. It contains approximately 50 per cent pebbles and is overlain by sandstone containing few pebbles.

Although glacial sand and gravel deposits are extensive (Figure 36) they are exploited only modestly, probably because of the proximity of more easily worked, river gravels in surrounding districts. Operators of the three active pits, at Heather, Cadeby and Huncote, report that the proportion of gravel to sand in their workings varies between 50 and 5 per cent, with average gravel yields of 15 to 25 per cent. About 20 to 25 per cent of the extracted material is unsaleable silt, which has to be removed by washing.

Extraction methods are conventional. Heather Pit, in an elevated position, is dry, whereas the others work below the water table and pump to keep the workings free of water. The

Table 13   Proportion of conglomerate in the Polesworth Formation, from selected borehole data

	Netherseal No.1	Netherseal No.2	Netherseal No.2 (Plant)	Netherseal No.4	Chilcote	Appleby No.1	Appleby No.2	Snarestone	Chilcote Trial	Chilcote P.S. No.1	Chilcote P.S. No.2
Grid reference	2873 1378	2853 1393	2678 1471	2757 1543	2828 1145	3205 1058	3176 1166	3389 1011	2844 1039	2841 1041	2840 1040
Total thickness of formation m	71	31	123	33	119	77	45	36	131	128	127
Total thickness of conglomerate beds m	7.6	6.1	8.8	14	30	0	17	2.6	13	20	20
Thickest conglomerate bed m	3.7	2.0	8.8	7.3	17	0	6.7	2.4	5.6	14	12

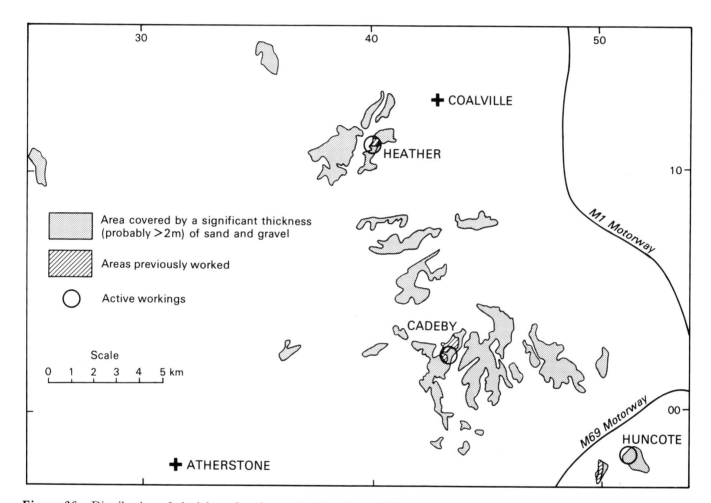

Figure 36   Distribution of glacial sand and gravel and active workings

gravel fractions consist mainly of hard 'Bunter' quartzite pebbles together with smaller quantities of porous sandstone. Boulders of igneous rock from the Croft outcrops nearby are found in the pit at Huncote, together with small quantities of coal which are removed by means of a jig.

At all the pits, the gravel is sold for coarse concrete aggregate and the washed sands for asphalt, building or concreting sand.

IRONSTONE

Ironstone (siderite) is present at several horizons in the Coal Measures and was of particular importance in the War-wickshire Coalfield, although the main centres of former exploitation lie beyond the present district. The aptly named Smithy Seam is said to have been worked mainly for ironstone at Amington Colliery in about 1880 and Birch Coppice Colliery may have once produced ironstone. In the South Derbyshire Coalfield, the Moira collieries operated an iron furnace in the last century at Furnace [314 152], for smelting ores from local coal mines. Iron mining was never more than a minor industry locally, and the deposits are unlikely to be important in the future.

SLATE

Slates from the Charnian Swithland Greywacke Formation, locally known as Swithland Slates, were used for roofing material in Roman Leicester. Production was first recorded officially in 1858 when Groby and Swithland Wood slate quarries produced 2000 and 1000 tons respectively. Systematic recording of annual production did not begin until 1881, when the quarries in Leicestershire appear to have been mainly idle. However, the two quarries were worked again between 1883 and 1887, when 2543 tons were produced, and between 1894 and 1906 when 12 678 tons were produced, with a maximum annual output of 4936 tons or 0.9 per cent of the national output in 1895. The Swithland Slate generally is purple-blue-grey with pale yellow and green, chlorite veins running through it. It does not cleave into slabs as thin as those of Welsh slate and consequently requires stronger timber roof supports. This factor is probably responsible for its previously limited exploitation, even in times of high slate demand.

NON-FERROUS METALS

No non-ferrous metals have been worked within the district, but some of the igneous rocks exhibit traces of mineralisation, and there are widespread signs of copper mineralisation associated with the sub-Triassic unconformity and occasional mineralisation in Triassic sandstones (King, 1968).

At Bardon Hill and Whitwick quarries, occasional specimens of gold have been found in small, quartz veins, and at Croft Quarry, small amounts of molybdenite are found in analcime-bearing veins.

Copper mineralisation at the sub-Triassic unconformity is usually malachite or other oxidised copper minerals, sometimes with chalcocite, cuprite and native copper. Occurrences of this type are exposed at Bradgate Quarry [514 088], Sheethedges Wood Quarry [525 084] and Bardon Hill Quarry [458 135]. Malachite-stained sandstone occurs in the Mercia Mudstone near Orton-on-the-Hill.

Galena occurs in small amounts in quartz veins cutting Namurian sandstones in the Rotherwood Borehole. It also occurs in the Coal Measures of the South Derbyshire Coalfield (Fox-Strangways, 1907, p.110) and was formerly worked in Dinantian limestones at Dimminsdale [377 215], in the Loughborough district (Fox-Strangways, 1905, p.17).

LIMESTONE

Thin seams of 'Spirorbis' limestone in the Coal Measures of the Warwickshire Coalfield are the only limestone outcrops within the present district. They have not been worked since official records began, although there are traces of shallow pits in places, which at one time presumably satisfied the local demand for agricultural lime. Dinantian limestones have been proved at depth in a number of borings.

The limestone sequence in the Rotherwood Borehole is about 71 m thick and its upper surface approaches within 75 m of the ground surface. It appears to offer the best prospect for limestone exploitation in the district, although it would have to be mined. The limestones are mainly dolomitic and impure with thin mudstone partings. There are traces of pyrite mineralization. Leaching tests show that the average content of insolubles, mostly clay and quartz, is 16.9 per cent with a standard deviation of 14.8. Strength measurements defined by 'aggregate impact values' indicate an average of 17.1, with a standard deviation of 3.5. Thus, the Rotherwood limestones seem unlikely to be suitable for chemical or metallurgical purposes because of their low carbonate content, but could produce adequate material for construction purposes.

WATER SUPPLY

Memoirs dealing with the groundwater supplies of Warwickshire, Derbyshire and Leicestershire (Richardson, 1928, 1929, 1931) contain valuable well and spring data, and discuss the properties of the principal aquifers.

The district can be divided into the catchments of the westerly-flowing River Anker and the north-easterly flowing River Soar, both tributaries of the River Trent. A review of the hydrogeology of the district was included in a study of the Trent Basin by Downing and others (1970).

Average annual rainfall is about 635 mm in the east, nearly 760 mm on the higher ground in the north-east and about 700 mm in the west. It is evenly distributed between the winter (October-March) and summer (April-September) months. Penman (1950) estimated actual evaporation at 460 mm/a. Since more than 80 per cent of the actual evaporation occurs in the summer, infiltration is mainly confined to the winter.

A number of surface reservoirs have been built around Charnwood Forest to supply water to Leicester. Thornton Reservoir [473 073] yields 2500 million litres a year whilst those at Cropston [557 149] and Swithland [550 111] together produce 7500 Ml/a. Two, further, impounding reservoirs in Charnwood, at Nanpantan [508 171] and Blackbrook [438 178], supply Loughborough.

The main aquifers of the district are the Triassic and Carboniferous sandstones. Small, but locally useful, supplies are obtained from the Mercia Mudstone and from glacial sands and gravels.

The hard, Precambrian rocks generally yield only small supplies of groundwater, which are restricted to joints and fissures. Small supplies can be obtained from wells which intersect these fractures. In Charnwood Forest, a 14 m bore [510 129] of 18 cm diameter produced 1.6 litres per minute (l/m).

Good yields are obtained from Carboniferous rocks at three localities. At Heather Pumping Station (P.S.) [394 107], a yield of 1.1 million litres per day (Ml/d) was obtained from the Wingfield Flags in the Lower Coal Measures. However, the impermeable cover of Mercia Mudstone and Coal Measures restricts recharge and thus seriously limits

Table 14 Representative physical properties of the Triassic sandstone aquifers

Well	Grid reference	Sample depth (m)	Permeability (millidarcys)	Porosity (%)	Aquifer
The Altons Bh. No.1	395 152	24.1	2165–4616	28.9	Bromsgrove Sandstone
Austrey Ho. No.1 Bh.	303 049	15.3–29.6	1–20.9	13.5–18.4	Bromsgrove Sandstone
		35.7–78.1	654–5952	27.1–32.0	
		88.7–135.9	3356–16278	26.6–34.0	
Newton Regis Bh.	282 073	22.8–24.4	2183–4746	25.1–31.6	Bromsgrove Sandstone
		75.8–87.9	422–3560	28.5–30.0	
		100–131	3594–18934	26.0–31.4	Polesworth Formation
		135.6–163.4	3124–15385	25.6–30.2	
Snarestone Lodge Bh.	343 101	59.3–62.8	78–616	13.9–20.8	Polesworth Formation or Moira Breccia
Appleby Parva Bh.	307 086	146.3–168.9	825–17283	22.4–30.2	Polesworth Formation

the extent to which these sandstones can be developed for water supply in the longer term.

At Holly Hayes P.S. [441 153], a yield of 2.1 Ml/d was obtained from three boreholes, each 100 m deep, drilled through Mercia Mudstone into Carboniferous beds. Most of the water probably comes from the latter rocks.

Sandstones in the Keele and Halesowen formations crop out in the Warwickshire Coalfield. At Bramcote P.S. [271 039], a 37 m borehole of 12.8 cm diameter produced 1.7 Ml/d from the Halesowen Formation. The water from these Carboniferous sandstones is frequently hard, with a high sulphate content.

Carboniferous Limestone underlies some of the northern part of the district and has been developed at outcrop in adjacent areas. However, under confined conditions, the water quality is generally too poor for water supply.

The Bromsgrove Sandstone, Polesworth Formation and Moira Breccia generally act as one aquifer, although, in places, interbedded marls may locally cause them to behave as multiaquifers. The aquifer as a whole is fairly porous, with values ranging from 14 to 32 per cent (Lovelock, 1972) and it has a high value for specific yield; in consequence, seasonal variations in water levels are small. Intergranular permeability is high; laboratory results are generally in the range 100 to 10 000 millidarcys (Table 14) although fissure flow almost certainly predominates in the field.

Where overlain by the Mercia Mudstone, these sandstones behave as a confined aquifer. Surface run-off is generally negligible, so that recharge at outcrop approximates to the difference between precipitation and actual evaporation, in practice about 200 mm/a. The aquifer crops out over an area of 24 km² in the north-west of the district, so that recharge can be roughly estimated to be about 13 Ml/d.

Groundwater at outcrop is of the calcium bicarbonate type with a chloride content of about 30 mg/l. Downdip, beneath the Mercia Mudstone, the quality deteriorates as both calcium and sulphate contents increase due to water/rock chemical interaction. A representative chemical analysis of

the water from the aquifer, determined by the Severn-Trent Water Authority, is given in Table 15.

At Warton P.S. [280 026], the sulphate and chloride concentrations are high, 770 mg/l and 116 mg/l respectively. This may be due to overpumping bringing up water from the underlying Carboniferous, although the borehole itself does not penetrate Carboniferous beds.

Important wells with their yields are listed in Table 16. Large quantities of water are also pumped from mine workings; these originate mainly from the overlying Bromsgrove Sandstone and Moira Breccia (Table 17). The mine-water is discharged into surface streams.

The total volume of water abstracted for public supply amounts to more than 9.6 Ml/d. This, together with the water abstracted from the mine workings, indicates that the potential for further abstraction from this aquifer is limited. As the total volume of water abstracted amounts to more than 17.5 Ml/d, and the recharge is only 13 Ml/d, some of the water must come from the underlying Carboniferous rocks.

Table 15 Analysis of water at Snarestone Pumping Station [339 101]

Aquifer	Bromsgrove Sandstone, Polesworth Formation and Moira Breccia
	mg/l
Calcium	107.8
Magnesium	32.4
Sodium	16.1
Bicarbonate	316.6
Chloride	35.1
Sulphate	139.7
Nitrate	3.7
Electrical conductivity (microsiemans/cm at 25°C)	780

The Mercia Mudstone is generally impermeable and is a poor aquifer. It restricts infiltration to the underlying Triassic sandstones and its thin bands of gypsum can reduce the quality of the water. However, where sandstones are present within the Mercia Mudstone, moderate supplies can be obtained, as at Upton, where a 33 m bore [363 995] of 15 cm diameter produces 41 l/min for a drawdown of 2.4 m.

In places, small domestic supplies of groundwater are obtained from glacial sands and gravels. At Sutton Cheney, for example, a 15 m bore [419 007] of 15 cm diameter yielded 41 l/min for a drawdown of 6 m.

Table 16 Yields from the main public water supply boreholes

Well	National Grid reference	Depth (m)	Yield (Ml/d)	Aquifer
Warton P.S.	280 026	(1) 92.7 (2) 91.4	2.4 Ml/d from 2 boreholes	Bromsgrove Sandstone Polesworth Formation Moira Breccia
Acresford P.S.	301 128	(1) 91.4 (2) 85.3	2.7 Ml/d from 2 boreholes	Bromsgrove Sandstone Polesworth Formation Moira Breccia
Broomleys P.S.	443 135	(1) 77.7 (2) 60.4	1.7 Ml/d from 2 boreholes drilled in 24.7 m shaft	Bromsgrove Sandstone Moira Breccia
Snarestone P.S.	339 101	(1) 264 (2) 122 (3) 122	2.6 Ml/d from 3 boreholes	Bromsgrove Sandstone Polesworth Formation Moira Breccia
Chilcote P.S.	284 104	(1) 246 (2) 212 (3) 212	These 3 boreholes together with 3 boreholes in the Lichfield district are licensed to abstract 8.1 Ml/d	Bromsgrove Sandstone Polesworth Formation Moira Breccia
Swepstone Road P.S.	382 109	79.6	0.218 Ml/d	Mercia Mudstone and Bromsgrove Sandstone (Lower Coal Measures)

Table 17 Volume of water pumped from mine workings

Name of works and abstraction shaft	National Grid reference	Strata penetrated	Estimated quantity of mine-water abstracted (Ml/d)
Bagworth No.1	444 087	Mercia Mudstone, Moira Breccia Middle Coal Measures	0.63
Desford No.2	460 069	Mercia Mudstone, Bromsgrove Sandstone, Moira Breccia, Middle and Lower Coal Measures	0.52
Merry Lees Sump	462 056	No record – but probably similar to Desford No.2	0.57
Donisthorpe No.1	313 145	Middle Coal Measures	0.70
Ellistown No.1 No.2	438 103 439 104	Mercia Mudstone, Moira Breccia, Middle and Lower Coal Measures	1.2
Minorca Shaft	355 114	Moira Breccia, Middle Coal Measures	1.1
South Leicester No.1	431 118	Mercia Mudstone, Middle Coal Measures (Whitwick dolerite)	0.52
Snibston No.2	419 145	Mercia Mudstone, Middle and Lower Coal Measures	0.87
Whitwick No.3 Whitwick No.6 Whitwick Pumping Shaft	430 147 431 145 429 148	Mercia Mudstone Whitwick Dolerite Middle Coal Measures	1.8
Total			7.91

REFERENCES

AITKENHEAD, N. 1977. The Institute of Geological Sciences Borehole at Duffield, Derbyshire. *Bull. Geol. Surv. G.B.* No. 59, 1–38

ALLEN, J. R. L. 1957. The Precambrian geology of Caldecote and Hartshill, Warwickshire. *Trans. Leicester Lit. Philos. Soc.*, Vol. 51, 16–31.

— 1965. Fining upwards cycles in alluvial successions. *Geol. J.*, Vol. 4, 229–246.

— 1968. Precambrian rocks. C. The Nuneaton district. 15–18 in *The geology of the East Midlands.* SYLVESTER-BRADLEY, P. C. and FORD, T. D. (editors). 400pp. (Leicester: Leicester University Press.)

— and RUSHTON, A. W. A. 1968. The Cambrian and Ordovician systems, in *The geology of the East Midlands.* SYLVESTER-BRADLEY, P. C. and FORD, T. D. (editors). 400pp. (Leicester: Leicester University Press.)

ALLPORT, S. 1879. On the diorites of the Warwickshire Coalfield. *Q. J. Geol. Soc. London*, Vol. 35, 637–642.

ALLSOP, J. M. and ARTHUR, M. J. 1983. A possible extension of the South Leicestershire Diorite complex. *Rep. Inst. Geol. Sci.*, No.83/10, 25–30.

ANON. 1882. *Reports of the Inspectors of Mines to the Secretary of State for the year 1881.* 679pp. (London: Her Majesty's Stationery Office.)

— Annually from 1883. *Lists of mines . . .* (London: Her Majesty's Stationery Office.)

— Annually from 1891. *List of the plans of abandoned mines deposited in the Home Office . . .* (London: Her Majesty's Stationery Office)

— 1920. Refractory materials: fireclays. Resources and geology. *Mem. Geol. Surv. Spec. Rep. Miner. Resour. G.B.*, Vol. 14, 243pp.

— Annually from 1948. *Guide to the coalfields.* (Redhill, Surrey: Colliery Guardian.)

— 1961. Croft granite. *Mine Quarry Eng.*, 1961, Vol. 27, No. 10, 436–447.

— 1974. Buddon Wood Quarry. A major new development in Leicestershire by Redland Roadstone. *Quarry Manage. Prod.*, 1974, Vol. 1, No. 5, 161–176.

ARTHURTON, R. S. 1980. Rhythmic sedimentary sequences in the Triassic Keuper Marl (Mercia Mudstone Group) of Cheshire, Northwest England. *Geol. J.* Vol. 15, 43–58.

AUDLEY-CHARLES, M. G. 1970. Triassic palaeogeography of the British Isles. *Q. J. Geol. Soc. London*, Vol. 126, 49–89.

BARROW, G., GIBSON, W., CANTRILL, T. C., DIXON, E. E. L., and CUNNINGTON, C. H. 1919. The geology of the country around Lichfield. *Mem. Geol. Surv. G.B.*

BENNETT, F. W., LOWE, E. E., GREGORY, H. H. and JONES, F. 1928. The geology of Charnwood Forest. *Proc. Geol. Assoc.*, Vol. 39, 241–298.

BERRY, E. E. 1882. Analyses of five rocks from the Charnwood-Forest district. Communicated with notes by Prof. T. G. Bonney. *Q. J. Geol. Soc. London*, Vol. 38, 197–199.

BISHOP, W. W. 1958. The Pleistocene geology and geomorphology of three gaps in the Midland Jurassic escarpment. *Philos. Trans. R. Soc. B.*, Vol. 241, 255–306.

BONNEY, T. G. 1895. Supplementary note on the Narborough district (Leicestershire). *Q. J. Geol. Soc. London*, Vol. 51, 24–34.

BOSWORTH, T. O. 1912. *The Keuper Marls around Charnwood.* 129pp. (Leicester Lit. and Philos. Soc.)

BOULTER, C. A. and YATES, M. G. 1987. Confirmation of the pre-cleavage emplacement of both northern and southern diorites into the Charnian Supergroup. *Mercian Geol.*, Vol. 10, 281–286.

BOULTON, W. S. 1934. The sequence and structure of the southeast portion of the Leicestershire Coalfield. *Geol. Mag.*, Vol. 71, 323–329.

BOYNTON, H. E. 1978. Fossils of the Pre-Cambrian of Charnwood Forest, Leicestershire. *Mercian Geol.*, Vol. 6, 291–296.

— and FORD, T. D. 1979. *Pseudovendia charnwoodensis*—A new Precambrian arthropod from Charnwood Forest, Leicestershire. *Mercian Geol.*, Vol. 7, 175–178.

BRIDGE, J. S. and LEEDER, M. R. 1979. A simulation model of alluvial stratigraphy. *Sedimentology*, Vol. 26, 617–644.

BRIDGER, J. F. D. 1968. Remarkable features in the Ulverscroft valley; a re-appraisal of the drainage history. *Trans. Leicester Lit. and Philos. Soc.*, Vol. 62, 73–77.

— 1975. The Pleistocene succession in the southern part of Charnwood Forest, Leicestershire. *Mercian Geol.*, Vol. 5, 189–203.

— 1981. The problem of discordant drainage in Charnwood Forest, Leicestershire. *Mercian Geol.*, Vol. 8, 217–223.

— and RICE, R. J. 1969. Disturbed Keuper Marl beneath the river gravels of the Soar basin. *Geol. Mag.*, Vol. 106, 554–561.

BROWN, G. M. 1957. Pyroxenes from the early and middle stages of fractionation of the Skaergaard intrusion, East Greenland. *Mineral Mag.*, Vol. 31, 511–543.

BROWN, H. T. 1889. The Permian rocks of the Leicestershire coalfield. *Q. J. Geol. Soc. London*, Vol. 45, 1–40.

BULMAN, O. M. B. and RUSHTON, A. W. A. 1973. Tremadoc faunas from boreholes in central England. *Bull. Geol. Surv. G.B.*, No. 43, 1–40.

BURGESS, I. C. and HOLLIDAY, D. W. 1974. The Permo-Triassic rocks of the Hilton Borehole, Westmoreland. *Bull. Geol. Surv. G.B.*, No. 46, 1–34.

BUTTERLEY, A. D. and MITCHELL, G. H. 1945. Driving of two drifts by the Desford Coal Co. Ltd., at Merry Lees, Leicestershire. *Trans. Inst. Min. Eng.*, Vol. 104, 703–713.

CALVER, M. A. 1968. Distribution of Westphalian marine faunas in northern England and adjoining areas. *Proc. Yorkshire Geol. Soc.*, Vol. 37, 1–72.

— 1969. Westphalian of Britain. *C. R. 6é Congr. Int. Strat. Géol. Carbonif.*, Sheffield 1967, Vol. 1, 236–264.

CATT, J. A. 1981. British pre-Devensian glaciations. 32–42 in *The Quaternary in Britain.* NEALE, J. and FLENLEY, J. (editors). *(Oxford: Pergamon Press.)*

CHRISTIANSEN, R. L. and LIPMAN, P. W. 1966. Emplacement and thermal history of a rhyolite lava flow near Fortymile Canyon, Southern Nevada. *Bull. Geol. Soc. Am.*, Vol.77, 671–684.

COOK, A. H., HOSPERS, J. and PARASNIS, D. S. 1951. The results of a gravity survey in the country between the Clee Hills and Nuneaton. *Q. J. Geol. Soc. London*, Vol. 107, 287–302.

COPE, J. C. W. 1977. An Ediacara-type fauna from South Wales. *Nature, London*, Vol. 268, No. 5621, 264.

COPE, K. G. and JONES, A. R. L. 1970. The Warwickshire Thick Coal and its mining development. *C. R. 6é Congr. Int. Strat. Géol. Carbonif. Sheffield 1967*, Vol. II. 585–598.

COWIE, J. W. and GLAESSNER, M. F. 1975. The Precambrian–Cambrian boundary: a symposium. *Earth Sci. Rev.*, Vol. 11, 209–251.

— RUSHTON, A. W. A. and STUBBLEFIELD, C. J. 1972. A correlation of Cambrian rocks in the British Isles. *Spec. Rep. Geol. Soc. London*, No. 2, 42pp.

CRIBB, S. J. 1975. Rubidium-strontium ages and strontium isotope ratios from the igneous rocks of Leicestershire. *J. Geol. Soc. London*, Vol. 131, 203–312.

DAVIES, D. and MATTHEWS, D. H. 1966. Geophysical investigations on the boundaries of the Mountsorrel Granite. *Geol. Mag.*, Vol. 103, 534–547.

DEELEY, R. M. 1886. The Pleistocene succession in the Trent Basin. *Q. J. Geol. Soc. London*, Vol. 42, 437–480.

DEER, W. A., HOWIE, R. A. and ZUSSMAN, J. 1962. *Rock-forming minerals*. (London: Longmans.)

DOUGLAS, T. D. 1974. The Pleistocene beds exposed at Cadeby, Leicestershire. *Trans. Leicester Lit. and Philos. Soc.*, Vol.68. 57–63.

— 1980. The Quaternary deposits of Western Leicestershire. *Philos. Trans. R. Soc. B.*, Vol. 288, 260–286.

DOWNING, R. A., LAND, D. H., ALLENDER, R., LOVELOCK, P. E. R. and BRIDGE, L. R. 1970. Hydrogeology of the Trent River Basin. Water Supply Papers. Inst. Geol. Sci., Hydrogeological Report No. 5.

DUFF, B. A. 1980. Palaeomagnetism of Late Precambrian or Cambrian diorites from Leicestershire, U.K. *Geol. Mag.*, Vol. 117, 479–483.

EASTWOOD, T., GIBSON, W., CANTRILL, T. C. and WHITEHEAD, T. H. 1923. The geology of the country around Coventry. *Mem. Geol. Surv. G.B.*

— WHITEHEAD, T. M. and ROBERTSON, T. 1925. The geology of the country around Birmingham. *Mem. Geol. Surv. G.B.*

EDEN, R. A. STEVENSON, I. P. and EDWARDS, W. 1957. Geology of the country around Sheffield. *Mem. Geol. Surv. G.B.*

— ELLIOTT, R. W., ELLIOTT, R. E. and YOUNG, B. R. 1963. Tonstein bands in the coalfields of the East Midlands. *Geol. Mag.*, Vol. 100, 47–58.

ELLIOTT, R. E. 1961. The stratigraphy of the Keuper Series in southern Nottinghamshire. *Proc. Yorkshire Geol. Soc.* Vol. 33, 197–234.

EVANS, A. M. 1963. Conical folding and oblique structures in Charnwood Forest, Leicestershire. *Proc. Yorkshire Geol. Soc.*, Vol. 34, 67–79.

— 1968. Precambrian rocks. A. Charnwood Forest. 1–12 in *The geology of the East Midlands*. SYLVESTER-BRADLEY, P. C. and FORD, T. D. (editors). 400pp. (Leicester: Leicester University Press.)

— 1979. The East Midlands aulacogen of Caledonian age. *Mercian Geol.*, Vol. 7, 31–42.

— and KING, R. J. 1962. Palygorskite in Leicestershire. *Nature, London*, Vol. 194, 860.

— and MAROOF, S. I. 1976. Basement controls on mineralisation in the British Isles. *Min. Mag.*, Vol. 134, 401–411.

FALCON, N. L. and KENT, P. E. 1960. Geological results of petroleum exploration in Britain 1945–1957. *Mem. Geol. Soc. London*, No. 2, 56pp.

FITCH, F. J., MILLER, J. A. EVANS, A. L. GRASTY, R. L. and MENEISY, M. Y. 1969. Isotopic age determinations on rocks from Wales and the Welsh Borders. 23–46 in *The Pre Cambrian and Lower Palaeozoic rocks of Wales*. WOOD, A. (editor). 461pp. (Cardiff: University of Wales Press.)

FORD, T. D. 1958. Pre-Cambrian fossils from Charnwood Forest. *Proc. Yorkshire Geol. Soc.*, Vol. 31, 211–217.

— 1968. Precambrian rocks. B. The Precambrian palaeontology of Charnwood Forest. 12–14 in *The geology of the East Midlands*. SYLVESTER-BRADLEY, P. C. and FORD, T. D. (editors). 400pp. (Leicester: Leicester University Press.)

— 1980. The Ediacaran fossils of Charnwood Forest, Leicestershire. *Proc. Geol. Assoc.*, Vol. 91, 81–83.

— and KING, R. J. 1968. Outliers of possible Tertiary age. B. Tertiary (?) deep weathering in Leicestershire. 329–339 in *The geology of the East Midlands*. SYLVESTER-BRADLEY, P. C. and FORD, T. D. (editors). 400pp. (Leicester: Leicester University Press.)

FOWLER A. and ROBBIE, J. A. 1961. The geology of the country around Dungannon. *Mem. Geol. Surv. N. Ireland.* 274pp.

FOX-STRANGWAYS, C. 1900. The geology of the country between Atherstone and Charnwood Forest. *Mem. Geol. Surv. G.B.*

— 1903. The geology of the country near Leicester. *Mem. Geol. Surv. G.B.*

— 1905. The geology of the country between Derby, Burton-on-Trent, Ashby-de-la-Zouch and Loughborough. *Mem. Geol. Surv. G.B.*

— 1907. The geology of the Leicestershire and South Derbyshire Coalfield. *Mem. Geol. Surv. G.B.*

FRANCIS, E. H. 1969. Les tonstein du Royaume-Uni. *Ann. Soc. Geol. Nord.* Vol. 99, 209–214.

FROST, D. V. and SMART, J. G. 1979. The geology of the country north of Derby. *Mem. Geol. Surv. G.B.*

GEORGE, T. N., JOHNSON, G. A. L., MITCHELL, M., PRENTICE, J. E., RAMSBOTTOM, W. H. C., SEVASTOPULO, G. D. and WILSON, R. B. 1976. A correlation of Dinantian rocks in the British Isles. *Spec. Rep. Geol. Soc. London*, No. 7, 87pp.

GIBSON, W. 1905. In *Summ. Prog. Geol. Surv.* for 1904, 151.

GREIG, D. C. ánd MITCHELL, G. H. 1955. The western extension of the Leicestershire and South Derbyshire Coalfield. *Bull. Geol. Surv. G.B.*, No. 7, 38–67.

— WRIGHT, J. E., HAINS, B. A. and MITCHELL, G. H. 1968. Geology of the country around Church Stretton, Craven Arms and Brown Clee. *Mem. Geol. Surv. G.B.*

GRIFFIN, C. P. 1969. The economic and social development of the Leicestershire and South Derbyshire Coalfield. Unpublished Ph.D. thesis. University of Nottingham.

HAGEDOORN, J. G. 1959. The plus-minus method of interpreting seismic refraction sections. *Geophys. Prospecting*, Vol. 7, 158–182.

HAINS, B. A. and HORTON, A. 1969. *British regional geology: central England.* (London: HMSO for Institute of Geological Sciences.)

HARRIS, A. L., SHACKLETON, R. M., WATSON, J., DOWNIE, C., HARLAND, W. B. and MOORBATH, S. 1975. A correlation of the Precambrian rocks in the British Isles. *Spec. Rep. Geol. Soc. London*, No. 6, 136pp.

HARRISON, W. J. 1884. The syenites of South Leicestershire. *Midland Nat.* Vol. 7, 7–11, 41–42.

— 1898. The ancient glaciers of the Midland Counties of England. *Proc. Geol. Assoc.*, Vol. 15, 400–408.

HATCH, F. H. 1909. *Text-book of petrology.* 219pp. (London: Swan Sonnenshein.)

HILL, E. and BONNEY, T. G. 1878. The Precarboniferous rocks of Charnwood Forest, Part II. *Q. J. Geol. Soc. London*, Vol. 34, 199–239.

HOLLAND, C. H., AUDLEY-CHARLES, M. G., BASSETT, M. G., COWIE, J. W., CURRY, D., FITCH, F. J., HANCOCK, J. M., HOUSE, M. R., INGHAM, J. K., KENT, P. E., MORTON, N., RAMSBOTTOM, W. H. C., RAWSON, P. F., SMITH, D. B., STUBBLEFIELD, C. J., TORRENS, H. S., WALLACE, P. and WOODLAND, A. W. 1978. A guide to stratigraphical procedure. *Spec. Rep. Geol. Soc. London*, No. 11.

HOLMES, W. D. 1960. The Leicestershire and South Derbyshire Coalfield (2). The clay industry. *E. Midland Geogr.*, Vol. 2, 9–13.

HORTON, A. 1963. In *Summ. Prog. Geol. Surv.* for 1962, 39.

— and HAINS, B. A. 1972. Development of porcellanous rocks and reddening of the Coal Measures in the South Derbyshire, Leicestershire and Warwickshire Coalfields. *Bull. Geol. Surv. G.B.*, No. 42, 51–77.

HOWELL, H. H. 1859. The geology of the Warwick Coalfield and the Permian rocks and Trias of the surrounding district. *Mem. Geol. Surv. G.B.*

HULL, E. 1860. The geology of the Leicestershire Coalfield and of the country around Ashby-de-la-Zouch. *Mem. Geol. Surv. G.B.*

— 1869. The Triassic and Permian rocks of the Midlands Counties of England. *Mem. Geol. Surv. G.B.*

HUNT, R. Annually 1854–1881. Mineral statistics of the United Kingdom of Great Britain and Ireland. *Mem. Geol. Surv. G.B.*

INSTITUTE OF GEOLOGICAL SCIENCES. 1976. IGS boreholes 1975. *Rep. Inst. Geol. Sci.* No. 76/10, 47pp.

— 1980a. IGS boreholes 1978. *Rep. Inst. Geol. Sci.* No. 79/12, 18pp.

— 1980b. 1:250 000 Series. Aeromagnetic anomaly map. (Provisional Edition) East Midlands Sheet. 52°N–2°W. (London: Institute of Geological Sciences.)

— 1982. 1:250 000 Series. Bouguer gravity map. (Provisional Edition) East Midlands Sheet. 52°N–2°W. (London: Institute of Geological Sciences.)

JENKINS, R. J. F. and GEHLING, J. G. 1978. A review of the frond-like fossils of the Ediacara assemblage. *Rec. S. Australia Mus.* Vol. 17, 347–359.

JONES, F. 1926. The petrology and structure of the Charnian rocks of Bardon Hill. *Geol. Mag.*, Vol. 63, 241–255.

— 1927. A structural study of the Charnian rocks and of the igneous intrusions associated with them. *Trans. Leicester Lit. and Philos. Soc.*, Vol. 28, 24–41.

JONES, P. A. 1981. National Coal Board exploration in Leicestershire. *Trans. Leicester Lit. and Philos. Soc.*, Vol. 75, 34–40.

JUKES, J. B. 1842. The geology of Charnwood Forest. Appendix pp.1–31 in *The history and antiquities of Charnwood Forest.* POTTER, T. R. 272pp. (London: Hamilton, Adams.)

KENT, P. E. 1966. The structure of the concealed Carboniferous rocks of North-eastern England. *Proc. Yorkshire Geol. Soc.*, Vol. 35, 323–352.

— 1968a. The buried floor of Eastern England. 138–148 in *The geology of the East Midlands.* SYLVESTER-BRADLEY, P. C. and FORD, T. D. (editors). 400pp. (Leicester: Leicester University Press.)

— 1968b. The Lower Carboniferous at Grace Dieu. 80–81 in *The geology of the East Midlands.* SYLVESTER-BRADLEY, P. C. and FORD, T. D. (editors) 400pp. (Leicester: Leicester University Press.)

KING, R. J. 1968. Mineralisation. 112–137 in *The geology of the East Midlands.* SYLVESTER-BRADLEY, P. C. and FORD, T. D. (editors). 400pp. (Leicester: Leicester University Press.)

LAPWORTH, C. 1898. A sketch of the geology of the Birmingham district. *Proc. Geol. Assoc.*, Vol. 15, 313–416.

LE BAS, M. J. 1968. Caledonian igneous rocks. 41–58, in *The geology of the East Midlands.* SYLVESTER-BRADLEY, P. C. and FORD, T. D. (editors). 400pp. (Leicester: Leicester University Press.)

— 1972. Caledonian igneous rocks beneath Central and Eastern England. *Proc. Yorkshire Geol. Soc.*, Vol. 39, 71–86.

— 1981. The igneous basement of southern Britain with particular reference to the geochemistry of the Pre-Devonian rocks of Leicestershire. *Trans. Leicester Lit. and Philos. Soc.*, Vol. 75, 41–57.

— 1982. Geological evidence from Leicestershire on the crust of southern Britain. *Trans. Leicester Lit. and Philos. Soc.*, Vol. 76, 54–67.

LLEWELLYN, P. G. and STABBINS, R. 1970. The Hathern Anhydrite Series, Lower Carboniferous, Leicestershire, England. *Trans. Inst. Min. Metall. Sect. B: Appl. Earth Sci.*, Vol. 79, 1–15.

LOVELOCK, P. E. R. 1972. Aquifer properties of the Permo-Triassic Sandstones of the United Kingdom. Unpublished PhD Thesis, University of London.

MACDONALD, G. A. 1972. *Volcanoes.* 510pp. (Englewood Cliffs: Prentice-Hall.)

MAGUIRE, P. K. H., WHITCOMBE, D. N. and FRANCIS, D. J. 1981. Seismic studies in the central Midlands of England 1975–1980. *Trans. Leicester Lit. and Philos. Soc.*, Vol. 75, 58–66.

MAROOF, S. I. 1973. Geophysical investigations of the Carboniferous and pre-Carboniferous formations of the east Midlands of England. Unpublished PhD Thesis, University of Leicester.

— 1976. The structure of the concealed pre-Carboniferous basement of the Derbyshire Dome from gravity data. *Proc. Yorkshire Geol. Soc.*, Vol. 41, No. 6, 59–69.

MASSON SMITH, D., HOWELL, P. M., ABERNETHY-CLARK, A. B. D. E. and PROCTOR, D. W. 1974. The National Gravity Reference Net, 1973. *Ordnance Surv. G.B. Prof. Pap.*, No. 26.

MENEISY, M. Y. and MILLER, J. A. 1963. A geochronological study of the crystalline rocks of Charnwood Forest, England. *Geol. Mag.*, Vol. 100, 507–523.

MITCHELL, G. F., PENNY, L. F., SHOTTON, F. W. and WEST, R. G. 1973. A correlation of the Quaternary rocks in the British Isles. *Spec. Rep. Geol. Soc.* No. 4.

MITCHELL, G. H. 1942. The geology of the Warwickshire Coalfield. *Wartime Pamph. Geol. Surv. G.B.*, No. 25.

— 1948. Recent geological work in the Leicestershire and South Derbyshire Coalfield. *Iron and Coal Trades Rev.*, Vol. 157, 501–505.

— and STUBBLEFIELD, C. J. 1941a. The Carboniferous Limestone of Breedon Cloud, Leicestershire and associated inliers. *Geol. Mag.*, Vol. 78, 201–219.

— and STUBBLEFIELD, C. J. 1941b. The geology of the Leicestershire and South Derbyshire Coalfield 1st edition. *Wartime Pamph. Geol. Surv. G.B.*, No. 22.

— 1948. The geology of the Leicestershire and South Derbyshire Coalfield 2nd edition. *Wartime Pamph. Geol. Surv. G.B.*, No. 22.

MORTIMER, M., CHALONER, W. G. and LLEWELLYN, P. G. 1970. Lower Carboniferous (Tournaisian) miospores and megaspores from Breedon Cloud Quarry, Leicestershire. *Mercian Geol.*, Vol.3, 373–385.

MOSELEY, J. 1979. The geology of the Late Precambrian rocks of Charnwood Forest, Leicestershire. Unpublished PhD Thesis, University of Leicester.

— and FORD, T. D. 1985. A stratigraphic revision of the Late Precambrian rocks of Charnwood Forest, Leicestershire. *Mercian Geol.*, Vol. 10, 1–18.

NATIONAL COAL BOARD. 1957. *Warwickshire Coalfield seam maps.* (NCB, Scientific Dept., Coal Survey.)

— 1962. *South Derbyshire and Leicestershire Coalfields seam maps.* (NCB, Scientific Dept., Coal Survey.)

PATCHETT, P. J., GALE, N. H., GOODWIN, R. and HUMM, M. J. 1980. Rb-Sr whole-rock isochron ages of late Precambrian to Cambrian igneous rocks from southern Britain. *J. Geol. Soc. London*, Vol. 137, 649–656.

PENMAN, H. L. 1950. Evaporation of the British Isles. *Q. J. R. Met. Soc.*, Vol. 76.

PERRIN, R. M. S., ROSE, J. and DAVIES, M. 1979. The distribution variation and origins of pre-Devensian tills in eastern England. *Philos. Trans. R. Soc. B.*, Vol. 287, 535–570.

PETTIJOHN, F. J. 1975. *Sedimentary rocks.* 628pp. (New York: Harper & Row.)

PHARAOH, T. C. and EVANS, C. J. 1987. Morley Quarry No.1 Geothermal Well: geological well completion report. (Keyworth: British Geological Survey.)

— MERRIMAN, R. J. WEBB, P. C. and BECKINSALE, R. D. 1987a. The concealed Caledonides of Eastern England: Preliminary results of a multidisciplinary study. *Proc. Yorkshire Geol. Soc.*

— WEBB, P. C., THORPE, R. S. and BECKINSALE, R. D. 1987b. 541–552 *in* Geochemistry and mineralisation of Proterozoic volcanic suites. *Spec. Publ. Geol. Soc.*, No.33. PHAROAH, T. C., BECKINSALE, R. D. and RICKARD, D. (editors).

PIDGEON, R. T. and AFTALION, M. 1978. Cogenetic and inherited zircon U-Pb systems in granites: Palaeozoic granites of Scotland and England. 183–220 *in Crustal evolution in Northwestern Britain and adjacent regions.* BOWES, D. R. and LEAKE, B. E. (editors). 492pp. (Liverpool: Seel House Press.)

POOLE, E. G. 1968. Some temporary sections seen during the construction of the M1 Motorway between Enderby and Shepshed, Leicestershire. *Bull. Geol. Surv. G.B.*, No. 28, 137–151.

POSNANSKY, M. 1960. The Pleistocene succession in the Middle Trent Basin. *Proc. Geol. Assoc. London*, Vol. 71, 285–311.

POTTER, T. R. 1842. *The history and antiquities of Charnwood Forest.* 272pp. (London: Hamilton, Adams.)

RAMSBOTTOM, W. H. C., CALVER, M. A., EAGER, R. M. C., HODSON, F., HOLLIDAY, D. W., STUBBLEFIELD, C. J. and WILSON, R. B. 1978. A correlation of Silesian rocks in the British Isles. *Spec. Rep. Geol. Soc. London,* No. 10, 81pp.

RICE, R. J. 1968. The Quaternary deposits of central Leicestershire. *Philos. Trans. R. Soc. A.*, Vol. 262, 459–509.

— 1981. The Pleistocene deposits of the area around Croft in south Leicestershire. *Philos. Trans. R. Soc. B.*, Vol. 293, 385–418.

RICHARDSON, L. 1928. Wells and springs of Warwickshire. *Mem. Geol. Surv. G.B.*

— 1929. Wells and springs of Derbyshire. *Mem. Geol. Surv. G.B.*

— 1931. Wells and springs of Leicestershire. *Mem. Geol. Surv. G.B.*

ROCK, N. M. S. and LEAKE, B. E. 1984. The International Mineralogical Association amphibole nomenclature scheme computerisation and its consequences. *Mineral. Mag.*, Vol. 48, 211–227.

RUSHTON, A. W. A. 1978. Fossils from the Middle–Upper Cambrian transition in the Nuneaton district. *Palaeontology*, Vol. 21, 245–283.

— 1979. A review of the Middle Cambrian Agnostida from the Abbey Shales, England. *Alcheringa*, Vol. 3, 43–61.

— 1983. Trilobites from the Upper Cambrian *Olenus* Zone in Central England. *Spec. Pap. in Palaeontol.*, No. 30, 107–139.

SEDGWICK, A. 1834. The rocks of Charnwood Forest (Abstract). *Philos. Mag. Ser. 3*, No. 4, 68–69.

SHEARMAN, D. J., MOSSOP, G., DUNSMORE, H. and MARTIN, M. 1972. Origin of gypsum veins by hydraulic fracture. *Trans. Inst. Min. Metall. Sect. B: Appl. Earth Sci*, Vol. 81, 149–155.

SHOTTON, F. W. 1927. The conglomerates of the Enville Series of the Warwickshire Coalfield. *Q. J. Geol. Soc. London*, Vol. 83, 604–621.

— 1953. The Pleistocene deposits of the area between Coventry, Rugby and Leamington, and their bearing upon the topographic development of the Midlands. *Philos. Trans. R. Soc. B*, Vol. 237, 209–260.

— 1976. Amplification of the Wolstonian Stage of the British Pleistocene. *Geol. Mag.*, Vol. 13, 241–250.

— 1977. *The English Midlands.* Guidebook for Excursion A2. INQUA (International Union for Quaternary Research), X Congress, 1977.

SMITH, D. B., BRUNSTOM, R. G. W., MANNING, P. I., SIMPSON, S. and SHOTTON, F. W. 1974. A correlation of Permian rocks in the British Isles. *Spec. Rep. Geol. Soc. London*, No. 5, 45pp.

SNOWBALL, G. T. 1952. The Leicestershire intrusions. *Leicester Museum and Art Gallery Bull.*, 3rd Ser., 1, 2–7.

SPEARS, D. A. 1970. A kaolinitic mudstone (tonstein) in the British Coal Measures. *J. Sediment. Petrol.*, Vol. 40, 386–394.

SPINK, K. 1965. Coalfield geology of Leicestershire and South Derbyshire: The exposed coalfield. *Trans. Leicester Lit and Philos. Soc.*, Vol. 59, 41–98.

— and STRAUSS, P. G. 1965. A possible coalfield east of Loughborough. *Trans. Inst. Mining Eng.*, Vol. 124, 581–590.

STEVENSON, I. P. and MITCHELL, G. H. 1955. Geology of the country between Burton upon Trent, Rugeley and Uttoxeter. *Mem. Geol. Surv. G.B.*

STRAUSS, P. G. 1971. Kaolin-rich rocks in the East Midlands coalfields of England. *C. R. 6é Congr. Int. Strat. Géol. Carbonif. Sheffield 1967*, IV, 1519–1531.

STREKEISEN, A. L. 1976. To each plutonic rock its proper name. *Earth Sci., Rev.*, Vol. 12, 1–33.

SUTHERLAND, D. S., BOYNTON, H. E., FORD, T. D., LE BAS, M. J. and MOSELEY, J. 1987. A guide to the geology of the Precambrian rocks of Bradgate Park in Charnwood Forest, Leicestershire. *Trans. Leicester Lit. and Philos. Soc.*, Vol. 81, 47–83.

TAYLOR, F. M. 1968. Permian and Triassic formations. 149–173 in *The geology of the East Midlands*. SYLVESTER-BRADLEY, P. C. and FORD, T. D. (editors). 400pp. (Leicester: Leicestershire University Press.)

TAYLOR, K. and RUSHTON, A. W. A. 1971. The pre-Westphalian geology of the Warwickshire Coalfield, with a description of three boreholes in the Merevale area. *Bull. Geol. Surv.*, G.B. No. 35, 152pp.

THORPE, R. S. 1972. The geochemistry and correlation of the Warren House, the Uriconian and the Charnian volcanic rocks from the English pre-Cambrian. *Proc. Geol. Assoc.*, Vol. 83, 269–286.

— 1974. Aspects of magmatism and plate tectonics in the Pre-Cambrian of England and Wales. *Geol. J.*, Vol. 9, 115136.

TIEN, PEI-LIN. 1973. Palygorskite from Warren Quarry, Enderby, Leicestershire, England. *Clay Miner.*, Vol. 10, 27–34.

WALSH, P. T., BOULTER, M. C., IJTABA, M. and URBANI, D. M. 1972. The preservation of the Neogene Brassington Formation of the southern Pennines and its bearing on the evolution of Upland Britain. *J. Geol. Soc. London.*, Vol. 128, 519–559.

WARRINGTON, G. 1970. The stratigraphy and palaeontology of the 'Keuper' Series of the central Midlands of England. *Q. J. Geol. Soc. Lond.*, Vol. 126, 183–223.

— AUDLEY-CHARLES, M. G., ELLIOTT, R. E., EVANS, W. B., IVIMEY-COOK, H. C., KENT, P., ROBINSON, P. L., SHOTTON, F. W. and TAYLOR, F. M. 1980. A correlation of Triassic rocks in the British Isles. *Spec. Rep. Geol. Soc. London*, No. 13, 78pp.

WATTS, W. W. 1903. A buried Triassic landscape. *Geogr. J.*, Vol. 21, 623–636.

— 1905. Buried landscape of Charnwood Forest. *Trans. Leicester Lit. and Philos. Soc.*, Vol. 9, 20–25.

— 1947. *Geology of the ancient rocks of Charnwood Forest, Leicestershire.* 160pp. (Leicester: Leicester Lit. and Philos. Soc.)

WHITCOMBE, D. N. and MAGUIRE, P. K. H. 1980. An analysis of the velocity structure of the Precambrian rocks of Charnwood Forest. *Geophys. J. R. Astron. Soc.*, Vol. 63, 405–416.

— 1981a. A seismic refraction investigation of the Charnian basement and granitic intrusions flanking Charnwood Forest. *J. Geol. Soc. London*, Vol. 138, 643–651.

— 1981b. Seismic refraction evidence for a basement ridge between the Derbyshire Dome and the W. of Charnwood Forest. *J. Geol. Soc. London*, Vol. 138, 653–659.

WILLIAMSON, I. A. 1970. Tonsteins—their nature, origins and uses. *Miner. Mag.*, Vol. 122, 119–125, 203–211.

WILLS, L. J. 1948. *The palaeogeography of the Midlands.* 147pp. (Liverpool: University Press of Liverpool.)

— 1951. *A palaeogeographical atlas.* (London: Blackie.)

— 1956. *Concealed coalfields.* 208pp. (London, Glasgow: Blackie.)

— 1970. The Triassic succession in the central Midlands in its regional setting. *Q. J. Geol. Soc. London*, Vol. 126, 225–285.

— 1973. A palaeogeological map of the palaeozoic floor below the Permian and Mesozoic formations of England and Wales. *Mem. Geol. Soc. London*, No. 7. 23pp.

— and SHOTTON, F. W. 1934. New sections showing the junction of the Cambrian and Pre-Cambrian at Nuneaton. *Geol. Mag.*, Vol. 71, 512–521.

WILSON, H. E. and MANNING, P. I. 1978. Geology of the Causeway Coast. *Mem. Geol. Surv. N. Ireland.*

WORSSAM, B. C. 1977. The Hanginghill Farm (1970) Borehole, South Derbyshire Coalfield. *Bull. Geol. Surv. G.B.*, No.63, 17pp.

— CALVER, M. A. and JAGO, G. 1971. Newly discovered marine bands and tonsteins in the South Derbyshire Coalfield. *Nature, London, Phys. Sci.*, Vol. 232, 121–122.

APPENDIX 1

List of boreholes and shafts

The appendix lists, by 1:10 000 quarter-sheet, those boreholes and shafts in the district that are of particular geological interest. The registered number of each borehole and shaft within each quarter sheet is given; this is its file number within the BGS Records System. This is followed in the lists by the name, date of drilling, National Grid Reference, surface level in metres above OD and the total depth in metres of the borehole or shaft. Shafts of which only the site is known are omitted, as are most water, motorway and site investigation boreholes.

The primary source of information on sinkings before 1907 is the Geological Survey Memoir 'Geology of the Leicestershire and South Derbyshire Coalfield' by C. Fox-Strangways (1907). The logs of many coalfield boreholes sunk between 1907 and 1955 were published in abstract by Greig and Mitchell (1955). The sites and brief abstract logs of most of the holes listed are shown on the six-inch geological maps.

The following abbreviations are used:

AM	Atherstone Memoir (Fox-Strangways 1900)
Fig	Number of text-figure in this memoir
FS	Fox-Strangways 1907
G & M	Greig and Mitchell 1955
LM	Lichfield Memoir (Barrow and others 1919)
Rep IGS	Report of the Institute of Geological Sciences
WSL	Wells and Springs of Leicestershire (Richardson 1931)
WSW	Wells and Springs of Warwickshire (Richardson 1928)

Registered number	Name	National Grid ref.	Surface level	Total depth	References
SK 20 SE					
1	Birch Coppice No. 4 Shaft	2500 0180	104.8	344.86	LM p.249; Figs. 5, 6
3	Pooley Hall Shaft	2582 0331	74.4	207.26	AM p.59; Figs. 5, 6
4	Butt Lane Pit	2632 0278	68.9	91.44	AM p.62–3
5	Kisses Barn No. 6	2775 0263	76	178.76	
6	Kisses Barn No. 7	2754 0256	c.70	117.8	
7	Lyndon Lodge No. 2	2738 0283	93.6	54.86	
8	Lyndon Lodge No. 1	2775 0315	90.7	203.91	
9	Lyndon Lodge No. 5	2759 0339	103.1	103.02	
10	Lyndon Lodge No. 4	2790 0376	75.5	147.52	
11	Lyndon Lodge No. 3	2790 0347	85.2	170.07	
12	Woodside Farm Trial Shaft	2617 0449	c.78	41.14	
14	Warton Pumping Station	2797 0259	79.9	91.44	LM pp.281–2
23	Bramcote Pumping Station	2711 0386	64	36.57	
36	Birch Coppice Colliery No. 2 Shaft	2526 0007	c.105	338.02	Figs. 5, 6
40	Birch Coppice Colliery Staple Pit	2527 0016	c.105	329.4	
44	Shaw & Co's Pit, Polesworth Colliery	2674 0289	76	59.6	AM p.63
SK 20 NE					
2	Shuttington Fields	2642 0610	71.5	297.4	
5	Newton Regis	2822 0728	78.3	163.3	
SK 21 SE					
1	Chilcote	2828 1142	c.82	292.3	FS pp.331–2; Fig. 17
2	Church Flats	2605 1452	97.7	946.6	Fis. 8, 9, 10
3	Netherseal No. 2 (Plant)	2678 1471	106.7	187.75	Binn's site of FS, pp.199–200
4	Rawdon West No. 1	2923 1477	88.0	309.98	
5	Rawdon West No. 2	2865 1471	97.9	262.74	Fig. 9
6	Gunby Lea No. 8	2912 1459	87.4	365.96	G & M p.61
7	Acresford 'L'	2943 1351	86.1	160.02	
8	Acresford 'M'	2980 1366	93.8	160.93	
9	Acresford No. 7	2960 1297	72.3	308.15	G & M p.61; Fig. 7
10	Netherseal No. 1	2873 1378	78.0	242.77	FS pp.193–5; Fig. 8
11	Netherseal No. 2	2853 1393	c.83.8	236.32	FS pp.195–7
12	Rawdon West No. 3	2928 1404	97.6	301.44	
13	Rawdon West No. 4	2851 1431	99.9	259.08	Fig. 8
14	Rawdon West No. 5	2818 1431	91.5	207.56	Fig. 8
15	Acresford Plantation	2997 1336	73.9	205.43	
16	Cadborough Farm	2966 1418	98.9	359.81	
17	Church Way	2964 1367	98.7	291.39	

Registered number	Name	National Grid ref.	Surface level	Total depth	References
18	Coalpit Lane	2525 1494	86.8	630.68	Figs 8. 9. 10
19	Chilcote Water Works Trial	2844 1039	72.7	245.36	
20	Chilcote Water Works No. 1	2841 1041	72.5	211.83	
21	Chilcote Water Works No. 2	2840 1040	72.2	211.83	
22	Broomfields	2894 1365	87.9	222.96	
25	Crickett's	2930 1327	80.9	136.0	
29	Hunts Lane	2834 1348	86.6	220.46	
30	Mill Farm	2924 1281	72.4	153.73	
31	Gunby Farm	2887 1413	88.4	113.0	
32	Clifton Road	2802 1291	77.2	190.33	
33	Hill	2834 1423	96.5	123.45	

SK 21 NE

27	Botany Bay	2558 1587	87.3	762	Fig. 10
28	Grange Wood	2652 1547	106.2	820.22	Figs. 7, 10
31	Wynne's BH	2715 1526	126.8	123.44	
37	Netherseal No. 3	2624 1551	105.2	256.46	
38	Netherseal 1912	2678 1521	121.6	399.20	G & M Fig. 10
39	Netherseal Colliery No. 1	2674 1590	103.3	106.68	
40	Netherseal Colliery No. 2	2789 1518	115.5	122.22	
41	Netherseal Colliery No. 3	2795 1581	114.0	145.08	
42	Netherseal Colliery No. 4	2817 1526	111.9	149.66	
47	Seal Wood	2798 1560	c.98	68.58	
59	Poplars Farm	2996 1514	101.9	130.76	
74	Green Lane	2959 1610	102.5	221.74	Fig. 9
79	Netherseal Colliery No. 4 (Plant)	2757 1543	114.9	340.6	FS pp.201 – 205; Fig. 8
107	Linton Lane	2562 1514	96.5	824.85	

SK 30 SW

6	Austrey House	3030 0488	c.74	152.4	

SK 30 SE

3	Bosworth Wharf	3916 0323	c.96	410.26	WSL pp.68 – 9; FS pp.342 – 3
4	Kingshill Spinney	3814 0154	c.85	321.25	FS p.344
5	Temple Hall Farm IGS	3597 0255	c.111	6	Rep. IGS No. 76/10
6	Freizeland No. 1 IGS	3821 0324	c.85	11.7	Rep. IGS No. 76/10
7	Freizeland No. 2 IGS	3845 0340	c.86	4.3	Rep. IGS No. 76/10
8	Freizeland No. 3 IGS	3835 0354	c.87	2.4	Rep. IGS No. 76/10
9	Hoo Hills No. 1 IGS	3721 0356	c.92.6	7.6	Rep. IGS No. 76/10
10	Hoo Hills No. 2 IGS	3709 0371	c.88.5	18.3	Rep. IGS No. 76/10

SK 30 NW

1	Appleby Parva	3067 0858	c.100	170.07	Fig. 17
2	Turnover	3423 0995	85.7	127.7	
6	Appleby Road	3365 0931	c.88	152.4	
8	Twycross, Gopsall House Farm IGS	3424 0580	c.101.7	4.9	Rep. IGS No. 76/10
9	Twycross, Lagos Farm IGS	3362 0572	c.123.0	7.4	Rep. IGS No. 76/10
10	Twycross Lodge	3215 0560	126.5	213.36	WSL pp.77 – 78
13	Twycross IGS	3387 0564	122.0	508.9	Rep. IGS No. 79/12 Figs. 17, 18

SK 30 NE

1	Measham No. 1	3670 0978	110.1	196.6	Fig. 4
2	Gopsall Hall	3526 0649	100.6	205.74	
3	Shackerstone	3802 0704	c.95.5	182.88	WSL pp.73 – 76
4	Gopsall Park	3573 0593	106.7	17.7	Rep. IGS No. 76/10
5	Twycross Lane B30/1	3552 0522	94.2	5.3	Rep. IGS No. 76/10
6	Twycross Lane B30/2	3556 0523	94.2	1.9	Rep. IGS No. 76/10
7	Twycross Lane B30/3	3559 0523	94.2	11.0	Rep. IGS No. 76/10

SK 31 SW

1	Measham Tramway No. 1	3488 1177	102.4	16.33	FS pp.85, 244
7	Measham Tramway No. 7	3475 1157	102.4	17.98	FS pp.85, 245

Registered number	Name	National Grid ref.	Surface level	Total depth	References
9	Measham Tramway No. 7	3470 1135	102.4	57.91	FS pp.85, 246
10	Measham Colliery No. 1	3382 1144	c.88	81.69	FS pp.254–5
11	Measham Colliery No. 2	3436 1166	c.100	35.66	FS p.255
12	Measham No. 2 (1954)	3446 1058	86.0	105.0	
13	Measham No. 3 (1954)	3458 1113	93.0	123.44	Fig. 8
14	Measham No. 4 (1955)	3469 1086	86.6	55.47	
15	Measham No. 10	3386 1120	90.1	174.8	
18	Hinckley U.D.C. (Snarestone)	3389 1011	82.3	121.92	
19	Snarestone Pumping Station	3475 1009	94.5	264.0	FS pp.335–8
20	Appleby (White House) No. 1	3205 1058	c.88	297.13	FS pp.333–4; Fig. 7
21	Appleby (Birdshill Gorse, Side Hollows) No. 2	3176 1166	83	219.2	FS p.334
54	Acresford Waterworks	3012 1288	79.2	91.44	
55	Acresford A	3165 1256	c.80	62.18	
56	Acresford B	3153 1266	c.80	68.26	
57	Acresford C	3146 1283	77.0	188.98	
58	Acresford D	3132 1276	77.1	215.8	
59	Acresford E	3114 1264	79.8	157.28	
60	Acresford F	3126 1312	92.1	224.94	
61	Acresford G	3082 1295	84.9	208.03	
62	Acresford H	3038 1331	90.1	268.22	
63	Acresford J	3049 1285	78.4	162.25	
64	Acresford K	3010 1263	77.4	227.08	
65	Acresford (Donisthorpe) No. 6	3096 1358	107.9	716.58	G & M p.61; Fig. 7
66	Stretton No. 1	3058 1235	c.73	211.88	Fig. 7
67	Stretton No. 2 (Bakewell's Close)	3101 1263	97.5	276.3	
68	Saltersford D	3096 1250	c.76	111.85	FS pp. 236–7
69	Oakthorpe B	3139 1212	c.74	35.95	FS p.239
70	Piddocke's BH	3170 1271	c.85	49.16	
71	Oakthorpe No. 1	3180 1283	c.90	92.96	
74	Springfield Pit No. 1 Shaft	3220 1275	c.98	85.95	FS p.240
77	Donisthorpe No. 1	3125 1450	85.3	51.2	FS p.234
78	Donisthorpe No. 2	3057 1432	76.2	44.18	FS p.235
79	Donisthorpe No. 3	3052 1395	c.73	29.95	FS pp.235–6
80	Donisthorpe No. 4	3058 1434	76.2	14.1	FS p.236
81	Donisthorpe Colliery Nos. 1 and 2 Shafts	3128 1430	c.99	272.86 286.51	FS pp.231–3; Figs. 8, 9
82	Donisthorpe Shaft	3189 1386	c.91	145.6	FS pp.230–1
106	Measham House No. 1 Shaft	3447 1251	c.120	54.9	FS pp.85, 247
107	Oakthorpe Colliery (Willesley Basin)	3317 1397	91.4	107.25	FS pp.241–2; Fig. 9
108	Oakthorpe Colliery (Daisy Pit)	3350 1372	107	97.84	FS p.242
109	Odd Barn	3406 1326	c112	30.78	
118	Measham House No. 4 Shaft	3482 1258	c.125	30.68	FS pp.85, 247, 249
119	Measham House No. 6 Shaft	3476 1239	c.114	41.59	FS pp.85, 247, 248–9
121	Measham No. 12	3495 1208	103.8	159.41	
122	Measham Colliery Upcast	3499 1201	102.4	115.79	FS pp.253–4
123	Measham Colliery Underground	3481 1201		27.21	FS p.254
124	Measham Hall No. 2 (Arlick Field)	3497 1250	c.119	46.32	FS pp.85, 251
125	Measham No. 11	3499 1109	c.87	203.3	Fig. 8
132	Pegg's Close	3357 1205	99.6	103.38	
133	Gallows Lane	3390 1177	96.0	120.45	Fig. 8
136	Horses Lane	3355 1166	93.3	183.03	Fig. 7
138	Coronet	3398 1080	87.9	101.49	
139	Snarestone Lodge	3432 1014	84.0	108.05	
140	Measham Sidings	3371 1139	86.3	58.97	
147	Park Farm (Willesley)	3404 1379	122.8	81.3	
148	Merryman's Farm	3298 1343	96.7	96.24	
149	Ashby Road	3403 1329	109.7	62.94	
150	Red Bank	3403 1060	85.3	82.29	
151	Gilwiskaw	3434 1029	85.7	110.49	
152	Canal (Measham)	3476 1046	93.9	115.67	
153	Odd House	3445 1047	92.2	110.62	
154	Bosworth Road	3474 1087	87.5	127.1	
155	Quarry Lane	3475 1025	92.2	98.15	
156	Brook Spinney	3402 1027	83.2	106.25	

Registered number	Name	National Grid ref.	Surface level	Total depth	References
157	Tow Path	3472 1054	93.8	113.0	
158	Barn Field	3463 1063	92.2	116.28	
169	Chapel Street No. 1	3308 1232	94.1	65.89	
170	New Street No. 1	3357 1307	110.2	96.31	
171	New Street No. 5	3354 1290	101.0	72.8	
247	Chapel Street No. 2	3276 1240	101.4	48.0	
248	Chapel Street No. 3	3239 1247	97.4	39.59	
270	Acresford Pumping Station No. 2	3012 1284	c.79	85.34	
277	New Street No. 3	3336 1296	c.108		
279	Richmond	3494 1011	96.7	97.4	
340	Buston Road	3018 1296	79.9	165.0	
341	Playing Fields	3277 1316	107.2	42.9	
SK 31 SE					
253	Measham No. 6	3586 1297	100.6	89.3	G & M
254	Measham No. 7	3614 1242	c.97	90.06	G & M
255	Measham No. 8	3698 1227	c.103.6	52.57	G & M
256	Measham No. 5	3689 1091	112.2	89.61	G & M Fig. 4
257	Measham Hall (First Arlick)	3529 1228	c.105	26.52	FS pp.249–50
279	Heather No. 5	3937 1212	116.7	107.0	Fig. 12
281	Measham No. 2 (Clock Mill)	3639 1192	c.98	182.88	Fig. 4
282	Measham No. 3	3602 1171	c.96	93.87	
283	Measham No. 4	3615 1187	94.5	91.74	
284	Measham Hall No. 1	3543 1182	c.94	52.59	FS p.250
285	Measham Hall No. 4	3577 1181	91.4	45.88	FS p.251
286	Measham Hall No. 9	3576 1168	c.93	18.46	FS p.253
287	Measham Hall No. 8	3578 1149	c.97	52.88	FS pp.252–3; Fig. 9
290	Minorca Colliery	3552 1141	95.1	148.52	Figs. 8, 9
291	Snarestone No. 1	3522 1065	c.99	145.08	FS pp.338–9
292	Snarestone No. 2	3505 1086	c.30	189.89	FS pp.339–42
293	Measham No. 1 (1954)	3525 1076	c.98	85.34	
294	Measham No. 9	3549 1049	100.4	185.92	
296	Normanton No.6	3836 1202	130.9	72.84	
297	Ashby R.D.C. No. 2 (Heather)	3820 1087	133.8	79.85	
298	Heather Old Colliery	3945 1147 3949 1141	c.113	84.8	FS p.292
299	Heather Colliery No. 1 (Meadow) Pit	3950 1091	c.111	41.68	FS p.292
300	Heather Colliery No. 2 (Brickyard) Pit	3969 1093	121.9	41.07	FS p.293
301	Heather Colliery No. 3 (Pumping Sh.)	3990 1112	c.123	55.77	
302	Heather Colliery No. 4 (Winding Sh.)	3997 1114	125.0	157.27	FS pp.293–4; Fig. 12
303	Ashby R.D.C. No. 1	3939 1073	109.4	42.67	
305	Old Shaft	3925 1004	c.112	83.66	
306	Ravenstone Mill B	3982 1204	118.9	108.9	FS pp.295–6
311	The Altons No. 2	3949 1460	146.6	70.71	
312	The Altons No. 3	3986 1422		67.88	Fig. 12
313	Ashby R.D.C. No. 3	3936 1070	c.110	39.32	
314	Ashby R.D.C. No. 4	3942 1073	c.110	47.24	
315	Colliery Sidings	3502 1216	105.4	16.15	
316	Measham Dirt Bank	3518 1208	98.9	22.55	
317	Coalfield Farm W1	3682 1163	112.2	100.58	
318	Coalfield Farm W2	3693 1140	107.2	90.0	
319	Coalfield Farm 2007	3694 1141	107.4	94.0	
320	Coalfield Farm W3	3739 1146	113.5	75.0	
321	Coalfield Farm 2006	3740 1146	113.2	88.0	
322	Coalfield Farm 2002	3755 1150	115.8	72.0	
323	Coalfield Farm 2003	3767 1154	118.0	78.0	
324	Coalfield Farm 2004	3779 1160	118.9	76.5	
325	Coalfield Farm 2005	3773 1179	127.4	90.5	
326	Coalfield Farm W5	3969 1280	152.4	110.0	
327	Quarry	3516 1058	95.0	130.75	
328	Smallholding	3514 1026	96.9	103.0	
329	Valley Bungalow A	3544 1026	113.7	113.23	
330	Valley Bungalow B	3544 1026	113.7	113.23	

Registered number	Name	National Grid ref.	Surface level	Total depth	References
400	Minorca Farm	3539 1080	95.8	148.13	
416	Dishley Farm	3623 1095	99.3	133.3	
417	St. Peter's	3567 1064	96.3	153.4	
418	Swepstone Road	3590 1151	95.8	76.0	
419	Hunters Lodge 'B'	3608 1008	116.3	147.2	
420	Snarestone Road	3640 1001	114.2	120.4	
421	Tie Barn	3682 1273	107.2	63.5	
422	Red Burrow Lane	3653 1233	99.0	76.4	

SK 31 NW

29	Rawdon East No. 1	3195 1730	129.0	304.8	
30	Rawdon Colliery	3127 1626	105.2	330.94	FS p.215 Fig. 9
34	Hastings and Grey Shaft	3199 1555	c.107	39.02	FS p.219; Figs. 9, 10
37	Bath Pit	3096 1562	c.91	18.44	FS p.226
38	Double Pits (Spinney Pit)	3104 1528	c.88	41.45	FS p.227
133	Barratt Mill (1968)	3076 1515	79.6	81.69	
134	Beehive	3286 1563	114.9	162.45	Fig. 9
136	Slackey Lane	3038 1576	91.6	78.64	Fig. 9
141	Hanging Hill Farm IGS	3135 1672	127.1	156.05	Worssam 1977; Fig. 10
260	Rotherwood IGS	3458 1559	109.5	199.0	Rep. IGS No. 78/21; Fig. 4
304	Shellbrook No. 1	3336 1559	126.0	187.8	
305	Shellbrook No. 2	3316 1571	123.9	196.1	
306	Shellbrook No. 3	3321 1589	124.2	164.1	

SK 31 NE

54	Farm Town No. 3	3899 1593	165.2	84.73	
55	The Altons No. 1	3946 1522	157.0	89.15	
56	The Altons Old Shaft No. 1	3941 1511	152.2	81.08	
57	The Altons Old Shaft No. 2	3842 1506	151.5	53.34	

SK 40 SW

1	The Fields No. 1	4482 0441	129.2	158.8	
2	The Fields No. 2	4387 0468	139.3	129.54	
3	Osbaston Field Farm (Desford No. 8)	4291 0483	118.7	131.06	Fig. 4
4	Newbold Verdon	4444 0414	133.4	139.9	
5	Heath Farm	4343 0456	134.6	151.33	
7	Cowpastures	4124 0383	c.116	166.72	

SK 40 SE

1	Desford No. 2	4545 0465	113.4	230 12	Fig. 4
2	Desford No. 4	4577 0492	114.0	185.31	
3	Desford No. 13	4665 0493	126.5	137.16	
4	Desford No. 14	4615 0489	124.9	216.68	Fig. 4
5	Desford No. 15	4586 0473	109.9	148.13	
6	Ibstock No. 2	4653 0453	104.4	216.4	Fig. 4
7	Lindridge No. 2	4645 0446	103.6	118.41	
8	Lindridge No. 1	4716 0484	99.7	117.04	FS pp.103, 344. WSL p.65
9	Ibstock No. 1	4732 0464	96.9	136.34	
10	Lindridge Colliery	4720 0454	99.1	154.53	FS p.345
11	Newtown Unthank	4947 0447	88.3	188.29	
12	Lindridge Hall Farm	4644 0485	124.4	133.50	
13	Hook House	4610 0448	106.0	136.55	
14	Newbold Road	4546 0404	131.6	162.0	
15	The Fields No. 3	4517 0425	132.5	159.41	
16	Stocks House	4680 0212	c.122	163.67	FS p.347; Fig. 4
17	Midland Tube Co.	4957 0444	84.7	125.57	

SK 40 NW

1	Ibstock No. 9	4091 0923	118.0	119.63	
2	Nailstone Colliery No. 1	4293 0852	152	138.99	
3	Nailstone Colliery No. 2	4294 0853	152	258.99	FS pp. 297–299; Fig. 4, 12
4	Bagworth Colliery Old Shaft	4438 0859	162.5	293.01	FS pp.304–6

Registered number	Name	National Grid ref.	Surface level	Total depth	References
5	Bagworth Colliery No. 1	4441 0866	162.5	289.76	FS pp.302–4; Figs. 12, 13
6	Bagworth Colliery No. 2	4440 0864	162.5	292.71	FS pp.303–4
8	Elms Farm No. 10	4118 0762	119.5	125.12	
9	Osbaston Hollow	4166 0635	114.3	213.36	Fig. 4
10	Bagworth Colliery (Desford No. 7)	4316 0728	128.9	66.45	
11	Desford No. 6	4397 0621	136.6	86.67	
12	Desford No. 1	4419 0532	143.9	169.67	
13	Desford No. 9	4353 0505	138.9	140.97	
14	Ibstock Manor	4062 0968	118.2	96.62	
15	Ibstock Grange No. 1	4107 0954	134.3	100.58	
16	Ibstock Lodge	4103 0894	125.3	115.67	
17	Cottage Farm	4046 0908	114.2	96.10	
18	Osbaston Gate	4333 0590	130.9	132.95	
19	Sefton House	4332 0551	132.1	156.97	
20	Barlestone Lodge	4255 0681	122.8	137.21	
21	Churchyard Farm	4293 0603	137.3	130.18	
22	Park Farm	4131 0895	137.1	65.30	
23	Nailstone Grange	4160 0843	141.9	86.79	
24	Manor House	4269 0525	121.0	121.03	
25	Orchard Street	4085 0996	125.0	86.34	
26	Grange Road	4233 0818	138.8	138.12	
27	Ibstock Grange No. 2	4155 0975	138.1	61.87	
28	Wigg Farm	4336 0917	163.7	161.54	
29	Nailstone Rectory	4210 0740	149.7	122.53	
30	Grange Farm	4180 0910	147.2	82.60	
31	Bagworth	4408 0791	167.4	179.22	
32	Bagworth Station	4422 0969	159.9	159.94	Fig. 13
33	The Hill	4390 0606	132.8	156.67	
34	Garland Lane	4323 0716	136.6	152.55	
35	Neville Arms	4355 0872	144.3	153.44	
36	Bagworth Moats	4485 0890	146.3	201.2	Fig. 13
45	Belcher's Bar	4090 0864	141.7	117.04	
46	Nailstone Hollows	4113 0806	136.3	29.0	
47	Stud Farm	4174 0700	141.9	103.94	
48	Tower House (Ellistown) No. 55	4436 0976	150.6	119.79	
49	Bungalow Farm (Ellistown) No. 54	4454 0944	149.4	131.47	
50	Barlestone Road No. 53	4370 0789	156.9	119.79	
51	Wood Road	4346 0960	162.8	104.85	
53	Battram	4316 0952	162.2	98.76	
55	Bagworth (Greenhouse)	4416 0728	150.6	152.04	
56	Bagworth (Oak Farm)	4442 0701	148.7	117.4	
57	Bagworth (Fox Covert)	4474 0673	145.1	122.05	
83	Bagworth (Poplar House)	4409 0813	156.7	119.33	
84	Pool House	4433 0505	135.9	117.55	
85	Gnarley House	4213 0613	131.2	83.97	
86	Hinckley Road	4178 0650	120.0	73.46	
93	Laurel Farm	4449 0785	170.1	131.67	
94	Craigmore Farm	.4485 0527	140.4	120.39	
95	Noah's Ark	4356 0551	135.3	103.63	
129	Kennels	4368 0707	140.6	182	
133	Powder Magazine	4394 0842	154.7	102.1	
134	Bagworth Tip	4371 0839	158.1	115.1	
135	Old House	4437 0751	161.4	119.2	
136	Paddock (Ellistown)	4357 0927	157.4	148.2	
137	Maynard Arms	4367 0904	152.3	147.5	

SK 40 NE

Registered number	Name	National Grid ref.	Surface level	Total depth	References
1	Desford Colliery No. 1 Shaft	4598 0685	c.122	219.25	FS pp.307–9 Fig. 12
2	Desford Colliery No. 2 Shaft	4596 0685	c.122	219.45	
3	Desford Colliery No. 12	4683 0504	118.8	143.26	
4	Desford Colliery No. 10	4636 0519	129.1	168.25	
5	Desford Colliery No. 5	4582 0587	131.1	99.67	
6	Desford Colliery No. 3	4509 0506	134.7	150.88	

Registered number	Name	National Grid ref.	Surface level	Total depth	References
7	Desford Colliery No. 11	4583 0505	113.4	154.38	
21	Heath Road Villas	4541 0698	124.9	162.92	
22	Little Bagworth	4539 0759	144.7	182.98	
32	Thornton Lane, Bagworth No. 28	4544 0791	151.8	144.48	
33	Thornton Mill	4581 0769	130.8	111.86	
34	Desford Tip	4593 0722	125.6	130.54	
35	Longmeadow	4556 0719	127.8	114.9	
36	Bagworth Church	4503 0753	152.7	132.28	
37	Desford Baths	4586 0656	139.6	136.8	
38	Little Fox, Desford	4537 0623	143.2	124.22	
45	Bagworth Heath	4516 0732	145.0	120.7	

SK 41 SW

Registered number	Name	National Grid ref.	Surface level	Total depth	References
24	Ravenstone Mill A	4002 1183	c.119	114.9	FS pp.294–5
26	Berryhills Farm	4091 1271	125.9	75.59	
27	Kelham Bridge	4110 1221	130.6	117.88	
28	Snibston No. 2 Pit	4187 1452	155.5	264.57	FS pp.273–6, 278
29	Snibston Colliery, Stephenson Shaft	4193 1447	155.8	280.87	Fig. 12
30	Snibston No. 1 Pit	4257 1460	155.5	272.62	
32	Hugglescote	4220 1333	156.7	142.93	
34	Battleflat Water Boring	4495 1163	169.9	100.74	
36	Whitwick Colliery No. 3 Pit	4302 1465	160.7	134.42	
38	Whitwick Colliery No. 6 Pit	4313 1450	161.5	299.39	Figs. 12, 13
44	Ibstock No. 11	4027 1116	141.1	74.14	
45	Blackberry Farm	4135 1162	145.5	126.8	
46	Ibstock Colliery Old Pit	4175 1108	143.3	117.12	FS pp.288–9
47	Ibstock Colliery No. 2 Pit	4161 1110	143.3	136.32	FS pp.289–92; Fig. 12
48	South Leicester Colliery No. 1 Pit	4307 1183	155.5	259.03	FS pp.282–5
49	South Leicester Colliery No. 2 Pit	4312 1181	157.3	269.14	FS pp.285–8; Figs. 12, 13
50	Ellistown Colliery Nos. 1 and 2 Sh.	4385 1033	170.7	303.63	FS pp.299–302; Figs. 12, 13
51	Ellistown Colliery BH	4390 1056	170.7	494.08	Boulton 1934; Figs. 4, 12
52	Valley Farm	4018 1008	128.8	78.03	
55	Ravenstone Road	4107 1439	153.8	134.72	
56	Leicester Road	4084 1359	149.2	97.84	
57	Standard Hill	4168 1323	147.6	152.4	
58	Ibstock Road	4253 1144	154.8	167.95	
59	Melbourne Road	4089 1157	134.8	55.78	
60	Victorial Terrace	4111 1122	141.5	65.0	
61	Donington Le Heath	4144 1226	134.6	82.91	
62	Meadow Row	4127 1019	124.6	85.19	
63	Wood Farm	4451 1057	174.4	227.97	
64	Battleflat	4356 1105	169.3	200.08	Fig. 13
65	Pickering Grange Farm	4305 1007	155.8	143.26	
66	Ibstock Railway	4294 1082	159.3	168.3	
75	Bardon Hill Station	4389 1256	154.0	275.23	Fig. 13
76	Upper Grange Farm	4411 1181	173.3	297.03	
77	Hugglescote Grange	4324 1270	140.0	87.86	Fig. 13
78	Spring Farm	4391 1362	155.5	269.64	Fig. 13
91	Glebe Farm	4358 1493	153.8	193.35	
92	Sharpley	4389 1467	161.7	173.74	
93	Broomleys	4375 1418	158.2	176.63	
94	St Mary's	4150 1296	149.2	47.40	
95	Ellistown Terrace	4331 1018	163.7	109.73	
96	Whitehill Farm	4313 1117	162.9	26.56	
97	Pretoria Road	4297 1050	158.6	92.41	
98	Beveridge Lane	4325 1151	161.7	92.71	
111	Station Road	4256 1241	133.8	163.7	
127	Green Lane (Whitwick)	4375 1480	156.4	267.2	
130	Broomleys Railway	4404 1438	160.9	274.2	
131	Ellistwon Church	4269 1104	158.2	153.65	
132	Ellistown Hotel	4264 1124	154.1	150.88	
149	King Edward's	4313 1316	150.4	43.28	
150	Granite Railway	4374 1433	158.6	81.07	

Registered number	Name	National Grid ref.	Surface level	Total depth	References
151	Yew's Farm No. 1	4140 1326	133.2	114.85	
153	Snibston Tip	4184 1397	154.5	153.7	Cores measured from 1.1 m above S.L.
154	Snibston Lodge	4122 1382	149.7	56.58	Cores measured from 1.1 m above S.L.
155	Ronlyn	4074 1132	141.7	62.75	
156	Leesons	4084 1121	142.0	59.00	
157	Avenue	4088 1115	143.6	47.9	
158	Yews Farm No. 2	4139 1327	132.0	37.5	
162	Richmond Bungalow No. 1	4177 1175	76.1	145.2	
163	Richmond Bungalow No. 2	4177 1175	100.0	145.2	
SK 41 NW					
21	New Swannington Colliery, Sinope Pit	4016 1532	158.5	96.09	FS p.330
23	Snibston Colliery No. 3 Pit	4203 1547	152.4	240.79	FS p.278
36	Coalville U.D.C. No. 1 (Holly Hayes Wood)	4405 1531	c.156		
39	Hoo Ash	4012 1525	154.5		
42	Spring Lane	4161 1579	128.9		
45	Thornborough Lane	4223 1543	148.7	205.74	Fig. 13
57	Hermitage Road	4303 1558	140.6		
58	Church Lane, Whitwick	4304 1586	139.0	131.19	
59	Whitwick Brickworks	4272 1511	130.8	15.3	
66	Raine House	4046 1606	141.2	103.02	
67	Talbot House	4234 1694	125.8	182.33	
71	Hough Hill	4100 1503	166.0	134.6	
72	West Hoo Ash	4046 1522	156.7	92.4	
SK 50 SW					
1	Barron Park Farm Shaft	5098 0454	80.2	35.97	FS pp.351–2; WSL p.65
71	Leicester Forest East IGS	5245 0283	104.3	178.94	Rep. IGS No. 79/12; Fig. 18
SP 29 NE					
3	Baddesley (Stratford Pit) Nos. 1 and 2	2790 9711	c.150	327.96	AM pp.47–9; Fig. 6
4	Baddesley (Speedwell Pit)	2663 9841	103.0	230.15	AM pp.50–3; Fig. 6
7	Woodend Shaft	2509 9830	136.6	426.11	LM p.253; Figs. 5,6
18	Baddesley Colliery No. 3 Shaft deepening	2790 9709	– 132.1	158.8	Fig. 5
SP 39 NE					
4	Dadlington	3993 9909	c.90	357.82	Fig. 17

APPENDIX 2

List of Geological Survey photographs

Copies of these photographs are deposited for reference in the British Geological Survey library, Keyworth, Nottingham NG12 5GG. Black and white prints and slides can be supplied at a fixed tariff, and in addition colour prints and transparencies are available for all the photographs.

All numbers belong to Series A. The National Grid references are those of the viewpoints.

PRECAMBRIAN

6535 High Sharpley. A crag of cleaved dacite.
6536 Pillar Rock, Benscliffe Wood. Benscliffe Agglomerate.
6537 Peldar Tor. Crag of agglomerate.
6538 Flat Hill, Charnwood. Hanging Stone, Benscliffe Agglomerate
6539 Charnwood Lodge Drive. Agglomerate.
6540 Hanging Rocks, Woodhouse Eaves. Bradgate Tuffs, bedding and cleavage.
6541 Hanging Rocks, Woodhouse Eaves. Crag of Sliding Stone Slump Breccia and Bradgate Tuffs above.
6542 Hanging Rocks, Woodhouse Eaves. Triassic deposits resting unconformably on Swithland Greywacke Formation.
6543 Hanging Rocks, Woodhouse Eaves, Charnwood. Bradgate Tuffs.
6546 Sheet Hedge's Quarry, Nr. Groby, Charnwood. Triassic deposits overlying diorite.
6547 Beacon Hill. Beacon Hill Tuffs.
6548 Bradgate Park. Sliding Stone Slump Breccia followed by basal Bradgate Tuffs.
6549 Bradgate Park. Sliding Stone Slump Breccia.
6550 Bradgate Park. Near Old John: Crag and fragments of Sliding Stone Slump Breccia in foreground.
10253 Beacon Hill. Beacon Hill Tuffs [509 148].
10254 Beacon Hill. Crags of Beacon Hill Tuffs [509 148].
10255 Beacon Hill. Crags of Beacon Hill Tuffs [509 148].
10264 West bank of M1 cutting, c.1.6 km NW of Markfield village. Beacon Hill Tuffs overlain by Mercia Mudstone breccia and marl [4787 1183].
10265 West bank of M1 cutting at E end of Birch Hill. Thrusted syncline in Beacon Hill Tuffs [4820 1346].
10267 West bank of M1 cutting at E end of Birch Hill. Cleaved Beacon Hill Tuffs [4820 1346].
12328 Warren Hill. Banded acid tuffs of Beacon Hill Formation [4577 1515].
12329 Hanging Stone, Charnwood Lodge. Benscliffe Agglomerate [467 160].
12330 800 m S. 25°W of Mount St Bernard's Abbey. Charnwood Lodge Agglomerate [4545 1543].
12332 Ives Head. Charnian tuffs of Blackbrook Group [4764 1704].
12338 Cutting on A50 at Raunscliffe. Park Breccia [4863 1085].
12239 Raunscliffe. Sliding Stone Slump Breccia [485 109].
12340 Cliffe Hill Quarry. Tuffs of Maplewell Group. Diorite forms extreme left of face [476 105].

12342 Billa Barra Quarry. Shallow syncline in tuffs of Maplewell Group [467 114].
12349 Swithland Quarry. Slates of Swithland Greywacke Formation [539 122].
12350 Bradgate Park. Type locality of Sliding Stone Slump Breccia [531 113].
12351 Sliding Stone, Bradgate Park. Contorted clast of fine-grained Charnian tuff [531 113].
12352 Bradgate Park. Park Breccia – type locality [531 115].
12353 Bradgate Park. Bradgate Tuffs [5245 1095].
12355 Pillar Rock, Benscliffe Wood. Benscliffe Agglomerate – type locality [515 124].
12356 Beacon Hill. Tuffs of Beacon Hill Formation – type locality [513 145].

CAMBRIAN

10409 Merevale Park. A steep bank of Cambrian shale. This forms the (faulted) junction with the Trias in the foreground [3000 9762].

WESTPHALIAN

1546 Old clay pit, Dordon. Wedge of sandstone at the base of the Halesowen Formation.
1547 Old clay pit, Dordon. The upper part of the Etruria Marl.
1548 Old sandstone quarry 90 m E of Polesworth Water Works. Typical band of pellet rock in the Keele Formation.
7974 Red Burrow opencast site, 1.5 km SE of Packington Church. Subsoil removal by scraper.
7982 Orchard St Leonards opencast site, 800 m ESE of St Leonards Church, Dordon. Subsoil excavation by dragline.
7981 Orchard St Leonards opencast site. Subsoil removal by dragline.
7986 Red Burrow opencast site. Excavating coal and loading on lorry. Nether Lount Coal.
8005 Red Burrow opencast site. Backfilling a cut on an opencast coal site.
8020 Red Burrow opencast site. Folding in New Red Sandstone and Coal Measures.
8021 Red Burrow opencast site. Sharp folding in New Road Sandstone and Coal Measures.
8022 Red Burrow opencast site. Folding and unconformity affecting New Red Sandstone and Coal Measures.
8023 Orchard St Leonards opencast site. Faulting in coal seam.
10381 Dordon Clay Pit (disused). Greenish grey massive sandstone on Middle Coal Measures with thin seams [2625 0018].
10382 Disused clay pit at Polesworth. Middle Coal Measures. Massive yellow-brown sandstone above grey Coal Measures. [2571 0107].
10386 400 m E of Baddesley Colliery. Old quarry in massive sandstone of Halesowen Formation [2823 9710].

11086   Bramborough Farm Quarry, Donisthorpe. Coarse sandstones are faulted against yellow-weathered clays near the Aegiranum Maine Band horizon [3194 1497].

11085   Bramborough Farm Quarry, Donisthorpe. General view of disused quarry [3200 1487].

11087   Willesley Quarry. Fault in Pottery Clays [3245 1535].

11088   Willesley Quarry. 'Moira Grits' on Middle Coal Measures [3240 1540].

11089   Willesley Quarry. 'Moira Grits' on Middle Coal Measures. A more distant view [3240 1540].

11090   Willesley Quarry. 'Moira Grits' on Middle Coal Measures mudstones with coal seams [3227 1548].

11091   Willesley Quarry. Cross-bedding in 'Moira Grits' [3242 1556].

11092   Willesley Quarry. Cross-bedding in 'Moira Grits' [3245 1561].

11094   Willesley Quarry. Section in Aegiranum Marine Band [3252 1528].

11097   Robinson & Dowlers Pit, Overseal. Large opencast working for coal and fireclay [2977 1605].

11729   Willesley Quarry. Large scale cross-bedding in Coal Measures [3255 1550].

11725   Robinson & Dowlers pit, Overseal. View of outcrop of Pottery Clays [2911 1636].

TRIASSIC

6532   Bardon Hill Quarry. Triassic deposits resting unconformably on Pre-Cambrian rocks.

6533   Broombriggs, Charnwood. Beacon Hill Tuffs; to show 'desert weathering' in Trias times.

6534   Ulverscroft Abbey. Scenery associated with the outcrop of Triassic rocks.

10268   M1 cutting at E end of Birch Hill. Head on boulder clay (uppermost 3 m) on Mercia Mudstone overlying Beacon Hill Tuffs and Benscliffe Agglomerate [480 135].

10378   Hill top about 730 m west Warton. Basal Trias scarp [276 037].

10380   Bramcote Hall, near Warton. The Warton Fault. View of the Warton Fault which runs under Bramcote Hall in the background [2728 0428].

10383   Old gravel pit 550 m west of Warton. Polesworth Formation pebble beds dipping south east at about 20° [2766 0381].

10384   Old gravel pit 550 m west of Warton. Polesworth Formation pebble beds [2766 0381].

11072   Small quarry 30 m NE of Chilcote Church. Honeycomb weathering of Bromsgrove Sandstone [2849 1141].

11073   Small quarry 30 m NE of Chilcote Church. Lamination in Bromsgrove Sandstone [2849 1141].

11074   Road junction 90 m W of Netherseal church. Pebble sandstone bed near base of Bromsgrove Sandstone [2876 1287].

11075   View southwards from bridge over River Mease at Netherseal. View of Bromsgrove Sandstone escarpment [2867 1276].

11076   Acresford Sand Pit. Pit in Polesworth Formation [3022 1335].

11077   Acresford Sand Pit. Sand in Polesworth Formation [3020 1350].

11078   Acresford Sand Pit. Pebbles in Polesworth Formation [3025 1325].

11079   Acresford Sand Pit. Polesworth Formation [3025 1325].

11080   Old sand pit 45 m south of Measham Station. Sandstone bed in Polesworth Formation [3326 1184].

11081   Quarry of Redbank Brick and Terra Cotta Works, near

Measham. Marl in Bromsgrove Sandstone [3340 1085].

11082   Quarry of Redbank Brick and Terra Cotta Works, Measham. Sandstone and marl in Polesworth Sandstone [3335 1077].

11083   Quarry of Redbank Brick and Terra Cotta Works, Measham. Marl and sandstone in Bromsgrove Sandstone [3355 1115].

11084   Old quarry at road junction 350 m SSE of Redbank Brick and Terra Cotta Works, Measham. Marl in Bromsgrove Sandstone [3190 1057].

12325   Foan Hill from Swannington. Basal Triassic scarp [4150 1615].

12326   Hermitage Brickworks, Whitwick. Mercia Mudstone and Bromsgrove Sandstone [429 153].

12327   Peldar Tor Quarry, Whitwick. Triassic land surface [444 159].

12336   Bardon Hill Quarry. Mercia Mudstone with coarse breccias infilling hollows in Charnian [4561 1303].

12337   Bardon Hill Quarry. Mercia Mudstone overlain by coarse head [4583 1333].

12341   Cliffe Hill Quarry. Trias unconformity [476 105].

12344   Ibstock Brickworks (east pit). Trias worked in series of benches defined by sandstone units [415 109].

12346   Ibstock Brickworks (west pit). Mercia Mudstone [412 109].

12358   Desford Colliery Brickworks. Horizontally bedded Mercia Mudstone with thin siltstone skerries, overlain by glacial gravel [461 066].

QUATERNARY

6545   Near Woodhouse Eaves. Cuckoo Hill, Brand. Superimposed drainage. A longitudinal stream has turned to its left and cut through the rib of rocks made by the highest Bradgate Tuffs and the lower part of the Swithland Greywacke.

10266   E Bank of M1 cutting 1.5 km NNW of Markfield. Mercia Mudstone-rich boulder clay with Charnian rock fragments [4796 1192].

10379   730 m S 20°W of Bramcote Hall near Warton. Basal Trias scarp and pre-glacial valley. The valley which opens into the foreground is partly filled with glacial till [2710 0377].

11098   House in Swepstone Lane, Measham. Effects of mining subsidence [3403 1220].

12331   Blackbrook Valley looking west towards Moult Hill. Fault-controlled and probably a glacial drainage channel [4712 1695].

12333   Cadeby Gravel Pits. Intebedded chalky till and Triassic till [4314 0278].

12334   Cadeby Gravel Pits. Triassic till on sand and gravel [4313 0282].

12335   Cadeby Gravel Pits. Interbedded tills on gravel [4314 0278].

12343   Spring Farm, Coalville. Flooded depression due to mining subsidence [439 135].

12345   Ibstock Brickworks (east pit). Chalky boulder clay [4175 1077].

12347   Ellistown Brickworks. Cryoturbation structures in Mercia Mudstone, overlain by red boulder clay [4345 1046].

12348   Heather Gravel Pit. Fluvio-glacial sand and gravel [4005 1095].

12354   Entrance to Bradgate Park. Fault-line in valley [524 099].

12357   M69 motorway, Forest Road bridge, Huncote. Deformed glacial sands [5165 9899].

FOSSIL INDEX

No distinction is made here between a positively determined genus or species and examples doubtfully referred to them (i.e. with the qualifications aff., cf or ?)

GENERAL INDEX

BRITISH GEOLOGICAL SURVEY

Keyworth, Nottingham NG12 5GG
Plumtree (060 77) 6111

Murchison House, West Mains Road,
Edinburgh EH9 3LA   (031) 667 1000

The full range of Survey publications is available through the Sales Desks at Keyworth and Murchison House. Selected items are stocked by the Geological Museum Bookshop, Exhibition Road, London SW7 2DE; all other items may be obtained through the BGS London Information Office in the Geological Museum ((01) 589 4090). All the books are listed in HMSO's Sectional List 45. Maps are listed in the BGS Map Catalogue and Ordnance Survey's Trade Catalogue. They can be bought from Ordnance Survey Agents as well as from BGS.

The British Geological Survey carries out the geological survey of Great Britain and Northern Ireland (the latter as an agency service for the government of Northern Ireland), and of the surrounding continental shelf, as well as its basic research projects. It also undertakes programmes of British technical aid in geology in developing countries as arranged by the Overseas Development Administration.

The British Geological Survey is a component body of the Natural Environment Research Council.

Maps and diagrams in this book use topography based on Ordnance Survey mapping

HER MAJESTY'S STATIONERY OFFICE

HMSO publications are available from:

HMSO Publications Centre
(Mail and telephone orders)
PO Box 276, London SW8 5DT
Telephone orders (01) 622 3316
General enquiries (01) 211 5656
Queueing system in operation for both numbers

HMSO Bookshops
49 High Holborn, London WC1V 6HB
   (01) 211 5656 (Counter service only)
258 Broad Street, Birmingham B1 2HE
   (021) 643 3740
Southey House, 33 Wine Street, Bristol BS1 2BQ
   (0272) 264306
9 Princess Street, Manchester M60 8AS
   (061) 834 7201
80 Chichester Street, Belfast BT1 4JY
   (0232) 238451
71 Lothian Road, Edinburgh EH3 9AZ
   (031) 228 4181

HMSO's Accredited Agents
(see Yellow Pages)

And through good booksellers